D0688182

Image Modeling of the Human Eye

Artech House Series
Bioinformatics & Biomedical Imaging

Series Editors
Stephen T. C. Wong, The Methodist Hospital Research Institute
Guang-Zhong Yang, Imperial College

Advances in Diagnostic and Therapeutic Ultrasound Imaging, Jasjit S. Suri, Chirinjeev Kathuria, Ruey-Feng Chang, Filippo Molinari, and Aaron Fenster, editors

Biological Database Modeling, Jake Chen and Amandeep S. Sidhu, editors

Biomedical Informatics in Translational Research, Hai Hu, Michael Liebman, and Richard Mural

Genome Sequencing Technology and Algorithms, Sun Kim, Haixu Tang, and Elaine R. Mardis, editors

Life Science Automation Fundamentals and Applications, Mingjun Zhang, Bradley Nelson, and Robin Felder, editors

Microscopic Image Analysis for Life Science Applications, Jens Rittscher, Stephen T. C. Wong, and Raghu Machiraju, editors

Next Generation Artificial Vision Systems: Reverse Engineering the Human Visual System, Maria Petrou and Anil Bharath, editors

Systems Bioinformatics: An Engineering Case-Based Approach, Gil Alterovitz and Marco F. Ramoni, editors

Image Modeling of the Human Eye

Rajendra Acharya U
Eddie Y. K. Ng
Jasjit S. Suri
Editors

ARTECH
HOUSE

BOSTON | LONDON
artechhouse.com

Library of Congress Cataloging-in-Publication Data
A catalog record for this book is available from the U.S. Library of Congress.

British Library Cataloguing in Publication Data
A catalogue record for this book is available from the British Library.

ISBN-13: 978-1-59693-208-1

Cover design by Igor Valdman

© 2008 ARTECH HOUSE, INC.
685 Canton Street
Norwood, MA 02062

All rights reserved. Printed and bound in the United States of America. No part of this book may be reproduced or utilized in any form or by any means, electronic or mechanical, including photocopying, recording, or by any information storage and retrieval system, without permission in writing from the publisher.

All terms mentioned in this book that are known to be trademarks or service marks have been appropriately capitalized. Artech House cannot attest to the accuracy of this information. Use of a term in this book should not be regarded as affecting the validity of any trademark or service mark.

10 9 8 7 6 5 4 3 2 1

Contents

CHAPTER 4

Automatic Identification of Anterior Segment Eye Abnormalities in Optical Images

Preface

Various disciplines have benefited from the advent of high-performance computing and the area of health care is no exception. Image processing, computer simulation tools, and infrared imaging have been developed to enhance computational capabilities, which helps physicians achieve practical solutions to their problems and helps with diagnosis and treatment.

The human eye is a complex sensory organ that perceives directly a tiny fraction of the world and is designed to optimize vision under conditions of varying light. A number of eye disorders, however, can influence vision. Eye disorders among the elderly are a major health problem. With advancing age, the normal function of eye tissues decreases and there is an increased incidence of ocular pathology. The most common causes of age-related eye disorders and visual impairment in the elderly are cataracts, diabetic retinopathy, glaucoma, iridocyclitis, and corneal haze. However, the human observer cannot directly monitor these subtle details. For proper care and management of eyes, we need an automatic system by which we can classify these eye diseases. Such a system would extract specific features using image processing techniques and analyze them using computers.

Imaging processing of optical and fundus images will aid clinicians with the accurate and fast diagnosis of eye abnormalities. One *potential* method is the measurement of the ocular surface temperature (OST), where thermal variations in the eyeball can be compared to existing models and clinical data to predict problems. Computer simulations with multiphysics approaches in conjunction with artificial intelligence systems can offer a possible solution to predicting eye abnormalities at an early stage. The bioheat transfer phenomena in the human eye have been of major interest for the past century. The earliest methods of eye temperature measurements were invasive and involved direct contact, but such methods are now confined to animal experiments. Most recently, efforts at noncontact temperature measurements have been focused on infrared (IR) imaging. Using IR imaging, however, provides only temperatures on the corneal surface. With the use of numerical computer simulations, researchers can create a model of the human eye and, using principles of heat transfer and the appropriate bioheat equation, to predict interior temperatures based on the surface temperature.

Chapter 1 of the book explains the structure and the function of the human eye and how they are related, often using functional issues as a guide to the most meaningful and important features of the anatomy. Different types of eye abnormalities are discussed, and diagrams showing the necessary level of detail are provided.

Chapter 2 deals with the concepts of light behavior and the nature of the light, analyzing several phenomena such as reflection and refraction. Different types of and prisms, their applications, and the changes that they produce based on the direction of light are discussed in detail, as are optical instruments.

It is best to have a device that can detect eye disease in its early stages. Physicians together with the engineers have designed equipment that allows them to view the eye for a quicker diagnosis of disease, often with greater accuracy and minimal invasion. So, Chapter 3 explains in detail the different techniques of image acquisition, namely, computed tomography, confocal laser scanning microscopy, magnetic resonance imaging, optical coherence tomography, and ultrasound imaging.

The most common anterior segment eye abnormalities are cataract, corneal haze, and iridocyclitis. Chapter 4 presents a method of automatic classification of four types of optical eye images (three different kinds of eye diseases and a *normal* class). Features are extracted from these raw images and fed to a neural network classifier.

Diabetic retinopathy is a complication of that affects the blood vessels in the. In Chapter 5, three groups are identified: normal, nonproliferative diabetic retinopathy, and proliferative diabetic retinopathy. The parameters used were extracted from raw fundus images using image processing techniques and fed to the Gaussian mixture model (GMM) classifier for identification of the unknown class.

During diabetic maculopathy, fluid rich in fat and cholesterol leaks out of damaged vessels. The fluid accumulates near the center of the retina (the macula) and leads to the distortion of central vision. Chapter 6 presents a computer-based intelligent system for the identification of *clinically significant* and *nonclinically significant* maculopathy fundus eye images. A Sugeno fuzzy model–based classifier was used for classification of these two stages.

In Chapter 7, the pathophysiology of glaucoma is described, and the use of confocal scanning laser ophthalmoscopes (CSLOs) for examining the optic nerve head (ONH) in glaucoma is introduced. The computational and statistical techniques currently available for detecting glaucoma and describing glaucomatous structural changes over time from CSLO-generated ONH topographs using ONH parameters and pixel-level topograph measurements are discussed in detail.

Chapter 8 presents a new technique for characterizing and detecting filamentous fungal infiltrates in the cornea of an eye for diagnosing fungal keratitis. The new technique locates any fungal infiltrates in the white-light confocal microscope (WLCM) optical section images of the cornea, characterizes the morphology of the fungal infiltrates using fractal dimension measures, and differentiates fungal infiltrates from any corneal nerves present in the WLCM images using a maximum likelihood classifier. A set of WLCM images of the cornea with culture-positive fungal infiltrates was used to demonstrate the performance of the new technique.

Chapter 9 discusses a Gaussian matched filter used in combination with different parameters to detect the vessels of retinal images. Filter parameters include the standard deviation, the filter rotation parameter, and filter width and height. The comparisons are performed using standard retinal image benchmarks.

During diabetic retinopathy, pathological changes may modify the color and shape of retinal vessels. For instance, some vessels may narrow or widen, new vessels may appear, or irregular edges may occur. Chapter 10 proposes a new method for using image processing techniques for the detection of blood vessels in fundus images of the retina. The technique includes the design of a bank of directionally sensitive Gabor filters with variable scale and elongation parameters.

Chapter 11 discusses the development of two-dimensional and three-dimensional models of the human eye using the finite element method (FEM). The models are used to simulate the steady-state temperature distribution inside the human eye, and results are compared with experimental measurements found in the literature.

Chapter 12 is dedicated to the application of the two-dimensional human eye model to simulate the changes in the ocular temperature distribution during contact lens wear. In addition, the thermal effects of different contact lens storage temperatures are investigated.

A boundary element method (BEM) is applied for the numerical solution of a boundary value problem for a two-dimensional, steady-state bioheat transfer model of the human eye in Chapter 13. It suggests that the BEM calculates the normal heat flux more accurately than the FEM on the corneal surface, near its edges in particular.

Chapter 14 illustrates the effects of aqueous humor (AH) flow on the temperature distribution inside the eye. The role of AH flow in the heat transfer at the eye anterior region is further evident when an artificial heat source is introduced.

Temperature changes across the ocular surface are small but occur rapidly. Ocular thermography is a technique that allows for noninvasive temperature assessment of the anterior eye and its surroundings at both high spatial and temporal resolutions. Chapter 15 provides a brief history of clinical ocular thermography and its clinical implications.

Chapter 16 presents the results of a clinical study that utilized dynamic ocular thermography. The temperature changes of the anterior eye during and immediately after wearing hydrogel contact lenses were studied by means of noncontact IR thermography.

Chapter 17 presents a statistical study on changes in the ocular surface temperature (OST) with age using IR thermography. Variations between the left and right OST are also discussed.

It is our earnest desire that this book will assist those who seek to enrich their lives and those of others with the wonderful powers of image processing and modeling. Electrical, computer, and biomedical engineering are important fields that contribute immensely to the service of humanity.

Rajendra Acharya
Eddie Y. K. Ng
Jasjit S. Suri
Editors
April 2008

CHAPTER 1

The Human Eye

Rajendra Acharya, William Tan, Wong Li Yun, Eddie Y. K. Ng, Lim Choo Min, Caroline Chee, Manjunath Gupta, Jagadish Nayak, and Jasjit S. Suri

The eyes are undoubtedly the most sensitive and delicate organs we possess and perhaps the most amazing. They present us with the window through which we view the world and are responsible for four-fifths of all of the information our brain receives—which is probably why we rely on our eyesight more than any other sense.

Being so small (about 24 to 25 mm in sagittal diameter) and so intricate, the way in which the eye functions is incredibly complex (Figure 1.1). In this chapter, we explain the important parts of the eye and describe different eye diseases in detail.

1.1 Basic Ocular Anatomy and Physiology

The human eye consists mainly of six regions: the cornea, aqueous humor, iris, lens, vitreous humor, and sclera. Other ocular domains include the retina and the choroid. The retina is the layer that is light sensitive and is the region where light energy is transformed into neural signals. The choroid is a highly vascularized structure in the human eye that accounts for 85% of the total ocular blood flow [1]. The human eye has a structure very close to that of a sphere. A typical human eye has a radius of 12 mm, and the length of the pupillary axis (the distance between the cornea and the posterior region of the eye) ranges from 23 to 25 mm [2]. Figure 1.2 shows the anatomical structure of the human eye.

1.1.1 The Cornea

The cornea is the only part of the human eye that is exposed to the environment. It is a transparent medium with a water content of 78% [3]. The cornea is elliptical in nature with a vertical and horizontal diameter of 11 and 12 mm, respectively [2, 4]. The thickness of the cornea is not uniform. The central thickness is about $530\,\mu m$, diverging to the periphery, which becomes $710\,\mu m$. The function of the cornea is to refract and transmit light. The cornea is avascular, which means that there are no blood vessels in it. Although it is avascular, the cornea is metabolically active. The cornea requires an oxygen supply, which is derived primarily from the atmosphere and absorption through the tear film.

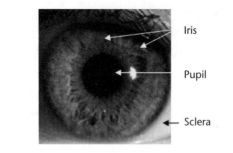

Figure 1.1   The human eye.

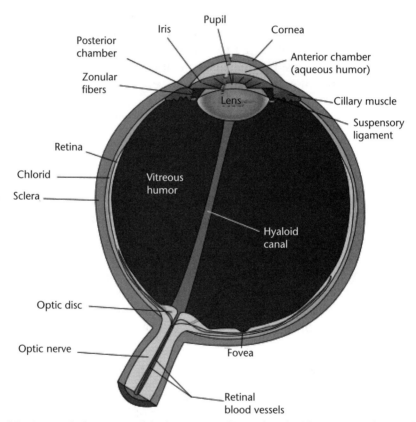

Figure 1.2   Anatomical structure of the human eye. (Reproduced with permission from Wikipedia, http://en.wikipedia.org/wiki/Aqueous_humor.)

The tear film is a layer of tears covering the surface of the cornea. It consists of three layers: the aqueous layer, the mucoid layer, and the lipid layer. It is approximately 10 μm thick [3]. The tear film serves as a lubricator for the smooth movement of the eyelids during blinking and is important in protecting the corneal surface. It is also responsible for the smooth optical properties of the corneal surface.

1.1.2   The Aqueous Humor

The aqueous humor is a chamber containing aqueous fluid. This fluid provides nutrients to the avascular cornea and lens. The aqueous is made up primarily of water, which is secreted into the chamber through the ciliary processes. The flow of aqueous has its own hydrodynamic features. This flow generates an intraocular pressure (IOP) of about 15 mmHg for normal eyes [5].

1.1.3   The Iris

The iris is made up of pigmented fibrovascular tissues known as the stroma, which connects a set of constricting muscles and dilator muscles named the sphincter pupillae and dilator pupillae, respectively, to control the size of the pupil. The function of the iris is to control the size of the pupil by adjusting it to the intensity of the lighting conditions. It is also responsible for determining the color of an individual's eye. The diameter of the pupil is dependent on the lighting conditions. At luminous conditions, the diameter of the pupil shrinks to a minimum of 1 mm and the opposite is observed when light intensity is low (9 mm). The division of the eye structure into anterior and posterior regions is based on the iris. Regions at the front of the iris are classified as the anterior ocular region, whereas regions at the back of the iris are classified as the posterior region. The iris is a vascular structure with the arteries originating from the ciliary body (near the iris roots).

1.1.4   The Lens

The human lens is biconvex with a central thickness between 3.5 and 5 mm depending on the age of the individual [6]. The lens is held in place by zonular fibers, and its function is to focus light that enters the eye onto the retina. The lens is made up of 65% water and 35% protein and is avascular. The radii curvature of the lens was found to decrease with age, whereas the thickness was found to increase [7]. Water content inside the lens has also been reported to increase with age [8].

1.1.5   The Vitreous Humor

The vitreous humor is the largest domain of the human eye. It fills up the space between the lens and retina with a clear aqueous solution that is made up of 98.5% to 99.7% water. The vitreous humor gives the spherical structure to the human eye. In an adult, the length of the vitreous is approximately 16.5 mm [9]. The vitreous serves as a storage area for metabolites of the retina and lens.

1.1.6   The Sclera

The sclera is usually referred to as the white outer layer of the eye and wraps around five-sixths of the posterior eye globe. The thickness of the sclera is nonuniform with the thickest part located at the posterior pole (1 mm). The blood supply in the sclera is minimal with only a few blood vessels passing through it. Without capillary beds, the sclera is assumed to be avascular. Beneath the sclera is the choroidal and retinal layer. The choroid contains a layer of capillary bed known as the choriocapillaries that make it highly vascular [10].

1.2 Cornea

The cornea (Figure 1.3) is a clear, dome-shaped surface that covers the front of the eye. It has two primary functions: to refract and transmit light. Factors that affect the amount of corneal refraction include the following:

- The curvature of the anterior corneal surface;
- The change in refractive index from air to cornea (actually the tear film);
- Corneal thickness;
- The curvature of the posterior corneal surface, and the change in refractive index from cornea to aqueous humor.

The clear cornea is actually a highly organized group of cells and proteins. Unlike most tissues in the body, the cornea contains no blood vessels to nourish or protect it against infection. Instead, the cornea receives its nourishment from the tears and aqueous humor that fills the chamber behind it. The cornea must remain transparent to refract light properly, and the presence of even the tiniest blood vessels can interfere with this process. To see well, all layers of the cornea must be free of any cloudiness or opaque areas.

1.2.1 Structure of the Cornea

The corneal tissue is arranged in five essential layers, each having an important function. The epithelium is the cornea's outermost region, comprising about 10% of the tissue's thickness. The epithelium functions primarily to:

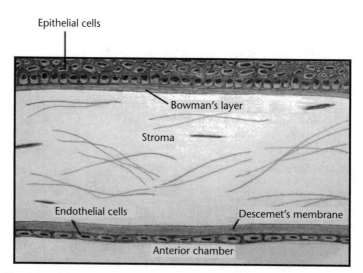

Figure 1.3 Structure of the cornea.

- Block the passage of foreign material, such as dust, water, and bacteria, into the eye and other layers of the cornea;
- Provide a smooth surface that absorbs oxygen and cell nutrients from tears, to distribute these nutrients to the rest of the cornea.

The epithelium is filled with thousands of tiny nerve endings that make the cornea extremely sensitive to pain when rubbed or scratched. The part of the epithelium that serves as the foundation on which the epithelial cells anchor and organize themselves is called the basement membrane.

Directly below the basement membrane of the epithelium is a transparent sheet of tissue known as Bowman's layer. It is composed of strong layered protein fibers called collagen. Once injured, Bowman's layer may form a scar as it heals. If these scars are large and centrally located, some vision loss may occur.

Beneath Bowman's layer is the stroma, which makes up about 90% of the cornea's thickness. It is primarily water (78%) and collagen (16%) and does not contain any blood vessels. The collagen fibrils are equally spaced in between and neatly packed together, forming parallel sheets of fibrils, also known as lamellar sheets. Each sheet is placed orthogonal to each other and to the path of light through the cornea. The uniformly arranged collagen fibril structure gives the cornea its strength, elasticity, and form. Essentially, the collagen fibrils' unique shape, arrangement, and spacing are essential in producing the cornea's light-conducting transparency.

Under the stroma is Descemet's membrane, a thin but strong sheet of tissue that serves as a protective barrier against infection and injuries. Descemet's membrane is comprised of collagen fibers (different from those of the stroma) and is supported by the endothelial cells that lie below it. Descemet's membrane is regenerated readily after injury.

The endothelium is the extremely thin, innermost layer of the cornea. Endothelial cells are essential in keeping the cornea clear by maintaining the equilibrium of the liquid in the cornea. A functioning endothelium's primary task is to pump excess fluid that leaks slowly from inside the eye into the middle corneal layer (stroma), out of the stroma. Without this pumping action, the stroma would swell with water, become hazy, and ultimately opaque. In a healthy eye, a perfect balance is maintained between the fluid moving into the cornea and fluid being pumped out of the cornea by means of the endothelium. Once endothelial cells are destroyed by disease or trauma, they are lost forever. If too many endothelial cells are destroyed, corneal edema and blindness ensue, with corneal transplantation being the only available therapy.

1.3   Retina

The retina is a thin layer of neural cells that lines the back of the eyeball of vertebrates and some cephalopods. In vertebrate embryonic development, the retina and the optic nerve originate as outgrowths of the developing brain. Hence, the retina is part of the central nervous system (CNS). It is the only part of the CNS that can be imaged directly [11].

The retina is the site of transformation of light energy into a neural signal and contains the first three types of cells in the visual pathway, the route by which visual information from the environment reaches the CNS for interpretation. Those three types of cells are the photoreceptor (rods and cones), the bipolar, and the ganglion cells (Figure 1.4). Photoreceptor cells respond and transform photons of light into a neural signal and then transfer this signal to bipolar cells, which in turn synapse into ganglion cells that transmit the signal from the eye. Other retinal cells—horizontal cells, amacrine cells, and the interplexiform neurons—modify and integrate the signal before it leaves the eye. The retinal output takes the form of action potentials in retinal ganglion cells whose axons form the optic nerve. Several important features of visual perception can be traced to the retinal encoding and processing of light.

The retina often is described as consisting of two regions; peripheral and central. The periphery is designated for detecting gross forms of motion, whereas the central area is specialized for visual activity. In terms of area occupied, the periphery makes up most of the retina, and rods dominate. The central retina is rich in cones, has more ganglion cells per area than elsewhere, and is a relatively small portion of the entire retina [11].

1.3.1 Structure of the Retina

The vertebrate retina has 10 distinct layers, listed from innermost to outermost:

- Inner limiting membrane (Mûller cell footplates);
- Nerve fiber layer;
- Ganglion cell layer (layer that contains nuclei of ganglion cells and gives rise to optic nerve fibers);
- Inner plexiform layer;
- Inner nuclear layer;
- Outer plexiform layer (in the macular region, this is known as the fiber layer of Henle);
- Outer nuclear layer;
- External limiting membrane (layer that separates the inner segment portions of the photoreceptors from their cell nuclei);
- Photoreceptor layer (rods and cones);

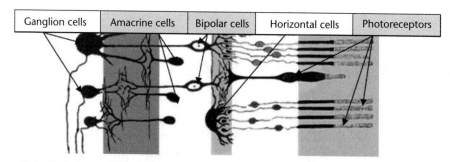

Figure 1.4 Retina's simplified axial organization. (Reproduced with permission from Wikipedia, http://en.wikipedia.org/wiki/Retina.)

- Retinal pigment epithelium (neuroretinal interface, which has a strong attachment to the choroid; has a part to play in retinal detachment, which is discussed later).

In adult humans the entire retina is 72% of the surface of a sphere, about 22 mm in diameter. A portion of the visual field is sometimes known as the "blind spot" because it lacks photoreceptors. It appears as an oval white area of 3 mm^2. Temporal (in the direction of the temples) to this disc is the macula. At the center of the macula is the fovea, a pit that is most sensitive to light and is responsible for our sharp central vision. Human and nonhuman primates possess one fovea as opposed to certain bird species such as hawks, which are actually bifoviate, and dogs and cats who possess no fovea but a central band known as the visual streak. Around the fovea, the central retina extends for about 6 mm and then the peripheral retina. The edge of the retina is defined by the ora serrata. The length from one ora to the other (or macula), the most sensitive area along the horizontal meridian, is about 3.2 mm.

1.3.2   Organization of Neuronal Cell Layers

The retina is a stack of several neuronal layers (see Figure 1.4). Light is concentrated from the eye and passes across these layers (from left to right) to hit the photoreceptors (right layer). This elicits chemical transformation, mediating a propagation of signal to the bipolar and horizontal cells (middle yellow layer). The signal is then propagated to the amacrine and ganglion cells. These neurons ultimately may produce action potentials on their axons. This spatiotemporal pattern of spikes determines the raw input from the eyes to the brain [11].

In cross section, the retina is no more than 0.5 mm thick. It has three layers of nerve cells and two of synapses. The optic nerve carries the ganglion cell axons to the brain and the blood vessels that open into the retina. As a by-product of evolution, the ganglion cells lie innermost in the retina, whereas the photoreceptive cells lie at the outermost point. Because of this arrangement, light must first pass through the thickness of the retina before reaching the rods and cones. However, it does not pass through the epithelium or the choroid (both of which are opaque).

The white blood cells in the capillaries in front of the photoreceptors can be perceived as tiny bright moving dots when looking into blue light. This is known as the blue field entoptic phenomenon (or Scheerer's phenomenon).

Between the ganglion cell layer and the rods and cones are two layers of plexus where synaptic contacts are made. The neuropil layers are the outer plexiform layer and the inner plexiform layer. In the outer, the rod and cones connect to the vertically running bipolar cells and the horizontally oriented horizontal cells connect to ganglion cells.

The central retina is cone-dominated and the peripheral retina is rod-dominated. In total there are about 7 million cones and 100 million rods in photoreceptors. The macula has an oval "yellow spot" known as the macula lutea, defined by having two or more layers of ganglion cells, thus attributing to the yellowish color. It contains the fovea centralis at which visual perception is at its highest acuity, where only cones are present and lack the presence of blood vessels. At the center of the macula is the foveal pit where the cones are the smallest and in a hexagonal mosaic and, hence, the most efficient and in the highest density. Below the pit the other retina layers are displaced, before building up along the foveal slope until reaching the rim of the fovea or parafovea, which is the thickest portion of the retina.

1.3.3   Path of Visual Signals

The difference between vertebrate and cephalopod retinas is that the vertebrate retina is *inverted* in the sense that the lightsensing cells sit at the back side of the retina, so that light has to pass through a layer of neurons before it reaches the photoreceptors. By contrast, the cephalopod retina is *everted*: The photoreceptors are located at the front side of the retina, with processing neurons behind them. Because of this, cephalopods do not have a blind spot.

The cephalopod retina does not originate as an outgrowth of the brain, as the vertebrate one does. This shows that vertebrate and cephalopod eyes are not homologous but have evolved separately.

An image is produced by the "patterned excitation" of the retinal receptors, the cones and rods. The excitation is processed by the neuronal system and various parts of the brain working in parallel to form a representation of the external environment in the brain.

The cones respond to bright light and mediate high-resolution vision and color vision. The rods respond to dim light and mediate lower-resolution, black-and-white, night vision. It is a lack of cones, a reduced sensitivity to red, blue, or green light, that causes individuals to have deficiencies in color vision or various kinds of color blindness. Humans and Old World monkeys have three different types of cones (trichromatic vision), whereas other mammals lack cones with red sensitive pigment and therefore have poorer (dichromatic) color vision [11].

When light falls on a receptor, it sends a proportional response synaptically to bipolar cells, which in turn signal the retinal ganglion cells. The receptors are also cross-linked by horizontal cells and amacrine cells, which modify the synaptic signal before reaching the ganglion cells (see Figure 1.4). Rod and cone signals are intermixed and combined, although rods are mostly active in very poorly lit conditions and saturate in broad daylight, whereas cones function in brighter lighting because they are not sensitive enough to work at very low light levels.

Despite the fact that all are nerve cells, only the retinal ganglion cells and few amacrine cells create action potentials. In the photoreceptors, exposure to light hyperpolarizes the membrane in a series of graded shifts. The outer cell segment contains a photo pigment. Inside the cell the normal levels of cGMP keep the Na channel open and, thus, while in the resting state, the cell is depolarized. The photon causes the retinal bound to the receptor protein to isomerize to transretinal. This causes receptors to activate multiple G-proteins. This in turn causes the Ga-subunit of the protein to bind and degrade cGMP inside the cell, which then cannot bind to the Na channels. Thus, the cell is hyperpolarized. The amount of neurotransmitter released is reduced in bright light and increases as light levels fall. The actual photo pigment is bleached away in bright light and only replaced as a chemical process, so in a transition from bright light to darkness the eye can take up to 30 minutes to reach full sensitivity.

The retinal ganglion cells make two types of responses, depending on the receptive field of the cell. The receptive fields of retinal ganglion cells comprise a central approximately circular area, where light has one effect on the firing of the cell, and an annular surround, where light has the opposite effect on the firing of the cell. In ON cells, an increment in light intensity in the center of the receptive field causes the firing rate to increase. In OFF cells, it makes it decrease. Beyond this simple difference, ganglion cells are also differentiated by chromatic sensitivity and the type of spatial summation. Cells showing linear spatial summation are termed *X cells* (also called *parvocellular, P,* or *midget* ganglion cells), and those showing nonlinear summation are *Y cells* (also called *magnocellular, M,* or *parasol* retinal ganglion cells), although the correspondence between X and Y cells (in the cat retina) and P and M cells (in the primate retina) is not as simple as it once seemed.

During the transfer of the signal to the brain, the visual pathway, the retina is vertically divided into two: a temporal half and a nasal half. The axons from the nasal half cross the brain at the optic chiasma to join with axons from the temporal half of the other eye before passing into the lateral geniculate body.

Although there are more than 130 million retinal receptors, there are only approximately 1.2 million fibers (axons) in the optic nerve, so a large amount of preprocessing is performed within the retina. The fovea produces the most accurate information. Despite occupying about 0.01% of the visual field (less than 2° of visual angle), about 10% of axons in the optic nerve are devoted to the fovea. The resolution limit of the fovea has been determined at around 10,000 points. The information capacity is estimated at 500,000 bits per second (bps) without color or around 600,000 bps including color.

1.4 Anterior Segment Abnormalities

The following types of corneal disorders are discussed in this section: keratoconus, corneal ulcer, Fuchs' dystrophy, pterygium, pingueculae, scleritis, uveitis (iridocyclitis), corneal haze, and cataract.

1.4.1 Keratoconus

Keratoconus (Figure 1.5) is the most common dystrophy of the cornea. It is the progressive thinning of the cornea in which focal disruptions of basement membrane

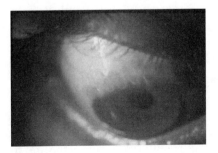

Figure 1.5 Patient with keratoconus.

and Bowman's layer occur. Metabolic or nutritional disturbances are among the possible causes. The degenerative process usually begins in the central area of the cornea; the stroma eventually degenerates and thins, and the affected area projects outward in a cone shape due to force exerted on the weakened areas by intraocular pressure [9]. Folds occur in the posterior stroma and endothelium. This abnormal curvature changes the cornea's refractive power, producing moderate to severe distortion (astigmatism) and blurriness (nearsightedness) of vision. Keratoconus may also cause swelling (hydration) and sight-impairing scarring of the tissue.

People with early keratoconus typically notice a minor blurring of their vision and come to their clinician seeking corrective lenses for reading or driving. At early stages, the symptoms of keratoconus may be no different from those of any other refractive defect of the eye. As the disease progresses, vision deteriorates, sometimes rapidly. Visual perception becomes impaired, and vision at night is often quite poor. Some individuals have vision in one eye that is markedly worse than that in the other eye. Some develop photophobia (sensitivity to bright light), eye strain from squinting in order to read, or itching in the eye. There is usually little or no sensation of pain.

The classic symptom of keratoconus is the perception of multiple "ghost" images, known as monocular polyopia. This effect is most clearly seen with a high-contrast field, such as a point of light on a dark background. Instead of seeing just one point, a person with keratoconus sees many images of the point, spread out in a chaotic pattern. This pattern does not typically change from day to day, but over time it often takes on new forms. Patients also commonly observe streaking and flaring distortion around light sources. Some even notice images moving relative to one another in time with their heartbeat.

1.4.1.1 Treatment for Keratoconus

In the early stages of keratoconus, spectacles or contact lenses can be used to correct the cornea's refractive problem. However, these only serve as temporary vision improvement solutions because the cornea continues to thin and changes shape. Rigid gas permeable (RGP) contact lenses would be the later option because they are made of firmer material than soft contact lenses, which vaults around the irregular shape of the affected cornea better, rectifying the refractive problem. The disadvantage of RGP contact lenses is that the rigidity might cause discomfort to the wearer, and repetitive adjustments to the fit have to be made.

Corneal inserts, or Intacs, are clear semicircular plastic rings used to help patients obtain a functional vision beyond the use of spectacles and contact lenses. The Intacs are inserted under the eye's surface in the periphery of the cornea, flattening out the central part of the cornea and improving its structural integrity. Its presence would not be felt once the corneal wound heals after insertion, and it is suited for exchanging because it is easily removed.

C3-R treatment, or corneal collagen cross-linking with riboflavin treatment, improves the adhesion of the Intacs in the cornea, by strengthening the cornea through inducing more cross-linking of the collagen fibrils, which are responsible for the prevention of irregularity in the shape and form of the cornea. The process of inducing more cross-linking between collagen fibrils present in the cornea is done by simple application of specially formulated riboflavin eyedrops, which are then activated by ultraviolet light. Cross-linking treatments together with Intacs flattens the cornea better than just application of the latter; it better stabilizes the condition and could well prevent keratoconus from worsening further.

For cases in which the therapies just discussed do not suffice or sustain functional vision for the patient, a corneal transplant can be made on the availability of a existing donor. However, even after a corneal transplant, the cornea might not be able to return to its original state before getting keratoconus, and patients may still require spectacles and contact lenses.

1.4.2 Corneal Ulcer

A corneal ulcer forms when the surface of the cornea is damaged or compromised. Ulcers may be sterile (no infecting organisms) or infectious. The term *infiltrate*, which refers to an immune response causing an accumulation of cells or fluids in an area of the body where they do not normally belong, is commonly used along with *ulcer*.

Extreme pain is one symptom of bacterial ulcers and is typically caused by a break in the epithelium, the superficial layer of the cornea. In some cases, the inflammatory response affects the anterior chamber along with the cornea. Certain types of bacteria, such as *Pseudomonas,* are extremely aggressive and can cause severe damage and even blindness within 24 to 48 hours if left untreated. Sterile infiltrates on the other hand, cause little if any pain. They are often found near the peripheral edge of the cornea and are not necessarily accompanied by a break in the epithelial layer of the cornea.

Corneal ulcers are most commonly caused by an infection with bacteria, viruses, fungi, or amoebas. Other causes are abrasions (scratches) or foreign bodies, inadequate eyelid closure, severely dry eyes, severe allergic eye disease, and various inflammatory disorders (these are predisposing factors) [9, 10].

Bacterial infections that cause corneal ulcers (Figures 1.6 and 1.7) are common in people who wear contact lenses. Viral infections are also possible causes of corneal ulcers. Such viruses include the herpes simplex virus (the virus that causes cold sores) or the varicella virus (the virus that causes chickenpox and shingles).

Fungal infections can cause corneal ulcers and may be induced by the overuse of eyedrops containing steroids. Disorders that cause dry eyes can leave your eye without the germ-fighting protection of tears and, hence, can result in ulcers. Similarly, disorders that affect the eyelid and prevent your eye from closing completely, such

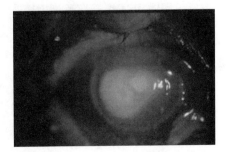

Figure 1.6 Bacterial corneal ulcer.

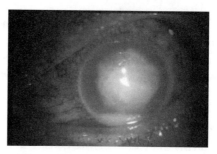

Figure 1.7 A bacterial corneal ulcer with an accumulation of pus (white blood cells) in anterior chamber of the eye.

as Bell's palsy, can dry your cornea and make it more vulnerable to ulcers. Furthermore, chemical burns or other caustic solution splashes can injure the cornea.

People who wear contact lenses are at an increased risk of corneal ulcers (Figure 1.8). In fact, the risk of corneal ulcerations increases tenfold when using extended-wear soft contact lenses. Extended-wear contact lenses are those contact lenses that are worn for several days without removing them at night. Contact lenses may damage the cornea in many ways. Scratches on the edge of the lens and tiny particles of dirt trapped underneath the contact lens may scrape the cornea's surface, opening it to bacterial infections.

Bacteria may thrive on the lens or in lens cleaning solutions. When the lenses are left in the eyes for long periods of time, the bacteria multiply and damage the cornea. In addition, wearing lenses for extended periods of time blocks oxygen to the cornea, making it more susceptible to infections.

Some symptoms of cornea ulcers are as follows:

- Red eye;
- Severe pain;
- Feeling that something is in the eye;
- Tears;
- Pus or thick discharge draining from the eye;
- Blurry vision;
- Pain when looking at bright lights;
- Swollen eyelids;

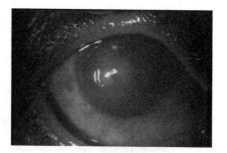

Figure 1.8 Corneal ulcer due to misuse of contact lens.

• A white round spot on the cornea that is visible with the naked eye if the ulcer is very large.

1.4.3 Fuchs' Dystrophy

Fuchs' dystrophy (Figure 1.9) is an inherited disease that affects the cornea. The exact cause is unknown, but research has shown that it results from a DNA coding error. In Fuchs' dystrophy, the cells that line the back surface (endothelium) of the cornea slowly deteriorate. The primary task of the endothelium is to pump and remove excess fluid from the cornea to maintain its transparency. Deterioration of the endothelial cells impairs this function and fluids build up in the cornea. The lost endothelial cells do not grow back; instead they spread out to fill empty spaces. This causes swelling in the corneal and stroma region, which reduces the transparency of the cornea resulting in blurred vision.

The excess fluids that build up in the cornea causes the epithelium to swell, which damages vision in two ways: (1) by changing the cornea's normal curvature, and (2) by causing a sight-impairing *haze* to appear in the tissue [10]. Epithelial swelling will also produce tiny blisters on the corneal surface, which might cause extreme pain when burst.

Signs and symptoms of Fuchs' dystrophy usually appear after age 50. In the early stages of the disease, one wakes up with blurred vision that gradually improves during the day. As the disease progresses, there are longer periods of impaired vision. This happens because the internal layers of the cornea tend to retain more moisture during sleep that evaporates when the eyes are open. As the dystrophy worsens, the vision becomes continuously blurred. There might be pain in the eye caused by the eruption of tiny blisters on the surface of the cornea. Eventually, blindness can result. In addition, a sandy and gritty sensation may be experienced in the eye [9].

1.4.4 Pterygium and Pingueculae

A pterygium (Figure 1.10) is a growth of fibrovascular tissue, that is, fibrous tissue with noticeable blood vessels. It forms on the conjunctiva and spreads across the eye surface, covering the sclera [9]. This triangular growth commonly appears at the corner of the eye closest to the nose, but may also grow from the opposite corner. Occasionally the growth extends as far as the pupil of the eye, obscuring the vision.

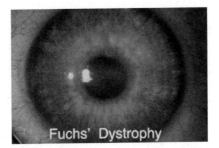

Figure 1.9 Fuchs' dystrophy. (Reproduced with permission from http://www.fuchssupport.info.)

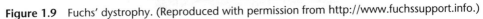

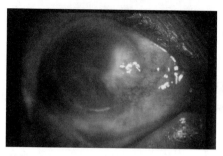

Figure 1.10   Pterygium growths on the conjunctiva.

Alternatively, it may cause the cornea to alter in shape and "warp," resulting in blurred vision.

Pterygia are benign (noncancerous) growths. They are sometimes difficult to distinguish from the less problematic eye condition called pingueculae. Pingueculae do not have the distinctive triangular shape of pterygia, are more yellow in appearance, and do not encroach far across the eye surface. They may, however, be a precursor to pterygium.

Pingueculae are yellowish, slightly raised lesions that form on the surface tissue of the white part of your eye (sclera) close to the edge of the cornea. They are typically found in the open space between your eyelids (palpebral fissure), which happens to be the area exposed to the Sun.

While pingueculae are more common in middle-aged or older people who spend significant amounts of time in the sun, they can also be found in younger people and even children, especially those who spend a lot of time in the sun without protection such as sunglasses or hats. It is a degeneration of normal tissue resulting in a deposit of protein and fat.

Although there may be a hereditary aspect to the development of pterygia, it is thought that chronic exposure to some environmental elements damages the tissue of the conjunctiva and cornea. Ultraviolet light, wind, and dust irritate the eye surface. This leads to the replacement of healthy conjunctival tissue with an overgrowth of thickened fibrous tissue containing blood vessels—a pterygium. The growth is usually progressive and eventually the corneal cells may be affected. The eye surface is roughened and this may interfere with the natural lubrication process of the eye.

Large and advanced pterygia can distort the surface of the cornea and induce astigmatism, create a feeling that there is something inside the eye, blur and obscure vision, and cause eye redness or inflammation on infected areas.

1.4.5   Scleritis

Scleritis is an inflammatory condition that can affect the conjunctiva, sclera, and episclera. The conjunctiva and episclera, however, are bystanders; they show some congestion, but are not affected primarily. The sclera is the white part of the eye. Most patients with scleritis will experience severe and teary eye pain, which may radiate to the forehead, cheek, or behind the eye. This is usually associated with a red eye, light sensitivity, and in some cases, reduced or poor vision. The affected eye often has a light bright hue at resolution of scleritis (Figure 1.11), which is best observed with the naked eye under natural light. However, scleritis can also occur without inflammation, which is indicated by grayish patches on the sclera, but this possibility is usually rare.

About 50% of patients will have an associated systemic autoimmune disease, such as rheumatoid arthritis, systemic lupus erythematosus, Wegener's granulomatosis, polyarteritis nodosa, or ankylosing spondylitis. Other associated disorders include herpes zoster infection (shingles), syphilis, tuberculosis, and gout.

Inflammation of the sclera can progress to ischemia and necrosis, eventually leading to scleral thinning and perforation of the globe. Necrotizing anterior scleritis, in particular, represents a destructive form of scleritis [9, 10].

Scleritis may cause inflammation of the anterior and/or posterior segments of the eye and manifests as severe eye pain. The four types of anterior scleritis are as follows:

- Diffuse anterior scleritis (Figure 1.12) is characterized by widespread inflammation of the anterior portion of the sclera. This is the most common form of anterior scleritis, as well as the most benign and responsive to therapy.
- Nodular anterior scleritis (Figure 1.13) occurs when the anterior sclera becomes erythematous, immovable, with tender and inflamed nodules. Approximately 20% of these cases progress to necrotizing scleritis.
- The inflammation associated with necrotizing anterior scleritis (Figure 1.14) frequently accompanies serious systemic collagen vascular disorders, including rheumatoid arthritis. An associated type of vascular inflammation, called

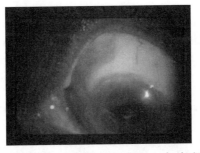

Figure 1.11 A light bright hue is often noticed at resolution of scleritis.

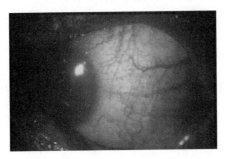

Figure 1.12 Diffuse anterior scleritis.

Figure 1.13 Nodular anterior scleritis.

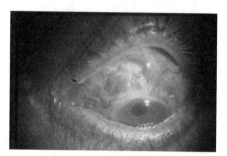

Figure 1.14 Necrotizing scleritis—arrow shows the vascular area with edema.

vasculitis, may threaten the lives of those patients with the disorder. Pain with this condition is usually extreme, and damage to the sclera is often marked.

- Scleromalacia perforans, also known as necrotizing anterior scleritis without inflammation, most frequently occurs in patients with long-standing rheumatoid arthritis and is notable for its absence of symptoms. Scleromalacia perforans is characterized by severe thinning of the sclera of the eye. The condition occurs in an otherwise white and "quiet" eye, without pain.

Posterior scleritis occurs much less frequently than anterior scleritis, but it may extend into the anterior segment of the eye. Its symptoms are poor or double vision, severe pain, proptosis (forward displacement of the eye), uveitis (inflammation inside the eye), and limitation of eye movement. An exudative retinal detachment (fluid under the retina) may occur, causing severe visual loss.

Scleritis affects women more frequently than men. It most frequently occurs in those who are in their forties and fifties. The problem is usually confined to one eye, but may affect both.

Some of the signs and symptoms that appear include the following:

- Severe, boring pain that wakes the patient from sleep;
- Local or general redness of the sclera and conjunctiva;
- Extreme tenderness to the eye;
- Light sensitivity and tearing (in some cases);
- Decreased vision (if other ocular tissues are involved).

1.4.6   Uveitis

Uveitis is defined as the inflammation of one or all parts of the uveal tract. Components of the uveal tract include the iris, the ciliary body, and the choroid. Uveitis may involve all areas of the uveal tract (pan-uveitis); however, involvement most often is subcategorized into one of three main diagnoses: iritis or anterior uveitis (anterior, confined to the iris and the anterior chamber), iridocyclitis or intermediate uveitis (confined to the iris, the anterior chamber, and the ciliary body), and choroiditis or posterior uveitis (Figure 1.15).

The most commonly seen form of uveitis is the anterior uveitis, or iritis. An entire class of uveitis is referred to as *idiopathic*, which simply means that the causes are unknown or cannot be related to other systemic disorders at the point of evaluation. However, uveitis may result from association with autoimmune disorders such as rheumatoid arthritis or ankylosing spondylitis and other diseases such as Reiter syndrome, sarcoidosis, and psoriasis. The importance of genetics in the role of causing uveitis remains uncertain.

Direct trauma to the eye in the case of iritis can cause the white blood cells to shed into the anterior chamber of the eye, causing accumulation of white blood cells and, ultimately, adhesion between components in the uveal tract, particularly the iris and the lens. Infectious causes can also come from viruses such as herpes, certain parasites such as toxoplasmosis, and even funguses.

The following is a list of the different disorders that might cause uveitis:

- Acute posterior multifocal placoid pigment epitheliopathy;
- Behçet's disease;
- Birdshot retinochoroidopathy;
- Brucellosis;
- Herpes simplex;
- Herpes zoster;
- Inflammatory bowel disease;
- Juvenile rheumatoid arthritis;
- Kawasaki's disease;
- Leptospirosis;
- Lyme disease;
- Multiple sclerosis;

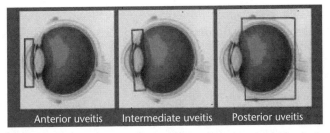

Figure 1.15   Classification of uveitis. (Reproduced with permission from Ocular Immunology and Uveitis Foundation, Massachusetts Eye Research and Surgery Institute, Cambridge, Massachusetts, http://www.uveitis.org/patient/articles/articles/background.html.)

- Presumed ocular histoplasmosis syndrome;
- Psoriatic arthritis;
- Systemic lupus erythematosus;
- Toxocariasis;
- Toxoplasmosis;
- Tuberculosis.

Anterior uveitis, or iritis, usually occurs in only one eye, with symptoms such as blurred vision; a small and constricted pupil due to increased intraocular pressure caused by the adhesion of the iris and lens; slight pain and a reddened eye, particularly the part adjacent to the iris; and sensitivity to light, which occurs mostly in acute cases. Acute iritis normally heals independently within a short period of time, usually a few weeks if enhanced with treatments. Chronic iritis can persist for months and even up to years before full recovery is achieved due to the fact that treatments for the condition are not well received, thus it may cause serious visual injury to the eye.

Intermediate uveitis, or iridocyclitis, is diagnosed when inflammation involves the anterior vitreous, peripheral retina, or the pars plana ciliaris (part of the ciliary body), and it can persist for a very long period of time [12]. It has been often related to diseases such as Lyme disease and multiple sclerosis [13, 14]. Many names have been used to describe this entity, such as chronic cyclitis, vitritis, peripheral uveitis, but only pars planitis has been preserved and applied. Pars planitis is considered a subset of intermediate uveitis because it describes the presence of white collagen exudates (snowbanks) or aggregates of inflammatory cells (snowballs) over the pars plana.

Patients with intermediate uveitis usually encounter blurry vision and floaters, with a few exceptional cases feeling pain and sensitivity to light. Severe complications of intermediate uveitis can lead to ocular hypertension and eventually glaucoma. Furthermore, cataracts develop in most patients. Also, they become more prone to posterior vitreous and retinal detachments [15].

Posterior uveitis is inflammation of the choroid. If the adjacent retina is involved, then it is known as choroiditis or chorioretinitis. Posterior uevitis is hard to detect because it is not accompanied by pain, and floaters would commonly be seen and vision become impaired suddenly or it might reduce gradually. Uveitis can be acute or chronic, with acute being the more commonly observed type. However, many cases of uveitis are chronic, and they can produce numerous possible complications, including clouding of the lens (cataract) or cornea, elevated intraocular pressure (IOP), glaucoma, and retinal problems (such as swelling of the retina or retinal detachment). These complications can result in vision loss.

1.4.7 Corneal Haze

Corneal haze (Figure 1.16) is a condition that can develop after corneal surface ablation procedures for refractive corrections such as excimer laser photorefractive keratectomy (PRK) or laser in situ keratomileusis (LASIK). In PRK, for example, the excimer laser creates a smooth stromal wound with minimal collateral damage. However, the cornea responds to the wound by reepithelialization through the

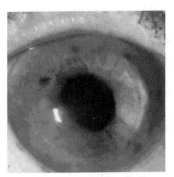

Figure 1.16 Corneal haze.

migration of adjacent epithelial cells over a newly formed substratum rich in fibronectin and fibrin [16]. This postoperative corneal wound healing response may cause a subepithelial haze to form, especially on patients with sensitive conjunctivitis. Histopathological studies revealed that subepithelial haze contains newly synthesized collagen such as type III collagen, type IV collagen, fibronectin, laminin, and proteoglycans [17].

These synthesized collagens diffuse in a random manner, and present as a white haze, which decreases the opacity of the cornea, restricting light from passing through and thus reducing vision clarity. The severity of the corneal haze corresponds to the aggressiveness of the healing procedure, in which the worst case causes corneal scars, which leads to haze formation, directly affecting the visual acuity and contrast sensitivity. Due to the haze, some patients may complain of glare or disturbance of contrast visual functions [17], vision distortion or fluctuation, and even development of irregular astigmatism.

In a commonly applied postoperative measure to prevent formation of the corneal haze, eyedrops containing corneal scar depressants such as local corticosteroids are administered to reduce scarring and increase corneal clarity [18–20], but this does not serve as a long-term solution. Because the haze is a natural healing process, it would subside gradually as the cornea epithelial cells heal themselves. However, excessive or persistent haze formation may result in permanent blurring of vision, which may result in additional laser ablation treatments.

1.4.8 Cataracts

A cataract is a clouding of the natural lens, the part of the eye responsible for focusing light and producing clear, sharp images. The lens is made mostly of water and protein fibers. The protein fibers are arranged in a precise manner that makes the lens clear and allows light to pass through without interference. With aging, the composition of the lens undergoes changes and the structure of the protein fibers breaks down. Some of the fibers begin to clump together, clouding small areas of the lens. As the cataract continues to develop, the clouding becomes denser and involves a greater part of the lens [9–11]. Age causes the lenses in the eyes to become less flexible, less transparent, and thicker.

A cataract can develop in one or both eyes. However, in most cases—except for those caused by injury or trauma—cataracts tend to develop symmetrically in both eyes. A cataract may or may not affect the entire lens. The lens is located just behind the iris and the pupil. It has a globular shape that is thicker in the middle and thinner near the edges and is held in place by bands of tough zonule fibers that connect the ciliary body to the lens.

When light passes through the cornea and the pupil to the lens, the lens focuses this light, producing clear, sharp images on the retina; the light-sensitive membrane on the back wall of the eyeball functions like the film of a camera. As a cataract develops (Figure 1.17), the lens becomes clouded; this scatters the light and prevents a sharply defined image from reaching your retina.

The lens consists of three layers. The outer layer (capsule) is a thin, clear membrane. It surrounds a soft, clear material (cortex). The hard center of the lens is the nucleus. Cataracts can form in any part of the lens.

1.4.8.1 Types of Cataracts by Location of Opacity within the Lens Structure

Nuclear

A nuclear cataract (Figure 1.18) occurs in the center of the lens. In its early stages, as the lens changes the way it focuses light, one may become more nearsighted or even experience a temporary improvement in reading vision. Some people actually stop needing their glasses. Unfortunately, this so-called second sight disappears as the lens gradually turns more densely yellow and further clouds your vision.

As the cataract progresses, the lens may even turn brown. Seeing in dim light and driving at night may be difficult. Advanced discoloration may lead to the inability to distinguish between shades of blue and purple [9–11].

Cortical

A cortical cataract (Figure 1.19) begins as whitish, wedge-shaped opacities or streaks on the outer edge of the lens cortex. As it slowly progresses, the streaks extend to the center and interfere with light passing through the center of the lens. Both distance and near vision can be impaired. Focusing problems and distortion are common, together with problems of glare and loss of contrast.

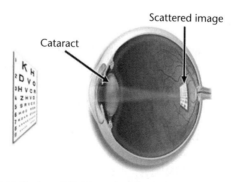

Figure 1.17 Effect of the scattering of light passing through the lens. (Reproduced with permission from St. Luke's Cataract and Laser Institute, http://www.stlukeseye.com.)

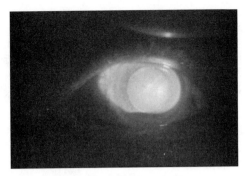

Figure 1.18 Nuclear cataract as seen in a human eye.

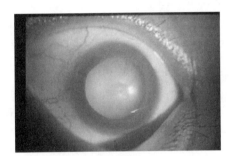

Figure 1.19 Anterior cortical cataract as seen in a human eye.

Subcapsular

A subcapsular cataract (Figure 1.20) starts as a small, opaque area just under the capsule of the lens. It usually forms near the back of the lens, closer to the posterior part of the eye, and it interferes with the path of light on its way to the retina. This type of cataract may occur in both eyes, but tends to be more advanced in one eye than the other. A subcapsular cataract often interferes with reading vision, reduces vision in bright light, and causes glare or halos around lights at night.

Lenses change with age for several possible reasons. One possibility is the damage caused by unstable atoms or molecules known as free radicals. Smoking and exposure to UV light are two primary sources of free radicals. General wear and tear on the lens over the years also may cause the changes in protein fibers.

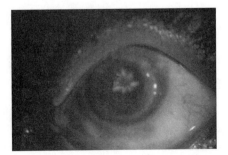

Figure 1.20 Anterior subcapsular cataract as seen in a human eye.

Age-related changes in the lens are not the only cause of cataracts. Some people are born with cataracts or develop them during childhood. Such cataracts may be the result of the mother contracting German measles (rubella) during pregnancy or due to metabolic disorders. Congenital cataracts, as they are called, do not always affect vision, but if they do they are usually removed soon after detection.

1.5 Posterior Segment Abnormalities

The following sections cover different types of retinal disorders: diabetes retinopathy, macular degeneration, retinal detachment, floaters, and glaucoma.

1.5.1 Diabetic Retinopathy

Diabetic retinopathy is a condition caused by complications of diabetes mellitus (DM). DM is also the main cause for several other diseases such as glaucoma and cataract. Diabetic retinopathy can ultimately lead to blindness. It is caused by destructive changes in the blood vessels of the retina. After 10 or 15 years, most people have signs of mild damage to the back of the eye called retinopathy. Diabetes causes damage to the blood vessels that nourish the retina [21–24].

The severity of the damage to the retina is highly correlated with the length of time the patient has had diabetes. In people with diabetes the retinal blood vessels may expand and leak fluid. In addition, abnormal new blood vessels may grow; these blood vessels may break and cause bleeding.

For some profound reasons that are not well identified, the blood vessels of the retina lack their normal oxygen load. Capillaries tend to close off, further depleting the oxygen supply. DM also weakens the walls of these blood vessels, which tend to become enlarged and form microaneurysms. Often, the small blood vessels break, causing hemorrhage and contributing to the patient's clouded vision. These changes may result in vision loss or blindness. In the later phases of the disease, continued abnormal vessel growth and scar tissue may cause serious problems such as retinal detachment and glaucoma.

The earliest phases (Figure 1.21) of the disease are known as background (nonproliferative—mild, moderate, and severe) diabetic retinopathy. In this phase, the arteries in the retina become weakened and leak, forming small, dot-like hemorrhages [21–24]. These leaking vessels often lead to swelling or edema in the retina and decreased vision.

The next stage is known as proliferative diabetic retinopathy. In this stage, circulation problems cause areas of the retina to become oxygen deprived or ischemic. New, fragile vessels develop as the circulatory system attempts to maintain adequate oxygen levels within the retina. This is called neovascularization. Unfortunately, these delicate vessels hemorrhage easily. Blood may leak into the retina and vitreous, causing spots or floaters, along with decreased vision.

Diabetic retinopathy has four stages:

1. *Mild nonproliferative retinopathy* (Figure 1.22). At this earliest stage, microaneurysms occur. They are small areas of balloon-like swelling in the retina's tiny blood vessels.

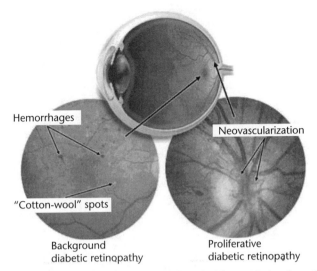

Hemorrhages

Neovascularization

"Cotton-wool" spots

Background
diabetic retinopathy

Proliferative
diabetic retinopathy

Figure 1.21 Stages of diabetic retinopathy. (Reproduced with permission from St. Luke's Cataract and Laser Institute, http://www.stlukeseye.com.)

2. *Moderate nonproliferative retinopathy* (Figure 1.23). As the disease progresses, some blood vessels that nourish the retina are blocked.

3. *Severe nonproliferative retinopathy* (Figure 1.24). Many more blood vessels are blocked, depriving several areas of the retina with their blood supply. These areas of the retina send signals to the body to grow new blood vessels for nourishment.

4. *Proliferative retinopathy* (Figures 1.25 and 1.26). At this advanced stage, the signals sent by the retina for nourishment trigger the growth of new blood vessels. This condition is called proliferative retinopathy. These new blood vessels are abnormal and fragile. They grow along the retina and along the surface of the clear, vitreous gel that fills the inside of the eye [1–3]. By themselves, these blood vessels do not cause symptoms or vision loss. However, they have thin, fragile walls. If they leak blood into the retina and vitreous, they can cause spots or floaters along with severe vision loss and even blindness [21–24].

With time, if the background diabetic retinopathy becomes more severe, the macula area may become involved. This is called maculopathy. In diabetic maculopathy (Figure 1.27), fluid rich in fat and cholesterol leaks out of damaged vessels. If the fluid accumulates near the center of the retina (the macula), central vision will become distorted. If too much fluid and cholesterol accumulate in the macula, permanent loss of central vision can occur.

1.5.2 Macular Degeneration

The retina's central portion, known as the macula, is responsible for focusing central vision in the eye, and it controls the ability to read, recognize faces or colors, and see objects in fine detail. The macula is made up of densely packed, light-sensitive cells called cones and rods. These cells, particularly the cones, are essential

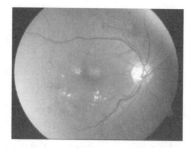

Figure 1.22 Mild nonproliferative diabetic retinopathy, showing microaneurysms and dot hemorrhages. Also demonstrates macular edema with a small amount of lipid exudate—not clinically significant.

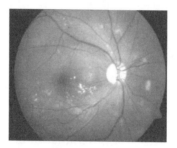

Figure 1.23 Moderate nonproliferative diabetic retinopathy, showing cotton wool spots, retinal hemorrhages, and microaneurysms.

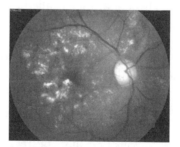

Figure 1.24 Severe nonproliferative diabetes retinopathy.

for central vision. The cones are responsible for color vision, and the rods are to see shades of gray.

Degenerative processes involving the choroid/retina interface in the macular area often manifest as age-related macular degeneration (AMD) [6]. AMD is caused by the deterioration of the central portion of the retina, the back layer of the eye that records the images we see and sends them via the optic nerve from the eye to the brain.

The choroid is an underlying layer of blood vessels that nourishes the cones and rods of the retina. A layer of tissue forming the outermost surface of the retina is called the retinal pigment epithelium (RPE). The RPE is a critical passageway for

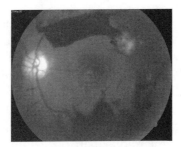

Figure 1.25   Proliferative diabetic retinopathy, showing peripheral new vessel, retinal hemorrhages, and no vitreous or preretinal hemorrhage. Note lack of other retinopathy features.

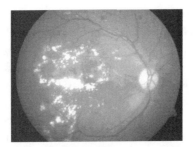

Figure 1.26   Proliferative diabetic retinopathy, showing subhyaloid hemorrhages. Intraretinal microvasular abnormalities, or IRMAs, can also be seen.

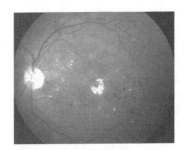

Figure 1.27   Diabetic maculopathy.

nutrients from the choroid to the retina and helps remove waste products from the retina to the choroid.

Age deteriorates the RPE to the point where it loses its pigment and becomes thin (a process known as atrophy), which sets off a chain of events. The nutritional and waste-removing cycles between the retina and the choroid are interrupted; thus, fatty waste deposits from the photoreceptors form on the macula. Lacking nutrients, the light-sensitive cells of the macula become damaged. The damaged cells can no longer send normal signals through the optic nerve to the brain, and vision becomes blurred. This is often the first symptom of macular degeneration.

Damaged parts of the macula often cause scotomas, or localized areas of vision loss. Macular degeneration varies widely in severity. In the worst cases, it causes a complete loss of central vision, making reading or driving impossible. For others, it may only cause slight distortion. Fortunately, macular degeneration does not cause total blindness, because it does not affect peripheral vision.

The two basic types of macular degeneration are referred to as "dry" or "wet." Approximately 85% to 90% of the cases of macular degeneration are the dry (atrophic) type. Dry AMD does not involve any leakage of blood or serum. Loss of vision may still occur. Patients with this dry form may have good central vision (20/40 or better), but substantial functional limitations, including fluctuating vision, difficulty in reading because of their limited area of central vision, and limited vision at night or under conditions of reduced illumination.

1.5.2.1 Dry AMD

In the "dry" type of macular degeneration, the deterioration of the retina is associated with the formation of small yellow deposits (basal laminar deposits), known as drusen, under the macula (Figure 1.28). This phenomenon leads to thinning and drying out of the macula, which causes the macula to lose its function. The amount of central vision loss is directly related to the location and amount of retinal thinning caused by the drusen.

The early stage of age-related macular degeneration is associated with minimal visual impairment and is characterized by large drusen and pigmentary abnormalities in the macula. Drusen are accumulations of acellular, amorphous debris subjacent to the basement membrane of the retinal pigment epithelium. Nearly all people over the age of 50 years have at least one small druse in one or both eyes. Only eyes with large drusen are at risk for late age-related macular degeneration.

This form of macular degeneration is much more common than the wet type of macular degeneration, and it tends to progress more slowly than the wet type. However, a certain percentage of the dry type of macular degeneration turns to wet with the passage of time. There is no known cure for the dry type of macular degeneration.

Dry AMD has three stages, all of which may occur in one or both eyes:

- *Early AMD*. People with early AMD have either several small drusen or a few medium-sized drusen. At this stage, there are no symptoms and no vision loss.

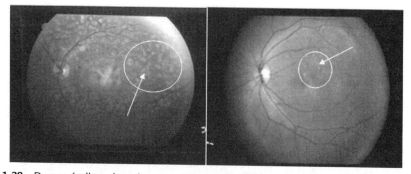

Figure 1.28 Drusen (yellow deposits or proteins and lipids) form under the macula, thinning and drying out the macula.

- *Intermediate AMD.* People with intermediate AMD have either many medium-sized drusen or one or more large drusen. Some people see a blurred spot in the center of their vision. More light may be needed for reading and other tasks.
- *Advanced dry AMD.* In addition to drusen, people with advanced dry AMD have a breakdown of light-sensitive cells and supporting tissue in the central retinal area. This breakdown can cause a blurred spot in the center of your vision. Over time, the blurred spot may get bigger and darker, taking more of your central vision. A person may have difficulty reading or recognizing faces until they are very close to him or her.

1.5.2.2 Wet AMD

Approximately 10% to 15% of all cases of macular degeneration are the wet (exudative) type. In the wet type of macular degeneration, abnormal blood vessels (known as choroidal neovascularization) grow under the retina and macula. These new blood vessels may then bleed and leak fluid, causing the macula to bulge or lift up, thus distorting or destroying central vision. Under these circumstances, vision loss may be rapid and severe.

With the wet type of AMD, the patient may see a dark spot (or spots) in the center of his or her vision due to blood or fluid under the macula. Straight lines may look wavy because the macula is no longer smooth. Side or "peripheral" vision is rarely affected. However, some patients do not notice any such changes, despite the onset of neovascularization.

In wet macular degeneration, new, weak blood vessels may grow in or under the retina (Figure 1.29) causing fluid and blood to leak into the space under the macula. As a result, wet macular degeneration is sometimes called exudative macular degeneration. (An *exudate* is material, such as fluid, that has escaped from blood vessels and has been deposited in tissues.)

Wet forms of macular degeneration are further classified into two general types:

- *Classic.* When blood vessel growth and scarring exhibit very clear, delineated outlines beneath the retina, this type of wet AMD is known as classic

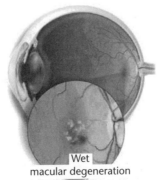

Wet
macular degeneration

Figure 1.29 In wet AMD, new blood vessels develop under the retina. This causes hemorrhage, swelling, and scar tissue. (Reproduced with permission from St. Luke's Cataract and Laser Institute, http://www.stlukeseye.com.)

choroidal neovascularization (CNV) and is usually associated with more severe vision loss.

- *Occult.* New blood vessel growth beneath the retina is not as pronounced and leakage is less evident in the occult CNV form of wet macular degeneration, which typically produces less severe vision loss.

Another form of wet macular degeneration, called retinal pigment epithelial detachment, occurs when fluid leaks from the choroid under the RPE even though it appears that no abnormal blood vessels have started to grow. The fluid collects under the retinal pigment epithelium, causing what looks like a blister or a bump under the macula.

This kind of macular degeneration causes similar symptoms to typical wet macular degeneration, but vision can remain relatively stable for many months or even years before it deteriorates. Eventually, however, this form of macular degeneration usually progresses to the more common wet form of macular degeneration, which includes newly growing abnormal blood vessels.

Macular degeneration usually develops gradually and painlessly. The signs and symptoms of the disease may vary, depending on which of the two types of macular degeneration one has—dry or wet—as discussed next.

Dry Macular Degeneration

With dry macular degeneration, the following symptoms may appear:

- The need for increasingly bright illumination when reading or doing close work;
- Increasing difficulty adapting to low levels of illumination, such as when entering a dimly lit restaurant;
- Increasing blurriness of printed words;
- A decrease in the intensity or brightness of colors;
- Difficulty recognizing faces;
- Gradual increase in the haziness of your overall vision;
- Blurred or blind spot in the center of your visual field combined with a profound drop in your central vision acuity.

Wet Macular Degeneration

With wet macular degeneration, the following symptoms may appear, and they may progress rapidly:

- Visual distortions, such as straight lines appearing wavy or crooked, a doorway or street sign that seems out of whack, or objects appearing smaller or farther away than normal;
- A decrease in or loss of central vision;
- A central blurry spot.

In either form of macular degeneration, vision may falter in one eye while the other remains fine for years. As the good eye compensates for the weak eye, one may not notice any changes. It is only when this condition develops in both eyes that vision and lifestyle begin to be dramatically affected [6].

In addition, some people with macular degeneration may experience visual hallucinations with increasing visual loss. These hallucinations may include unusual patterns, geometric figures, animals, or even grotesque-appearing faces. Many people who develop these symptoms are afraid to discuss them with their doctors or friends and families because others cannot relate to the situation. However, although these hallucinations may be frightening, they are not a sign of mental illness.

1.5.3 Retinal Detachment

Retinal detachments often develop in eyes with retinas weakened by a hole or tear. This allows fluid to seep underneath, weakening the attachment so that the retina becomes detached, rather like wallpaper peeling off a damp wall (Figure 1.30). When detached, the retina cannot compose a clear picture from the incoming rays, and vision becomes blurred and dim [11].

Retinal detachment (Figure 1.31) may be caused by trauma, advanced diabetes, or an inflammatory disorder (retinal thinning and degenerations). But it often occurs spontaneously, as a result of changes in the jelly-like vitreous that fills the vitreous cavity of the eye.

As one ages, the vitreous may change in consistency and partially liquefy or shrink. Eventually, the vitreous may sag and separate from the surface of the retina, a common condition called posterior vitreous detachment (PVD), or vitreous collapse. This occurs to some extent in most people's eyes as they age.

PVD usually does not cause serious problems, but it can cause visual symptoms. If the vitreous pulls on the retina as it shifts and sags, patients may see flashes of sparkling lights (photopsia) when their eyes are closed or in a darkened room. The shifting or sagging vitreous may also cause the appearance of new or different floaters in the field of vision. These spots, specks, hairs, and strings are actually small clumps of gel, fibers, and cells floating in the vitreous.

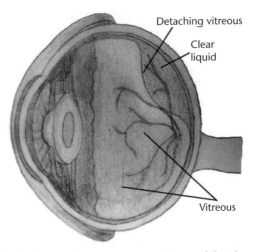

Detaching vitreous

Clear liquid

Vitreous

Figure 1.30 Changes in the vitreous humor result in shrinking of the vitreous chamber.

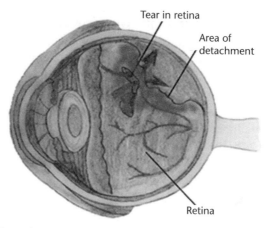

Tear in retina

Area of detachment

Retina

Figure 1.31   Tear in the retina.

Common floaters appear gradually over time and, although they are annoying, they are rarely a problem and hardly ever require treatment. However, the sudden onset of floaters can signal the development of a retinal tear, particularly when accompanied by flashes of light. This occurs when the pulling of the sagging vitreous becomes strong enough to tear the retina, leaving what looks like a small, jagged flap.

Retinal detachment (see Figure 1.31) occurs when vitreous liquid (vitreous humor) leaks through the tear and accumulates underneath the retina. Leakage can also occur through tiny holes where the retina has thinned due to aging or other retinal disorders. Less commonly, fluid can leak directly underneath the retina, without a tear or break.

As liquid collects, areas of the retina can peel away from the underlying choroids (pigment epithelium). Over time these detached areas may expand, like wallpaper that, once torn, slowly peels off a wall. The areas where the retina is detached lose their ability to see.

Most retinal tears caused by PVD lead to retinal detachment if left untreated. Detachments that go undetected and untreated can progress and eventually involve the entire retina, causing complete loss of vision.

The three types of retinal detachment are as follows:

- *Rhegmatogenous retinal detachment* happens when the detachment occurs due to a hole, tear, or break in the retina that allows fluid to pass into the subretinal space between the sensory retina and the retinal pigment epithelium.
- *Exudative, serous,* or *secondary retinal detachment* occurs due to inflammation, injury, or vascular abnormalities that result in fluid accumulating underneath the retina without the presence of a hole, tear, or break.
- *Tractional retinal detachment* occurs when the fibrovascular tissue (fibrous tissue with noticeable blood vessels), caused by an injury, inflammation, or neovascularization, pulls the sensory retina from the retinal pigment epithelium.

The early symptoms of a detachment may include flashes of light and the sudden appearance of floaters. Many people have floaters, which have been present for many years and are not a cause for concern. If a retinal detachment occurs, a visual field defect can usually be noticed as a "curtain" or "shadow" appearing in the peripheral vision and such a symptom should be taken seriously. Other signs and symptoms include these:

- Light flashes;
- "Wavy" or "watery" vision;
- Veil or curtain obstructing vision;
- The sudden appearance of many floaters—small bits of debris in the field of vision that look like spots, hairs, or strings and seem to float before your eyes;
- Sudden decrease of vision.

1.5.4   Floaters

The "dots" that one may sometimes see moving in the field of vision are called *floaters*. They are frequently visible when looking at a plain background (Figure 1.32), such as a blue sky. Floaters are actually tiny clumps of gel or cellular debris inside the vitreous chamber. They are quite common and are more likely to develop as the eyes get older.

As the vitreous humor gets older, strands of collagen start to denature and they become visible within it. These strands swirl gently when the eye moves. Rather than being solid blobs, floaters are actually shadows cast on the retina by these pieces of collagen (Figure 1.33). This is because light travels through the vitreous in order to reach the retina, so any objects in the vitreous reflect (shadow) on the retina. When the vitreous humor starts to shrink away from the retina because of aging, it thickens and clumps together, leading it to create floaters. This is the aforementioned posterior vitreous detachment (see retinal detachment discussion). The debris left at the site (not at the site of separation) where the vitreous separates from the retina becomes floaters.

Sometimes, as the vitreous humor pulls on the retina, it causes tiny blood vessels in the retina to burst, and slight bleeding occurs in the vitreous. Red blood cells in the blood appear as tiny black dots, or may look like a swarm of gnats, or like smoke. As the blood is reabsorbed, these sorts of floaters generally go away, although it can take a few months.

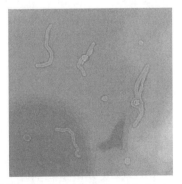

Figure 1.32   Floaters can be visible when looking at a plain background such as a blue sky. (Reproduced with permission from Wikipedia, http://en.wikipedia.org/wiki/Floater.)

Floating cells

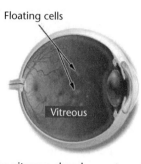

Vitreous

Figure 1.33 Floating particles in the vitreous chamber cast a reflection (shadow) on the retina. (Reproduced with permission from St. Luke's Cataract and Laser Institute, http://www.stlukeseye. com.)

Floaters by themselves are not dangerous, but when a floater develops it can sometimes cause a tear in the retina, which can later develop into a retinal detachment, usually indicated by a sudden onset of a new floater (shower of floaters) accompanied by flashing lights.

1.5.5 Glaucoma

Glaucoma is caused by a number of different eye diseases that, in most cases, produce an increase in pressure within the eye. This elevated pressure is caused by a buildup of fluid in the eye. Over time, it causes damage to the optic nerve. Through early detection, diagnosis, and treatment, vision can be preserved.

The eye is sphere shaped, divided into its anterior and posterior chambers (Figure 1.34). The aqueous humor is the liquid that fills and constantly circulates through both chambers. It is produced by a tiny gland, called the ciliary body, situated behind the iris. It flows between the iris and the lens and, after nourishing the cornea and lens, drains out into the venous circulation through a very tiny spongy tissue maze, only about 0.5 mm wide, called the trabecular meshwork. The trabecular meshwork is situated in the angle where the iris and cornea meet. Deficiencies of this outflow system imply that the aqueous humor cannot leave the eye as fast as it is produced, causing the fluid to build up. Because the eye is a closed compartment, the backed-up fluid causes increased intraocular pressure to build up within the eye. This is called open-angle, or wide-angle, glaucoma.

The eye is a very strong structure and could withstand this increasing pressure. However, when the pressure reaches the maximum amount the eye can withstand, it gives in at the weakest point, which is the site at the sclera where the optic nerve leaves the eye.

As mentioned earlier, the optic nerve is the part of the eye that carries visual information to the brain. It is made up of more than 1 million nerve cells and, although each cell is several centimeters long, they are extremely thin—about one 0.001 mm in diameter. When the pressure in the eye builds, the nerve cells become compressed, causing them to become damaged and eventually die. The death of these cells results in permanent visual loss. Early diagnosis and treatment of glaucoma can help prevent this from happening.

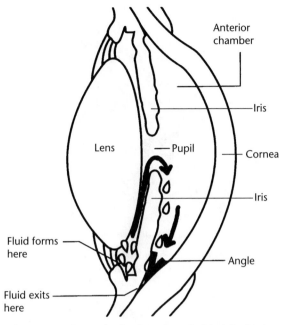

Figure 1.34 Aqueous humor continuously circulates from behind the iris into the anterior chamber. It exits the eye where the iris and the cornea meet. The fluid filters through the trabecular meshwork before passing into an open channel called Schlemm's canal. (Reproduced with permission from National Eye Institute, National Institutes of Health, http://www.nei.nih.gov/health.)

1.5.5.1   Types of Glaucoma

Primary Open-Angle Glaucoma

This form, also called chronic open-angle glaucoma, accounts for most cases of the disease [9, 10]. Although the drainage angle formed by the cornea and the iris remains open (Figure 1.35), the aqueous humor drains too slowly. This leads to fluid buildup and a gradual increase of pressure within the eye. Damage to the optic nerve is so slow and painless that a large portion of vision can be lost before one is even aware of the problem.

The exact cause of primary open-angle glaucoma remains unknown. It may be that the aqueous humor drains or is absorbed less efficiently with age, but not all older adults get this form of glaucoma. About 2% of Americans older than age 40 have elevated eye pressure. For Americans older than 70, the number is 8%.

Angle-Closure Glaucoma

Angle-closure glaucoma, also called closed-angle glaucoma, is a less common form of the disease. This type of glaucoma is a medical emergency that can cause vision loss within a day of its onset.

It occurs when the drainage angle formed by the cornea and the iris closes or becomes blocked (Figure 1.36). Many people with this type of glaucoma have a very narrow drainage angle, which may be an abnormality from birth. As one gets older, his o her lens becomes larger, pushing the iris forward and narrowing the space between the iris and the cornea.

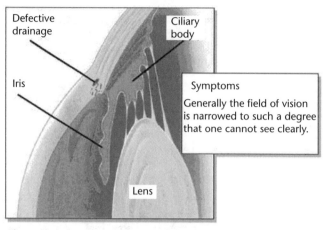

The angle is open but drainage is defective.

Figure 1.35   Blockage of the trabecular meshwork slows drainage of the aqueous humor, which increases intraocular pressure. (Reproduced with permission from the Pretoria Eye Institute, http://www.eyeinstitute.co.za/b_surgery_Glaucoma.asp.)

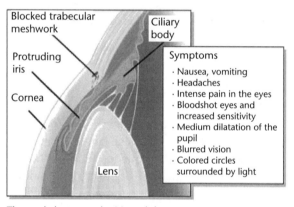

The angle between the iris and the cornea narrows or closes, blocking the drainage of the aqueous humor.

Figure 1.36   The angle formed by the cornea and the iris narrows, preventing the aqueous humor from draining normally out of the eye. This can lead to a rapid increase in intraocular pressure. (Reproduced with permission from the Pretoria Eye Institute, http://www.eyeinstitute.co.za/b_surgery_Glaucoma.asp.)

Whether the narrow drainage angle is an abnormality from birth or a result of aging, as this angle gets more narrow, the iris gets closer to the trabecular meshwork. If it gets too close, the aqueous humor cannot exit through the trabecular meshwork, resulting in a buildup of fluid and an increase in eye pressure.

Angle-closure glaucoma can be chronic (progressing gradually) or acute (appearing suddenly). The acute form occurs when the iris is forced up against the trabecular meshwork and completely blocks the drainage of the aqueous humor.

Angle-closure glaucoma is more common among farsighted people. Normal aging also may cause angle blockage.

For one with a narrow drainage angle, when the pupils dilate, the angle may close and cause a sudden increase in eye pressure. This attack of acute angle-closure glaucoma requires immediate treatment. Although an acute attack often affects only one eye, the other eye is at risk of an attack as well.

Several factors can cause the pupils to dilate; this includes darkness or dim light, stress or excitement, and certain medications. Medications, such as antihistamines, tricyclic antidepressants, and eyedrops can dilate the pupils. However, dilating eyedrops may not cause the angle to close until several hours after the drops are put into the eyes.

Secondary

Both open-angle and angle-closure glaucoma can be primary or secondary conditions. There are primary conditions when the cause of the condition is unknown, or secondary when the condition can be traced to a known cause, such as an injury or an eye disease. Secondary glaucoma may be caused by a variety of medical conditions, medications, physical injuries, and eye abnormalities or deformities. Infrequently, eye surgery can be associated with secondary glaucoma.

Low-Tension Glaucoma

Low-tension glaucoma is a poorly understood, though not uncommon, form of the disease. In this form, eye pressure remains within what is ordinarily thought to be the normal range, but the optic nerve is damaged nevertheless. The cause is still unknown.

Low-tension glaucoma may have an abnormally sensitive optic nerve or a reduced blood supply to the optic nerve caused by a condition such as atherosclerosis, a hardening of the arteries caused by accumulation of fatty deposits (plaques) and other substances. Under these circumstances even normal pressure on the optic nerve is enough to cause damage.

References

[1] Flammer, J., et al, "The Impact of Ocular Blood Flow in Glaucoma," *Prog. Retinal Eye Res.*, Vol. 21, No. 4, 2002, pp. 359–393.

[2] Warwick, R., *Eugene Wolff's Anatomy of the Eye and Orbit*, 7th ed., Philadelphia, PA: W. B. Saunders, 1976.

[3] Kaufman, P. L., and A. Alm, *Adler's Physiology of the Eye: Clinical Application*, 10th ed., St. Louis, MO: Mosby, 2003.

[4] van Buskirk, E. M., "The Anatomy of the Limbus," *Eye*, Vol. 3, No. 2, 1989, p. 101.

[5] Millar, C., and P. L. Kauffman, "Aqueous Humor: Secretion and Dynamics," in *Duane's Foundations of Clinical Ophthalmology*, W. Tasman and E. A. Jaeger, (eds.), Philadelphia, PA: Lippincott-Raven, 1995.

[6] Dubbelman, M., G. L. van der Heijde, and H. A. Weeber, "The Thickness of the Aging Human Lens Obtained from Corrected Scheimpflug Images," *Optometry Vis. Sci.*, Vol. 78, No. 6, 2001, pp. 411–416.

[7] Brown, N., "The Change in Lens Curvature with Age," *Experimental Eye Res.*, Vol. 19, No. 2, 1974, pp. 175–183.

[8]   Siebinga, I., et al., "Age-Related Changes in Local Water and Protein Content of Human Eye Lenses Measured by Raman Microspectroscopy," *Experimental Eye Res.*, Vol. 53, No. 2, 1991, pp. 233–239.

[9]   Oksala, A., "Ultrasonic Findings in the Vitreous at Various Ages," *Graefes Arch. Clin. Exper. Ophthalmol.*, Vol. 207, No. 4, 1978, pp. 275–280.

[10] Remington, L.A., *Clinical Anatomy of the Visual System,* Boston, MA: Butterworth-Heinemann, 1998.

[11] Snell, S. R., and A. M. Lemp, *Clinical Anatomy of the Eye,* 2nd ed., New York: Blackwell Science, 1998.

[12] Bloch-Michel, E., "Opening Address: Intermediate Uveitis," *Dev. Ophthalmol.,* Vol. 23, 1992, pp. 1–2.

[13] Zierhut, M., and C. S. Foster, "Multiple Sclerosis, Sarcoidosis and Other Diseases in Patients with Pars Planitis," *Dev. Ophthalmol.,* Vol. 23, 1992, pp. 41–47.

[14] Breeveld, J., A. Rothova, and H. Kuiper, "Intermediate Uveitis and Lyme Borreliosis," *Br. J. Ophthalmol.,* Vol. 77, 1993, pp. 480–481.

[15] Smith, R. E., W. A. Godfrey, and S. J. Kimura, "Complications of Chronic Cyclitis," *Am. J. Ophthalmol.,* Vol. 82, 1976, p. 277.

[16] Angela, F., et al., "Persistent Corneal Haze After Excimer Laser Photokeratectomy in Plasminogen-Deficient Mice," *Invest. Ophthalmol. Vis. Sci.,* Vol. 41, 2000, pp. 67–72.

[17] Nakamura, K., et al., "Intact Corneal Epithelium Is Essential for the Prevention of Stromal Haze After Laser Assisted *In Situ* Keratomileusis," *Br. J. Ophthalmol.,* Vol. 85, 2001, pp. 209–213.

[18] Gartry, D.S., et al., "The Effect of Topical Corticosteroids on Refractive Outcome and Corneal Haze After Photorefractive Keratectomy," *Arch. Ophthalmol.,* Vol. 110, 1992, pp. 944–952.

[19] O'Brart, D. P., C. P. Lohmann, and G. Klonos, "The Effects of Topical Corticosteroids and Plasmin Inhibitors on Refractive Outcome, Haze, and Visual Performance After Photorefractive Keratectomy: A Prospective, Randomized, Observer-Masked Study," *Ophthalmology,* Vol. 101, 1994, pp. 1565–1574.

[20] Baek, S.H., J. H. Chang, and S. Y. Choi, "The Effect of Topical Corticosteroids on Refractive Outcome and Corneal Haze After Photorefractive Keratectomy," *J. Refract. Surg.,* Vol. 13, 1997, pp. 644–652.

[21] Wu, G., *Retina: The Fundamentals,* Philadelphia, PA: W. B. Saunders, 1995.

[22] Hamilton, A. M. P., M. W. Ulbig, and P. Polkinghorne, *Management of Diabetic Retinopathy,* London, U.K.: BMJ Publishing Group, 1996.

[23] Olk, R. J., and C. M. Lee, *Diabetic Retinopathy: Practical Management,* Philadelphia, PA: Lippincott Williams & Wilkins, 1993.

[24] van Paul, O. B., *Diabetic Retinopathy,* London: Informa Healthcare, 2000.

CHAPTER 2

Introduction to Imaging Optics

Johnny Chee

The simplest way to create an image is to make a pinhole in the center of a large piece of cardboard and place the cardboard over a flat white surface (screen). The pinhole allows light rays from objects facing it to go through and fall onto the screen and form an image. The image, however, is low in intensity and therefore appears dim. To increase image brightness, we need to gather more light rays emitted by the objects and converge them into corresponding points (pixels) of the image. It turns out that a simple convex lens (such as that of a magnifying glass), when mounted at the aperture (the enlarged pinhole), will perform this action when the screen is placed at a distance equal to the focal length of the lens.

However, to control both the image size and image distance, a more complex arrangement involving more than one lens may be required. The goal of this chapter, therefore, is to introduce a fairly intuitive graphical approach that allows us to do this. The chapter starts with a description of the basic properties of light, especially in refraction through an optical medium, and then it discusses a range of prisms that are useful for managing the path of the rays that form the optical image. From there, the chapter introduces Gauss points (or cardinal points) and their use in determining image height (size) and image distance (from the lens). Paraxial optics (light rays close to the optical axis) are then introduced to approximate and simplify the analysis for forming images. The effects of physical dimensions of real optical compoenents on physical light rays (versus virtual or construction rays) are then considered. The use of virtual images is studied, and combined with real images to allow controlled sizing, and positioning of the final image is discussed. Finally, a design solution using multiple lenses is illustrated. The mathematics usually associated with optical design is not introduced in this chapter. Instead a wholly graphical approach is used.

2.1 Properties of Light

Light travels fastest through a vacuum at 2.99792498×10^8 m/s. In comparison, it travels slower through an optical medium such as air, and even slower through a denser medium such as transparent plastic and glass. Its speed in relation to that in a vacuum may be expressed as a refractive index η, where $\eta = c/v$, where c is the speed of light in vacuum and v is the speed of light in the optical medium of interest.

Table 2.1 lists the refractive index of some optical media for green light at a wavelength of 455 nm.

Light at different wavelengths exhibits varying properties such as color. Colors in relation to wavelength have not been precisely defined, but are approximated as shown in Table 2.2. The speed of light also varies slightly with wavelength. The shorter wavelength travels a little slower than the longer one, that is, blue light is slower than red light. Consequently, blue light exhibits a slightly higher refractive index than red.

2.1.1 Refraction

Refraction, which is the bending of light at the interface between two optical media, occurs as a result of different propagation speeds within the two media. Because light exhibits dual wave-particle properties, its direction of travel is influenced by the propagation of its wavefront [1]. Referring to Figure 2.1, when part of the wavefront hits the denser optical medium, the propagation speed of that part slows down. The rest continues to propagate at the original speed until it, in turn, strikes and enters the optical medium. Figure 2.1 clearly illustrates this phenomenon and shows the change in direction (refraction) in the path of the light ray. For a light ray entering a denser optical medium, the refraction results in the ray bending closer toward the imaginary normal interface at the point of entry.

Table 2.1 Refractive Indices of Light Through Different Optical Media

Medium	Refractive Index
Vacuum	1.000000
Air	1.000292
Water	1.333
Ordinary crown glass	1.516
Borosilicate crown glass	1.524
Light flint glass	1.571
Medium flint glass	1.627
Dense flint glass	1.754
Extra dense flint glass	1.963

Note that, for most purposes, the refractive index of air can be taken as 1.000.

Table 2.2 Refractive Index of Different Colors of Lights in a Vacuum

Color	Wavelength
Violet	380–450 nm
Blue	450–495 nm
Green	495–570 nm
Yellow	570–590 nm
Orange	590–620 nm
Red	620–750 nm

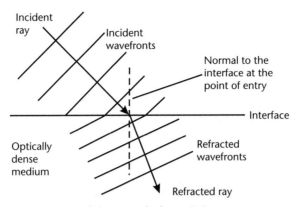

Figure 2.1 Refraction occurs toward the normal when a light ray enters an optically denser medium.

Similarly, it can be shown, as in Figure 2.2, that when a light ray exits an optically denser medium, its direction of propagation bends away from the normal.

2.1.2 Angle of Refraction

The relationship between the incident angle and the refracted angle is mathematically expressed in Snell's law as:

$$\eta_i \sin \theta_i = \eta_r \sin \theta_i$$

where

θ_i = incident angle measured with respect to the normal;

θ_r = refracted angle measured with respect to the normal;

η_i = refractive index of the optical medium for the incident ray;

η_r = refractive index of the optical medium for the refracted ray.

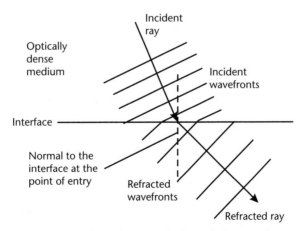

Figure 2.2 Refraction occurs away from the normal when a light ray exits an optically denser medium.

If the ray starts from an optical medium (say, air) and travels through another optical medium (say, glass) having two parallel surfaces, the light ray will exit in the same direction as that of the incident ray (by Snell's law). Its position, however, is displaced by an offset (Figure 2.3).

2.1.3 Critical Angle and Total Internal Reflection

Because a ray refracts away from the normal as it exits a dense optical medium, there is a critical incident angle at which the ray will exit at a right angle to the normal (i.e., along the interface). If the incident angle is greater than the critical angle, the incident ray is totally reflected and will not exit the dense optical medium (Figure 2.4). The ray is then said to be totally internally reflected.

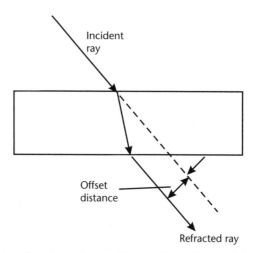

Figure 2.3 A light ray traveling through an optical medium with parallel surfaces will exit in the same direction as the incident direction. However, its original position is displaced by an offset distance that is determined by the incident angle and the thickness of the medium.

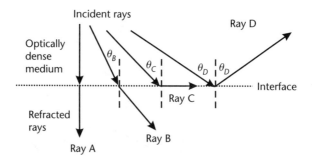

Figure 2.4 Different incident angles of light rays within a dense optical medium leads to different results: Ray A is unaffected (incident angle to the normal is 0°). Ray B is refracted away from the normal upon exit (incident angle θ_B is less than the critical angle). Ray C is refracted at 90°, that is, perpendicular to the normal (incident angle is equal to the critical angle). Ray D is totally internally reflected (its incident angle is greater than the critical angle). Note that the reflected angle is always equal to the incident angle.

This critical angle at the interface of a dense and less dense optical media is dependent on the media's refractive indices (Snell's law). Table 2.3 lists the critical angles for an incident ray in different dense optical media in contact with air.

2.1.4 Loss of Energy

When a light ray crosses the boundary between two optical media of different refractive indices, a portion of its energy is reflected back. An air-to-glass interface exhibits a minimum loss of about 4% on entry (air-to-glass) at a 0° incident angle (normal to the surface) and another 4% on exit (glass-to-air) (Figure 2.5). However, for total internal reflection, energy loss is negligible. Almost 100% of the incident ray is reflected. In comparison, the best performance obtainable from a polished silver surface is only 95%. The remaining 5% is absorbed.

2.1.5 Dispersion of Light into a Spectrum

A ray of white light comprises many colored components (rays with different wavelengths). Because propagation speed varies slightly with wavelength, the refracted angle also differs slightly for the different colored light rays [1]. Blue light, with the shortest wavelength, is refracted the most, whereas red, with the longest visible wavelength, is refracted the least.

Table 2.3 Critical Angles of Different Materials with Air as the Adjacent Medium

Material	Critical Angle
Water	48° 36′
Crown glass	41° 18′
Quartz	40° 22′
Flint glass	37° 34′
Diamond	24° 26′

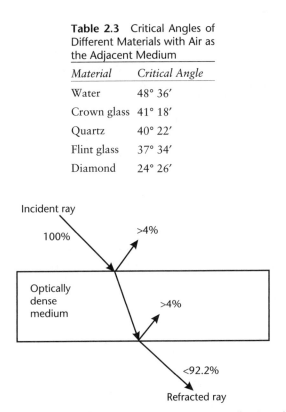

Figure 2.5 The different refractive indices of optical media cause reflections that appear as energy loss from the incident beam. For air/glass and glass/air interfaces, this energy loss amounts to more than 4% of the incident energy.

In an optical medium with nonparallel surfaces, such as a lens or prism (Figure 2.6), the double surface enhances the differences in refraction angles and makes the dispersion greater. Although color dispersion is a problem in optics, for the rest of the chapter, we will neglect its effect and assume that color dispersion is negligible. Dispersion is not a problem if the light is monochromatic, and it is a minimal problem if the optical elements selected have surfaces that are close to parallel or have chromatic compensation (e.g., achromat lenses).

2.2 Redirection of the Light Ray

There are many ways to redirect a light ray using reflection and refraction. The following subsections describe how this can be done using standard optical elements designed for this purpose.

2.2.1 Basic Prism

A prism has the ability to deflect light. Its ability to do so is measured in terms of diopters, where 1 diopter of power deflects the incident ray by 1 cm at a distance of 1m from the incident surface of the prism (Figure 2.7).

2.2.2 Optical Wedge

An optical wedge (Figure 2.8) has nonparallel surfaces. When directing a ray perpendicular to one surface, the ray will exit the second surface at an angle to the original path. Rotating the wedge will then cause the exit ray to trace the locus of a circle, providing a beam steering function.

Adding a second wedge (Figure 2.9) will add more functionality. When the second wedge is placed in a mirror image position [Figure 2.9(a)], the deflection of the

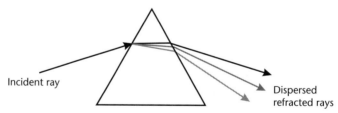

Figure 2.6 A ray of light is dispersed into its component colors by an optical medium with non-parallel surfaces. The shortest wavelength component, blue in this illustration, is deflected the most.

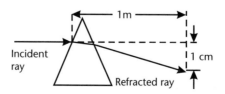

Figure 2.7 The power of a prism is measured in diopters, where 1 diopter indicates that the incident ray is deflected by 1 cm at a distance of 1m from the incident surface of the prism.

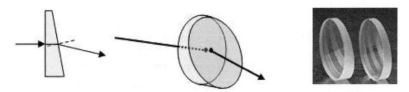

Figure 2.8   An optical wedge has nonparallel surfaces and serves to deviate the incident beam. (Reprinted with permission from Edmund Optics (http://www.edmundoptics.com/onlinecatalog/browse.cfm).)

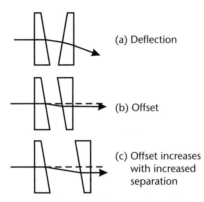

(a) Deflection

(b) Offset

(c) Offset increases with increased separation

Figure 2.9   A second optical wedge adds more functionality when placed as shown. (a) Increased deflection. (b) Exit ray is in the same direction as the incident ray but with an offset from the optical axis. (c) The offset is increased by increasing the separation between the wedges.

ray is enhanced. When rotated about its axis so that their faces are parallel to each other, the exit ray's direction is restored back to the original direction, but displaced off the axis (offset) [Figure 2.9(b)]. Increasing the separation of the wedge increases the amount of offset [Figure 2.9(c)]. Wedges with beam deviations that range from 1° to 10° are available commercially.

2.2.3   Reflecting Surfaces

When a ray is reflected at an optical surface, its reflected angle is identical to the incident angle. The angles are measured relative to the normal on the surface that receives the incident ray. Via triangulation, the image of the object is found to be at the same distance behind the surface as the object is in front of the surface (Figure 2.10).

2.2.4   The Reflected Image

The reflected image of an object will appear reverted (laterally inverted or mirror imaged). This effect can be understood by studying the position of different points on the surface of an object. For example, consider a person A facing a mirror as shown in Figure 2.11 as viewed from the top. An observer sees L′ as the right arm of

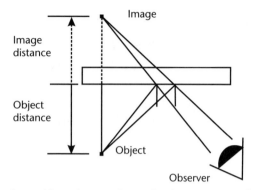

Figure 2.10 The image formed in a planar surface reflection appears at the back of the reflecting surface. Its distance from this reflecting surface is identical to the distance that the object is in front of the surface.

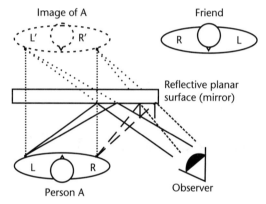

Figure 2.11 A single-reflected image is always reverted. In this illustration, we can see that the left side L of person A appears as L′ in the image of person A. But L′ is actually the right-hand side of image A. Thus the image of person A has been reverted (laterally inverted) or, as it is commonly described, presented as a mirror image.

the image, and R′ as the left arm of the image. This contrasts clearly when compared with his friend who is standing opposite him and adjacent to the mirror. The right arm of his friend is on the right side of his body, and the left on the left side, where they should be.

Similarly, we can deduce that a reflected image remains upright; that is, the top of the image corresponds to the top of the object, whereas the bottom of the image corresponds to the bottom of the object.

2.2.5 Double Reflection

When two reflective surfaces are present and placed to reflect the rays as shown in Figure 2.12, the image of the person is reverted twice. The final image now appears like the normal, or original, image. The left arm of the image is at the left-hand side

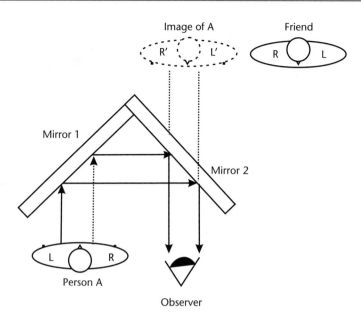

Figure 2.12   In double reflection, the image as viewed by the observer is reverted twice to become normal. The image is thus described as normal and upright.

of the image, and the right arm of the image is at the right-hand side of the image. The image is thus just like the original object (person A).

2.3   Practical Prisms

Reflecting prisms act like mirrors. Although they are more expensive than mirrors, they possess certain desirable characteristics. They are geometrically stable; that is, their reflecting surfaces remain in a fixed relationship to each other. No adjustments need to be made—as would be the case if plane mirrors were used. Multiple reflections within the prism experience no significant loss of energy when the reflections are totally internally reflected. Prisms without any silver or aluminium coating do not tarnish with age, and there is nothing to peel off (as would be the case with mirrors).

Prisms can be used to achieve a diversity of functions. When uncoated, they may be used to disperse light into its component spectrum (e.g., an equilateral prism, as shown later in Figure 2.15). Specifically, coated and uncoated prisms can be used in the following manner:

- To deviate the ray away from its original direction by 90°, 60°, and 30° (right angle and Littrow prisms; see Figure 2.13).
- To displace rays by a fixed distance from the original axis (a rhomboid; see Figure 2.14).
- To rotate the ray about its axis by a specific angle [dove prism; see Figure 2.16(a)].

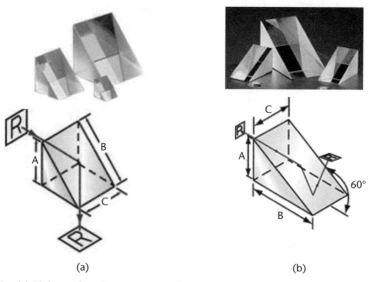

(a)   (b)

Figure 2.13 (a) Right angle prisms are generally used to achieve a 90° light path bend. (b) Littrow prisms at 30°, 60°, and 90° are coated prisms which deviate the line of sight by 60°. (Reprinted with permission from Edmund Optics (http://www.edmundoptics.com/onlinecatalog/browse.cfm).)

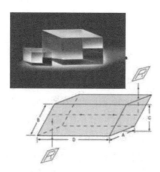

Figure 2.14 In a rhomboid prism, the ray is displaced by a distance D, which corresponds to the length of the prism. The direction remains unchanged. (Reprinted with permission from Edmund Optics (http://www.edmundoptics.com/onlinecatalog/browse.cfm).)

- To displace a ray and retroreflect it [reflective-coated dove prism; see Figure 2.16(b)].
- To revert, invert, and bend the line of sight by 90° (Amici prism; see Figure 2.17) or 45° (Schmidt prism; see Figure 2.18).
- To bend the line of sight by 90° (penta prism; see Figure 2.19).
- To split a beam differently depending on the beam's direction of propagation (beam splitter; see Figure 2.20).
- To generate an additional parallel beam from the original beam with a fixed displacement (lateral displacement beam splitter; see Figure 2.21).

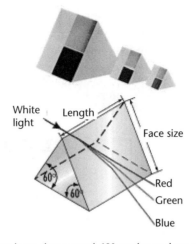

Figure 2.15   Equilateral prisms have three equal 60° angles and are generally used as dispersing prisms. (Reprinted with permission from Edmund Optics (http://www.edmundoptics.com/onlinecatalog/browse.cfm).)

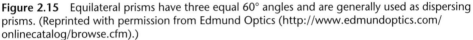

(a)   (b)

Figure 2.16   (a) An uncoated dove prism rotates the image through twice the angle that the prism rotates through around its longitudinal axis. The uncoated top surface is ground (roughened) for this type of application. (b) A coated dove prism has a protected aluminum coating to perform displaced retroreflection. The prism may be used in imaging systems to view around corners. (Reprinted with permission from Edmund Optics (http://www.edmundoptics.com/onlinecatalog/browse.cfm).)

2.3.1   Amici (Roof Edge) Prism

Amici prisms, also called *roof prisms* or *right angle roof prisms,* revert and invert the image and also bend the line of sight through a 90° angle (Figure 2.17). They serve as prism diagonals in optical systems to erect the inverted image. They are also ideal for use in spotting scopes, and in any optical instrument where an inverted image from an objective has to be turned right side up and bent through a 90° angle to maintain the correct visual orientation.

2.3.2   Schmidt Prism

Schmidt prisms are used to invert and revert an image while deviating it through an angle of 45° (Figure 2.18). Its function is similar to that of Amici prisms but the

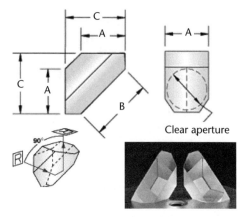

Figure 2.17 An Amici prism reverts, inverts, and bends the line of sight through 90°. (Reprinted with permission from Edmund Optics (http://www.edmundoptics.com/onlinecatalog/browse.cfm).)

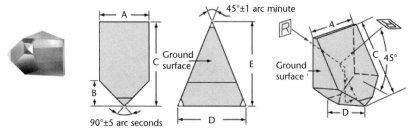

Figure 2.18 A Schmidt prism inverts and reverts the image. The line of sight is deviated by 45°. (Reprinted with permission from Edmund Optics (http://www.edmundoptics.com/onlinecatalog/browse.cfm).)

deviation is 45°. It is used in eyepiece assemblies and imaging systems that require a path bend.

2.3.3 Penta Prism

The five-sided penta prism changes the direction of the beam 90° while preserving the orientation of the image (i.e., the image is not inverted or reversed) (Figure 2.19). The image is reflected twice to achieve the preservation of the orientation, while the line of sight is deviated by a right angle turn. Slight movement of the prism does not affect the right angle at which the rays are reflected. It is also used for shortening the length of an instrument.

2.3.4 Cube Beam Splitters

Cube beam splitters split the beam differently depending on which direction the beam comes from. For Figure 2.20 a ray traveling from right to left goes through without any deviation, but a ray traveling from left to right is deviated downward by

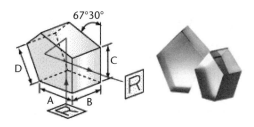

Figure 2.19 A penta prism deviates the line of sight by 90°, but does not invert or revert the image. (Reprinted with permission from Edmund Optics (http://www.edmundoptics.com/onlinecatalog/browse.cfm).)

Figure 2.20 A cube beam splitter allows light to go through in one direction (right to left), but in the opposite direction (left to right) a 90° deflection results. (Reprinted with permission from Edmund Optics (http://www.edmundoptics.com/onlinecatalog/browse.cfm).)

90° as shown. Cube beam splitters are constructed by cementing two precision right angle prisms together with the appropriate interference coating on the hypotenuse surface. The absorption loss to the coating is minimal. Transmission and reflection approach 50% (average), though the output is partially polarized.

2.3.5 Lateral Displacement Beam Splitters

Lateral displacement beam splitters output two parallel beams separated by a standard distance. Typical distances are 10 or 20 mm, depending on size of the prism (Figure 2.21). These beam splitters consist of a precision rhomboid prism cemented to a right angle prism.

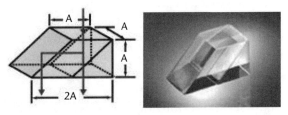

Figure 2.21 A lateral displacement beam splitter splits an incident beam into two displaced parallel beams. (Reprinted with permission from Edmund Optics (http://www.edmundoptics.com/onlinecatalog/browse.cfm).)

2.4 Basic Imaging (First-Order Paraxial Optics)

In first-order imaging optics, we make simplifications by assuming that aberrations are negligible when light rays travel through an optical element, and that the image resolves to a perfect dot (pixel) without the limitation of real-world diffraction. This simplification helps us to analyze and understand the basic design of an optical system. When aberrations and diffraction limits have to be considered, specialized optical design software is typically used. Many such software programs are commercially available. A directory Web site such as http://www.optics.net lists many of these software programs.

By constraining the light rays of an optical system close to its optical axis, the optical system is said to be operating in the paraxial region. In this region, the incident angles of all rays to every optical element are assumed small and Snell's law of refraction ($\eta_i \sin \theta_i = \eta_r \sin \theta_r$) simplifies to $\eta_i \theta_i = \eta_r \theta_r$ where the angles θ_i and θ_r are in radians.

The normal conventions used in optical system design are typically as follows:

1. Light rays are assumed to travel from left to right in an optical medium of positive index. When the optical medium is negative (e.g., reflecting surface), the rays travel from right to left.

2. A distance is positive in value if it is measured to the right of the reference point, but negative if measured to the left.

3. The radius of curvature r of a surface is positive if its center of curvature lies to the right of the surface, but negative if the center lies to the left. The curvature c is the reciprocal of the radius r, that is, $c = 1/r$.

4. Spacings between surfaces are positive if the next (following) surface is to the right; they are negative if the surface is to the left (as can happen after a reflection).

5. Heights, object sizes, and image sizes are measured normal to the optical axis and are positive above the axis, negative below.

6. The term *optical component* refers to one or more optical elements that collectively behave as a single optical unit. An optical element is a simple lens.

2.4.1 Cardinal (Gauss) Points and Focal Length

An optical system, whether a single element or a component, can be conveniently represented by four points on its optical axis called the *cardinal*, or *Gauss, points*. These points are the first and second focal points and the first and second principal points. The labeling of "first" and "second" indicates their physical locations in the optical diagram. Figure 2.22(a, b) illustrates the definitions of these points.

An incident light ray traveling from left to the right, parallel to the optical (principal) axis, is refracted as it enters the lens and refracted again as it exits the lens [Figure 2.22(a)]. The exiting ray cuts the optical axis at a point known as the second focal point, F_2. When this exiting ray is projected backward to meet the incident ray, the incident ray appears to have made a bend at the plane called the second principal plane. This plane cuts the optical axis at the second principal point, P_2. For this

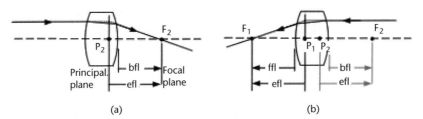

(a)   (b)

Figure 2.22 (a) An incident light ray traveling from left to right, parallel to the optical (principal) axis, appears to bend at the second principal plane and cuts the optical axis at the second focal point F_2. (b) A light ray traveling from right to left, parallel to the optical (principal) axis, appears to bend at the first principal plane and cuts the optical axis at the first focal point F_1.

biconvex lens, the plane lies within the lens. The effective focal length (efl) of the lens is the distance from P_2 to F_2 and has a positive value for the convex lens. The back focal length (bfl) is the distance from the back (outer surface) of the lens to F_2. The value of bfl is also positive.

When the light ray travels in the reverse direction, from right to left, parallel to the optical (principal) axis, it is refracted as it enters the lens and refracted again as it exits the lens, and it cuts the optical axis at the first focal point F_1. When the exiting ray is projected backward to meet the incident ray, it appears to have bent from the incident ray at the principal plane. This principal plane cuts the optical axis at the first principal point P_1. The efl of the lens in this direction is the distance from P_1 to F_1. The forward focal length (ffl) is the distance from the front outer surface of the lens to F_2.

Points P_1, F_1, P_2, and F_2 are used to simplify the effort involved in determining the location and size of the image formed from an object placed in front of an optical component.

2.4.2 Image Formation Using a Positive Lens

When an object O is placed before a convex lens, every point on its surface that faces the lens will emit or reflect light rays that fall incident onto the whole surface of the lens. Let's use Figure 2.23 to consider the paths of three strategic rays emitted from the top of such an object to see how its image is formed.

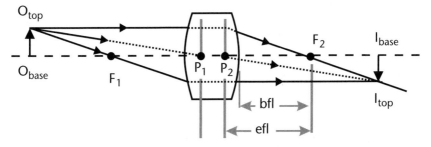

Figure 2.23 The paths of just two rays—the horizontal ray and the (chief) ray—through the lens center are sufficient to determine the formation of a pixel of any image.

1. The first ray we consider is the horizontally emitted ray that passes through the second focal point F_2. It travels horizontally to enter the lens, "bends" at the second principal plane, exits the lens, and travels through the second focal point F_2, to reach the image at point I_{top}. (Note that if the horizontal ray is from the base, it will travel undeviated through the lens along the optical axis. This ray is called the *principal ray*.)

2. The second ray we consider passes through the first focal point F_1. It is emitted from O_{top}, travels through the first focal point F_1 to the lens, "bends" at the first principal plane P_1, and exits horizontally to meet the first ray at I_{top}.

3. The third ray we consider is the chief ray. It travels midway between the preceding two rays. It enters the lens, "bends" at P_1, travels horizontally to P_2, "bends" again to regain its original incident direction, and meets the first two rays at I_{top}.

We need to use only two of the three rays just considered to determine where the image of a point on the object will be formed. Because object O rests on the optical axis, the ray from the base of the object is the principal ray. This ray travels through the lens undeviated to form the image I_{base} at a location on the optical axis directly below the image I_{top}. Thus, image size and location have been determined simply by analyzing the paths of strategic rays traveling through the cardinal points of the optical component.

Note the following interesting characteristics:

1. If the lens had been very thin, P_1 and P_2 would have coincided, and the chief ray would have appeared to pass through the lens without any deviation (i.e., the ray remains straight).

2. The whole conical bundle of emitted rays from the object's top, O_{top}, falls on the lens and is focussed to a pixel of the image, I_{top}. If a bigger sized lens is used, more rays from the object would be captured and the image would be brighter.

3. When object O is placed far from the lens, the image shifts closer toward the lens, approaching second focal plane F_2. When the object is at infinity, its image will be formed at this focal plane.

In first-order optics, we simplify this analytical procedure by assuming the lens to be very thin and represent it by a single vertical line. The two principal planes within the lens coincide together with the center of the lens ($P_1 = P_2 =$ lens center). The bfl, the efl, and the ffl are all the same length.

2.4.3 Optical Axis, Chief Rays, and Marginal Rays

We now define some optical terms (Figure 2.24).

- *Optical axis:* The optical axis is the axis of rotation for the lens. It lies perpendicular to the lens. A ray traveling along the optical axis is known as the principal ray. It goes through the lens undeviated.

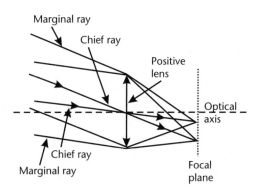

Figure 2.24   Each chief ray goes through the center of the lens and exits without deviation. The marginal rays are parallel to the chief ray and are the limiting rays that the lens can capture.

- *Chief rays:* These are rays from different points of an object that pass through the optical center of the lens undeviated. There is only one chief ray for every image pixel.
- *Marginal rays:* These are rays parallel to a chief ray or the principal ray. They fall at the edge of an optical element, and their positions mark the boundary within which rays are captured by an optical element to form an image pixel. Marginal rays indicate how much light is captured to form the image pixel and therefore how intense (bright) the pixel will be.

2.4.4   Aperture Stop

The aperture stop of an optical component is the surface where all of the chief rays from an object cuts its optical axis and thus appear to pivot about. For a single element, the aperture stop usually lies within the lens.

2.4.5   Iris

An iris is mechanical diaphragm with an opening for light to pass through. The size of the opening may be adjustable. The iris is usually placed at the aperture stop to control the brightness of the image. While the marginal rays are cut off, rays closer to the chief rays continue to pass through. Thus, every image pixel has less light, and the image appears dimmer, yet remains undistorted. For a single-element component, the iris may have to be placed either immediately at the front or back of the lens, because the aperture stop may lie within the lens.

2.4.6   Entrance and Exit Pupils

The entrance pupil is the image of the aperture stop when viewed from the front of an optical component. Conversely, the exit pupil is the image of the aperture stop when viewed from the back of an optical component. The locations of the entrance and exit pupils become important when multiple groups of lenses are used together, because the exit pupil of the front group of lenses must match the entrance pupil of the back group of lenses in order to capture all usable light rays.

2.4.7 Vignetting

Not all light that passes through the front element of an optical component will necessarily form an image. If the rays are at a steep angle to the optical axis, the rays near the edge of the front element may not be captured by subsequent elements to form the image pixel. Such off-axis pixels will be dimmer than those at the center, causing a gradual reduction in image brightness away from the center. This variation of intensity is known as *vignetting*. Telescopes, projectors, and other visual systems may have vignetting of about 30% to 40% of the energy from the center. The eye can tolerate this amount, and the user will not generally notice it.

2.5   Real Images

An object O located far from a positive (convex) lens, will form a real inverted image I_1 at the image plane [Figure 2.25(a)]. This image can be seen by placing a screen at the image plane. The image is formed by the lens focusing the light rays to form

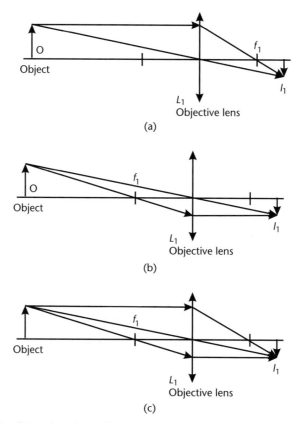

Figure 2.25   (a) An object placed at a distance (> focal length) from a positive lens, will generate a real inverted image of the object, as shown by this ray construction diagram. (b) This second form of construction yields the same results as the earlier form and may be used in place of the first form of part (a). (c) Both construction forms can be used together to determine the size and location of the image. If you use both forms together, however, redundancy and, hence, unnecessary work results.

bright spots (image pixels) at the image plane. Each image pixel directly corresponds to a surface on the object that fronts the lens.

To graphically determine the image size and image distance, we can use construction rays. These are the strategic rays that pass through the focal point and center of the lens as discussed earlier (see Figure 2.23). In drawing the ray construction diagram, we take liberties to extend the location of the rays and sizes of lenses as necessary. The real-world limits are not considered initially.

A ray construction diagram is formed using a few basic rules:

1. Draw the (objective) lens perpendicular to the optical axis.
2. Draw the object resting perpendicularly on the optical axis, to the left of the lens.
3. The base of the object sends a principal ray traveling along the optical axis and its image pixel will rest on the optical axis.
4. Mark the two focal points of the lens; they are on the lens' right and left. Usually only one will need to be used.
5. Draw the chief ray from the object's top to the center of the lens. This ray goes through the lens center without any deviation, that is, the ray remains a straight line.
6. Draw another ray from the object's top to go through one of the focal points. Either start with a horizontal ray [Figure 2.25(a)] and bend at the lens to pass through the second focal point (the first form of construction), or start with a diagonal ray through the first focal point [Figure 2.25(b)] and bend to form a horizontal line (second form of construction). The image of the object's top is formed at the point where the ray cuts the chief ray.
7. Draw a vertical line for the image starting from the top of the image to the optical axis. The base of the image is located at the point where the vertical line cuts the optical axis.

Both the first and second forms of construction will yield the image height and location, and their results will be identical [Figure 2.25(c)]. Notice that the construction rays do not necessarily form a parallelogram.

When the object is bigger than the lens, or vice versa (the lens is smaller than the object), the graphical construction can still proceed by "extending" the lens vertically as shown in Figure 2.26. The horizontal construction ray does not correspond to any real light ray from the object that strikes the lens. Nevertheless, the construction rays serve a useful purpose when we do not limit them to real physical sizes.

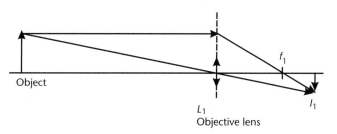

Figure 2.26   Notice that even though the physical lens is small, the graphical analysis can still proceed by extending the vertical line that represents the lens.

2.6 Physical Devices and Their Effects

So far we have used construction rays to determine image size and image distance. To consider other optical effects, we also need to consider the spread of all light paths that are captured by the lens to form the image.

Let us consider the bundle of rays that originates from the object's top and base. After drawing the construction rays to determine the image, we add the marginal rays to the lens from the object's top and bottom as shown in Figure 2.27(a).

The resulting diagram can be simplified by removing the construction rays as shown in Figure 2.27(b). Of course, once removed, it is difficult to verify whether the image was constructed correctly. Nevertheless, for the purpose of studying the optical effects we will continue to use the simplified diagrams. Because most objects are viewed symmetrically rather that resting on the optical axis, we modify the diagram as shown in Figure 2.27(c).

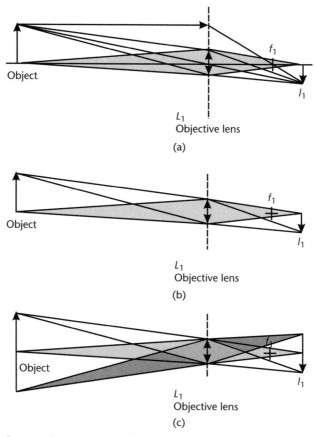

Figure 2.27 (a) Construction rays are used to determine image size and image location. Marginal rays are added to determine the bundle of rays captured by the lens. (b) Same as part (a) but simplified by excluding the construction rays. The bundle of rays captured by the lens from the top and bottom of the object is now more clearly seen, but verification of correctness of construction is a problem. (c) Typically objects are placed symmetrically about the optical axis.

2.6.1   Aperture and Lens Hood

An aperture is an opening (e.g., an iris) that allows rays of interest to pass through the lens to form the image (Figure 2.28). Ideally all other rays will be blocked by it so that interference with the image is reduced. Interference may be caused either directly by rays from an interfering source being focused onto the image (glare) or through diffused reflection smearing the image and reducing image contrast (image washout). These forms of interference can be eliminated to a large extent by the combination of a lens hood placed in front of the lens and an aperture placed just in front of the lens. To understand how this works, let us consider three unwanted rays, UR_1, UR_2, and UR_3, which fall on the lens hood:

1.   Unwanted ray UR_1 is totally blocked by the hood. It creates no impact on the image.
2.   Unwanted ray UR_2 goes through the hood opening, but is partially absorbed by the aperture stop and partially diffusely reflected. A second diffused reflection may occur on the inside of the lens hood, which may send rays through the lens. But with very much lower energy, its effect on the image will be negligible (very mild washout).
3.   Unwanted ray UR_3 falls on the inside of the lens hood and is diffusely reflected to pass through the lens and fall on the image, smearing it. Depending on how intense the ray was initially, and whether multiple reflections are involved, the image may or may not experience any washout or glare. Blackening the inside surface of the hood to absorb light reduces the interference to a negligible level.

2.6.2   Obstruction

What happens if there is an obstruction (unwanted blocking object) O_b in front of the lens? How will the image be affected?

2.6.2.1   Obstruction Close to the Lens

Let us consider an obstruction O_b close to the lens [Figure 2.29(a)]. Rays from every point of the object attempt to reach the lens, but some are blocked and do not reach the lens. The rest that do reach the lens are focused to form an image of the object. The image intensity is lower because there are fewer rays and hence less energy. The

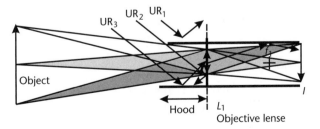

Figure 2.28   A lens hood and an aperture stop can protect the image from glare and also from smearing, which reduces image contrast.

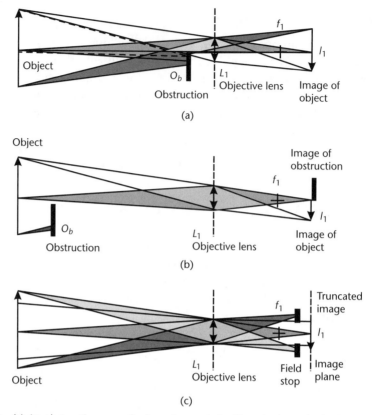

Figure 2.29 (a) An obstruction near the lens does not significantly degrade the image other than reducing image brightness and smearing the image with a very dim, defocused image of itself. (b) An obstruction at a distance far from the lens appears as another object and is imaged by the lens as such. (c) A field stop blocks part of the image. This is useful in eliminating glare coming from a bright part of an object, so that the dimmer parts of the image may be more clearly examined.

net result is that an obstruction close to the lens does not block the image, but merely reduces the brightness of the image unevenly [4, 5].

2.6.2.2 Obstruction Far from the Lens

If the obstruction O_b is far from the lens, then the effect will be very different. Referring to Figure 2.29(b), all rays from the base of the object are blocked, and thus there will be no image of the object's base. Instead, rays from the obstructor reach the lens and are focused by the lens, forming an image of the obstruction. Thus, an obstruction far from the lens appears as part of the image.

2.6.2.3 Field Stop

If an obstruction is placed after the lens as shown in Figure 2.29(c), a very different effect is obtained. All rays are captured by the lens and imaged, but not all image pixels fall onto the image plane. The obstruction effectively crops out a portion of the image. Thus, when placed close to the image plane, the obstruction behaves as a field stop cropping out a part of the object's image.

A useful application of the field stop is to limit the field of view and block out the bright part of an image (usually at the periphery). With the glare blocked, the eye is able to examine the dimmer parts of image.

2.7   Virtual Images

Virtual images are formed when the rays from an object diverge after going through the lens (Figure 2.30). The divergent rays emulate the emission of light rays from a more distant object. When projected backward, the rays appear to converge from a point in space farther from the object, where physically there is no object [4, 5]. This point lies on a virtual plane where the rest of the virtual image also appears to exist. Whereas physical light rays travel from left to right through the lens to form a real image, virtual rays travel from right to left from the lens to form the virtual image. The convergent point of virtual rays is to the left of the lens (while it is to the right for real images). The ray construction diagram of Figure 2.30 shows how the virtual image is formed. Note that the virtual image is upright, not inverted.

2.7.1   Converting a Virtual Image into a Real Image

Because the light rays do not converge, an image cannot be formed on a screen. Instead an additional lens L_4 (Figure 2.31) must be used to converge the rays and focus into a real image. Figure 2.31 shows a ray construction diagram of how this is achieved. The additional lens L_4 may be that of the eye or the camera, forming an image, respectively, on the retina or the CCD.

2.7.2   Minimum and Maximum Separation Between Lenses

In Figure 2.31, lens L_3 may be the eyepiece lens of an optical instrument. Lens L_3 generates a virtual image for the eye (lens L_4) to convert into a real image. When the distance between the eyepiece to the eye is too close, the user will be irritated every

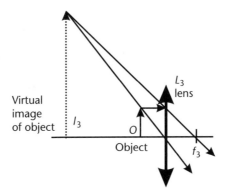

Figure 2.30   Divergent light rays from the lens give an appearance that the rays have been emitted from a point on the virtual image.

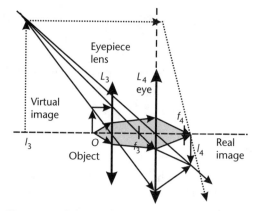

Figure 2.31   The virtual image can be made real by using an additional positive lens (L_4). The divergent rays will then converge and form a real image.

time his eyelashes brush against the eyepiece. Therefore, a minimum separation distance of about 15 to 20 mm will be needed for the user's comfort.

Figure 2.32 shows the eye at a farther distance from the eyepiece. The virtual image is formed at the same distance from the eyepiece lens L_3 but is, of course, now farther away from the eye. The ray diagram reveals that the real image derived from the virtual image has been reduced in size although the virtual image is still of the same size. The reason for this is that the chief ray from the top of the virtual image to the eye is now less steep (smaller subtended angle).

A virtual image, therefore, behaves just like a real object, because an object at a farther distance also results in a smaller image on the retina.

Notice that in Figure 2.32 the increased eyepiece-to-eye separation has resulted in some of the divergent rays going past the edges of the eye lens. These rays are not captured by the lens and are lost. With fewer rays, the image of the top of the object appears dimmer—the vignetting effect.

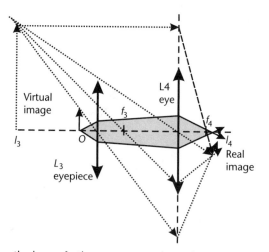

Figure 2.32   Separating the lenses further apart not only results in a smaller reel image (smaller subtended angle) but also a timer image of the top of the object (vignetting effect).

2.7.3   Virtual Image at Infinity

A virtual image may be placed at any desired distance from 20 cm to infinity—the comfortable range of viewing distances of the eye. Such an image can be used to eliminate the need for a real image to be formed on a screen—a desirable feature for some optical instruments.

When an object is placed at the focal plane of the lens, rays exiting the lens are parallel to each other (Figure 2.33). The location of the virtual image is therefore at an infinite distance in front (to the left) of the lens. The image size is also infinitely large. An infinitely far and infinitely large virtual image will produce a constant slope for the chief ray, meeting the eye even when the eye is moved slightly away from the lens. The image size on the retina, therefore, remains constant (unchanged) because the subtended angle is constant.

2.7.4   Vignetting Due to Eye Position

Practical optical systems have physical constraints and unwanted side effects. Consider the actual paths taken by the physical light rays from the object, through the eyepiece and eye, onto the retina, as shown in Figure 2.34. The bundles of rays from both the top and the base of the object take different paths to reach the retina. However, fewer rays from the top reach the retina, producing a dimmer image of the object's top as compared to that of the base—again, the vignetting effect.

If the eye is brought nearer to the eyepiece, the vignetting is less severe, because more rays from the object's top will be captured whereas those from the base remain the same. Thus, vignetting is less severe when the distance between the eye and eyepiece is small.

2.8   Controlled Size Imaging

The undilated eye will typically have a pupil size from 2 to 5 mm in diameter depending on the luminosity of the object being observed. Also unlike a man-made

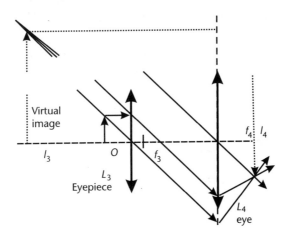

Figure 2.33   An object placed on the front focal plane of a convex lens produces rays that exit the lens that are parallel to each other.

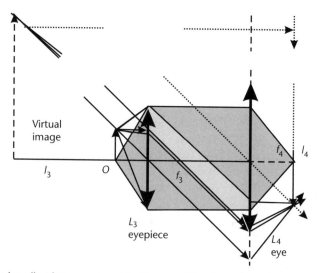

Figure 2.34 The bundle of rays captured by the eye differs in amount depending on the location from which the rays are emitted. The maximum bundle size is captured if rays are emitted from a point that rests on the optical axis. The minimum bundle size is captured for those rays emitted from the top of the object. The vignetting effect in this case is severe.

lens, the human eye lens changes its focal length to maintain a constant image distance while gazing at objects near or far. A practical optical system must therefore take care of these physical considerations and still ensure that the image formed is of acceptable size and has sufficient uniformity in brightness [4, 5].

Let us consider a distant object being imaged into the eye (Figure 2.35). Because the object is far, the chief ray has a gentle slope and the image formed is consequently small in size.

2.8.1 Adding an Objective (Convex) Lens to Increase Image Size

To make the image larger, a convex lens L may be placed to the right of the object such that the object is at a distance shorter than the lens' focal length (Figure 2.36). This creates a virtual image larger than the object. The chief ray from the virtual image to the eye is steeper, resulting in a larger image on the retina.

Note that with limited accommodation power, the human eye is unable to produce a sharp image if the object is too close to the eye. A virtual image can be used to

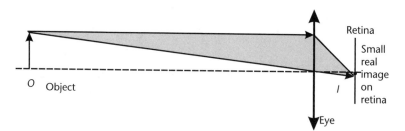

Figure 2.35 An object at a distance creates a small-sized image on the retina.

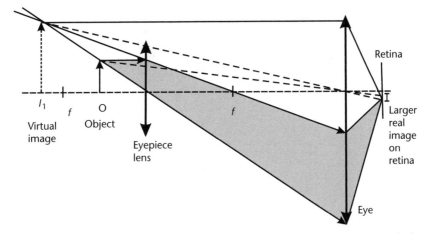

Figure 2.36 Adding a positive lens close to an object creates a larger virtual image, which results in a larger image size on the retina. However, vignetting appears as the rays toward the eye are deviated at steeper angles.

help the eye cope because the distance to formed is farther than that of the original object.

To get a larger image on the retina, the chief ray must have a steeper angle. In other words, the virtual image must be made larger. This can be achieved by using an eyepiece with a shorter focal length.

2.8.2 Multiple Lens Imaging

The limitation of this simple one-additional-lens scheme is that there is a conflicting requirement for the lens to be near the eye to minimize vignetting, yet also be near the object to generate the virtual image. These two conditions cannot always be met simultaneously, especially if the object is located at a distance. A solution to this dilemma is to use two positive lenses in different manners as follows:

1. Create a real image I_1 of the object using a positive lens (objective lens) [Figure 2.37(a)] with the object located farther than its focal length. Note that the image size decreases with distance.
2. Magnify the real image I_1 into a bigger real image I_2 using another positive lens (zoom lens) [Figure 2.37(b)]. The image I_1 behaves like a real object, and when placed within two focal lengths from the lens but more than the focal length (i.e., distance $< 2f_2$ but greater than f_2), a magnified image I_2 is formed.

The real image I_2 can now be used by the eyepiece to create a magnified virtual image for the eye. The final layout using the three lenses (objective, zoom, and eyepiece) is shown in Figure 2.38(a). For comparison, the ray diagram of Figure 2.38(b) has been drawn at the same scale to show clearly the magnification obtained using the three-lens design.

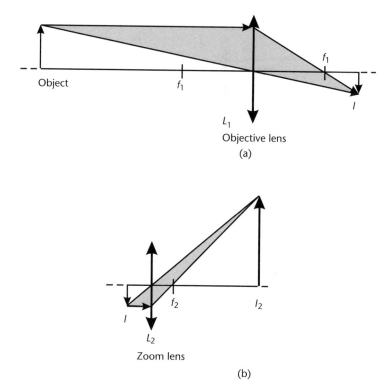

Figure 2.37 (a) An object may be "brought" near the observer by using a positive lens (objective lens L_1). If the object is farther than the focal length, its image I_1 will be inverted and real. If the object is very far, the image formed will be very small. (b) An object or a real image I_1 may be magnified by placing it within a distance of two focal lengths ($2 \times f_2$) from the lens. If the distance of I_1 from the lens is also greater than the focal length f_2, the image I_2 will be magnified, real, and inverted. If the distance is shorter than the focal length, a virtual image will be formed instead.

2.8.3 A Practical Three-Lens System

When we take into account the physical light rays, the final paraxial ray scaled diagram is as shown in Figure 2.39(b). Figure 2.39(a) shows how a distant object forms a very small image on the retina. The construction rays have been left in this diagram to show how the image was constructed and to assist with verification of the correctness of the construction.

Note the following points:

1. The vertical dimension is elongated relative to the horizontal dimension. This helps to show the angles more clearly.
2. The analysis and design are accurate only as a first-order approximation. The lenses are assumed to be thin, and the optical instrument is meant for paraxial rays (rays close to the optical axis). Aberrations and color dispersions have not been considered and are assumed noncritical (or easily compensated, e.g., by the use of acromat lenses) for this instrument.
3. It can be seen that the size of the lenses chosen (other than the objective lens) determines whether there is the extent of vignetting and to what extent. Note

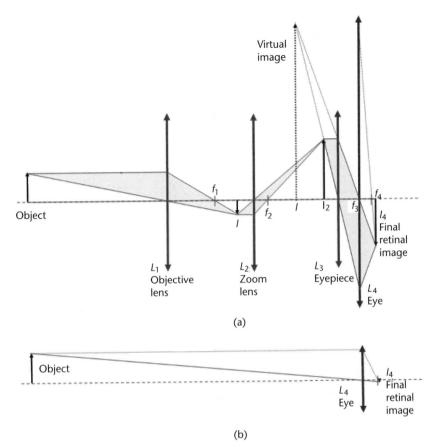

Figure 2.38   (a) The triple lens paraxial optics design generates a large image onto the retina. (b) An object far from the eye creates a very small image on the retina.

that a smaller diameter lens can be chosen for the zoom lens because the captured bundle of rays only goes through half of the diameter of the lens.

4. From the scale drawing shown in Figure 2.39(a), a magnification of 31 was obtained using standard lenses with focal lengths of 25 and 50 mm. The image distance is about 21 cm from the eyepiece. The separation between the eyepiece to the eye may reach up to 20 mm before vignetting makes the instrument unusable. These values are approximate. For accuracy, the design has to be reworked using optical design software.

5. The length of the instrument is fairly long at about 29 cm. To make it shorter, the zoom lens may be substituted for with one having a shorter focal length lens. Alternatively, or in addition, prisms may be introduced to fold back the optical path.

6. To control image brightness, an adjustable iris has been added after the objective lens. Varying the aperture size of the iris varies the total amount of light captured without blocking out the chief rays. Thus, adjusting to a small aperture size will still retain the full image (i.e., all pixels of the image are formed), but with image intensity reduced (i.e., dimmer image).

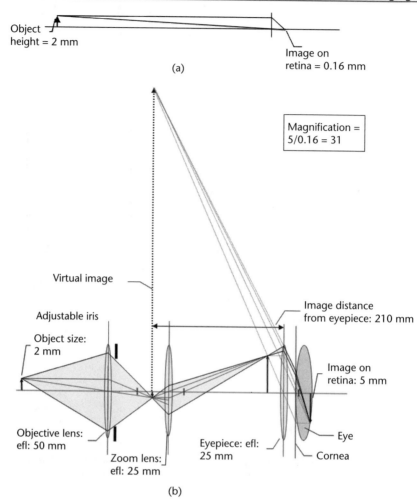

Object height = 2 mm

Image on retina = 0.16 mm

(a)

Magnification = 5/0.16 = 31

Virtual image

Adjustable iris

Object size: 2 mm

Image distance from eyepiece: 210 mm

Image on retina: 5 mm

Objective lens: efl: 50 mm

Zoom lens: efl: 25 mm

Eyepiece: efl: 25 mm

Eye

Cornea

(b)

Figure 2.39 (a) A distant object forms a very small image on the retina. (b) This three-lens paraxial optics diagram shows the physical ray tracings together with construction rays for a magnification of 31 for the same distant object as in part (a). A variable iris (aperture stop) after the objective lens allows control of image brightness.

2.9 Lasers

A laser is a device that emits light through optical amplification. The term *laser* is an acronym for "light amplification by stimulated emission of radiation." A typical laser emits light in a narrow, low-divergence beam and with well-defined specific wavelengths. The light beam is coherent; that is, the photons have identical amplitude, polarity, and time phase. In contrast, a light source such as the incandescent lightbulb emits light over a large solid angle and over a wide spectrum of wavelengths, and without coherence.

2.9.1 Generation of a Laser

Any light-emitting medium can be turned into a laser. The medium may be a gas, liquid, or solid, including a semiconductor. Figure 2.40 shows the different compo-

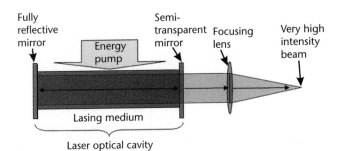

Figure 2.40 A typical arrangement for generating a laser beam. An energy pump excites the lasing medium by electrical or optical means. The laser optical cavity formed by two mirrors facing each other reflects to and fro the photons released by stimulated emission. Part of this intense beam escapes the cavity via the semitransparent mirror, appearing as an intense beam of light. A focusing lens placed in front of the beam can be used to concentrate the beam and increase the intensity of the beam over a tiny spot.

nents of the laser. For lasing to occur, there must first be an inversion population of a high number of atoms in an excited state. This excitation can be induced by pumping the medium with energy using sources such as flash tubes, an electrical field, or the light from another laser. Excited atoms will spontaneously decay to their unexcited state, releasing photons of identical characteristic energy. The atoms also instantly decay when triggered (hit) by photons of their characteristic energy (wavelength). Such photons are released with the same characteristic energy, phase, and direction as the triggering photon. By placing two mirrors at opposite ends of the optical medium to form an optical cavity, the stimulated-released photons bounce to and fro, triggering a further release, which leads to an avalanche of photons and forms an intense beam within the cavity. Some of this high-intensity beam of coherent light escapes the optical cavity through the semitransparent (partially silvered) mirror.

A lasing medium will normally emit photons at specific spectral lines (wavelengths). Spectral lines are determined by the different quantum levels, or energy states, of the material. The material chosen as the lasing medium typically has a metastable state at which its atom or molecule will remain at an excited state for some time before decaying to the ground state. When excited, the lasing medium stores energy in the form of electrons trapped in the metastable energy levels. Pumping therefore produces a population inversion (i.e., more atoms in the metastable state than the ground state), a necessary condition before laser action can take place. Many types of lasers are available for research, medical, industrial, and commercial uses. Lasers are often described by the kind of lasing medium they use: solid state, gas, excimer, dye, or semiconductor. Their emissions at different wavelengths determine their application (Figure 2.41).

- *Gas lasers.* The helium-neon (HeNe) laser is the most common gas laser. It has a primary output of a visible red light at 632.8 nm. CO_2 lasers emit photons of lower energy with a wavelength at 10.6 μm (far infrared). Photons at this wavelength are strongly absorbed by water, which makes up about 80% of

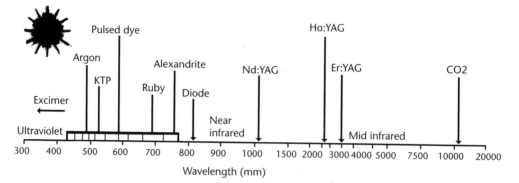

Figure 2.41 Different laser materials emit photons at different wavelengths. CO_2 lasers operate at the far infrared region. Yttrium aluminum garnet (YAG) lasers are doped with different rare Earth elements [neodynium (Nd), holmium (Ho), and erbium (Er)] to operate at different wavelengths in the near infrared region. Excimer lasers operate at very short wavelengths releasing photons with energy high enough to break molecular bonds.

the composition of soft tissue. A CO_2 laser, therefore, cuts like a scalpel on soft tissue. It can also be defocused to heat and coagulate.

- *Excimer lasers.* The name *excimer* is derived from the terms *excited* and *dimers (molecule with dual subunits or monomers)*. Excimer lasers use reactive gases such as chlorine and fluorine mixed with inert gases such as argon, krypton, or xenon. When electrically stimulated, a pseudomolecule or dimer is produced and, when lased, produces photons in the ultraviolet range. The high-energy photons are well absorbed by tissue and organic compounds, break molecular bonds, and disintegrate the surface tissue on which the beam falls. Very fine layers of surface material can be ablated with almost no heating to the underlying tissue, making it suitable for precision micromachining of organic materials and delicate surgery, for example, LASIK where the cornea of the eye is ablated about 1 nm deep per pulse. Wavelengths of excimer lasers range from 126 to 337 nm.

- *Dye lasers.* These lasers use complex organic dyes such as rhodamine 6G in liquid solution or suspension as lasing media. They are tunable over a broad range of wavelengths. Dye lasers can be used to treat vascular lesions and even up the tone of skins. This is achieved by tuning the wavelength to that which is strongly absorbed by hemoglobin and melanin and directing the beam to the area under treatment.

- *Semiconductor lasers.* In these lasers, also called *diode lasers*, photons are spontaneously emitted when electrons and holes in the *p-n* semiconductor material recombine. At low rates of recombination (below the lasing threshold), the light emitted is similar to that of a light-emitting diode (LED). Its output power increases with excitation (electrical current). But beyond the lasing threshold, the dominant mechanism for photon emission changes from spontaneous emission to stimulated emission. The output power shoots up dramatically by orders of magnitude, while its emission linewidth narrows dramatically also by orders of magnitude. Lasing has occurred. Laser diodes are generally very small and use low power. An advantage of semiconductor lasers is that they can be fabricated in arrays and may be applied as light sources for laser printers.

References

[1]   Hecht, E., *Optics,* 4th ed., Reading, MA: Addison-Wesley, 2002.

[2]   Navedtra, *Basic Optics and Optical Instruments,* rev ed., New York: Dover Publications, 1989.

[3]   Smith, W. J., *Practical Optical System Layout and the Use of Stock Lenses,* New York: McGraw-Hill, 1997.

[4]   Fischer, R. E., and T. -G. Biljana, *Optical System Design,* Bellingham, WA: SPIE Press, and New York: McGraw-Hill, 2000.

[5]   Elkington, A. R., H. J. Frank, and M. J. Greaney, *Clinical Optics,* 3rd ed., Boston, MA: Blackwell Science, 1999.

CHAPTER 3

Eye Imaging Systems

Rajendra Acharya, Wong Li Yun, Kenneth Er, Eddie Y. K. Ng, Lim Choo Min, Wenwei Yu, and Jasjit S. Suri

Various parts of the eye and the different kinds of diseases affecting the eye were described in detail in Chapter 1. Devices that can detect disease in its early stages are preferred over those that detect disease later. Physicians and engineers together have designed equipment that allows physicians to view the eye for a quick diagnosis of disease, often with great accuracy and minimal invasion. Because eye diseases afflict different parts of the eye, specialized equipment has to be designed to capture images at each location. This chapter explains in detail various techniques used for image acquisition: computed tomography, confocal laser scanning microscopy, magnetic resonance imaging, optical coherence tomography, and ultrasound imaging.

3.1 Computed Tomography

Computed tomography (CT) is a powerful nondestructive evaluation (NDE) technique for producing two-dimensional and three-dimensional cross-sectional images of an object from flat X-ray images. Characteristics of the internal structure of an object such as dimensions, shape, internal defects, and density are readily available from CT images. The CT scanner sends X-ray pulses through the body area being studied.

The limitation of X-rays alone is that the patient is subject to ionizing radiation and only dense structures such as bones are clearly visible. Computed tomography, previously called computed axial tomography or CAT scans, was a significant step toward overcoming these limitations. CT came about in the 1970s and makes use of a highly refined X-ray machine. Weak X-ray beams are focused simultaneously from various angles at a specific location in the body. A mathematical computer construction translates the X-ray absorption patterns into detailed three-dimensional cross-sectional pictures of the body regions that have been scanned.

CT allowed details of soft tissues such as the brain to become visible for the first time. The first CT scanner was installed in 1971 and the technique has found wide application in medicine, not only on living bodies but also in forensic science. CT scans are now commonly used to detect and assess size and position of structural changes in the head, brain tissues, and soft abdominal tissues and organs including the liver, pancreas, kidney, chest, spine, and eyes.

CT scan images represent the density and atomic number of human tissue just like general X-ray imaging, such as chest and bone X-rays. The denser the tissue, and the higher the atomic number, the whiter the CT image is. Bone and calcium appear white; air in the lungs appears black; water, blood, and internal organs such as liver, kidneys, and intestines appear gray; and fat tissue appears dark gray.

The basic principles of CT are shown in Figure 3.1(a). The X-ray source is tightly collimated to interrogate a thin "slice" through the patient. The source and detectors rotate together around the patient, producing a series of one-dimensional projections at a number of different angles. These data are reconstructed to give a two-dimensional image, as shown in Figure 3.1(b) [1]. CT images have a spatial resolution of approximately 1 mm and provide reasonable contrast between soft tissues.

Ever since its introduction, CT has become an important tool for medical imaging to enhance diagnoses made through X-ray and medical ultrasonography. Also, the CT scan can show the physicians small nodules or tumors that often cannot be seen on a conventional X-ray [2].

3.1.1 Principle

CT is based on the principles of X-ray production and is explained as follows:

X-rays result from the conversion of the kinetic energy attained by electrons accelerated under a potential difference—the magnitude of which is termed *voltage*

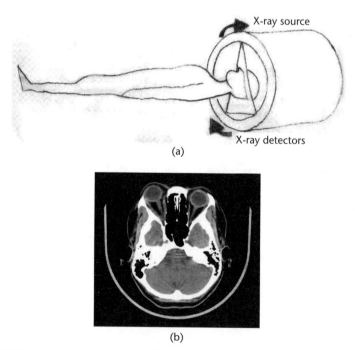

(a)

(b)

Figure 3.1 (a) X-ray source and detector unit rotating synchronously around the patient. (b) CT scan of the left eye, with a large mass filling the vitreous cavity. Areas of calcification appear as bright spots.

with units of volts (V)—into electromagnetic radiation, as a result of collisional and radiative interactions. An X-ray tube and X-ray generator are the necessary components for X-ray production and control. The X-ray tube provides the proper environment and components to produce X-rays, whereas the X-ray generator provides the source of electrical voltage and user controls to energize the X-ray tube. Basic components [3] of an X-ray system are illustrated in Figure 3.2.

The production of X-rays involves accelerating a beam of electrons to strike the surface of a metal target. The X-ray tube has two electrodes, a negatively charged cathode, which acts as the electron source, and a positively charged anode, which contains the metal target. A potential difference is applied between the cathode and the anode; the exact value depends on the particular application [4].

X-ray production occurs when the highly energetic electrons interact with the X-ray tube anode or target [5]. During the interaction of the accelerated incident electrons with the target nucleus, kinetic energy is lost and converted to electromagnetic radiation with equivalent energy in a process known as Bremsstrahlung (a German term meaning "braking radiation").

In an X-ray tube the electrons emitted from the cathode are accelerated toward the metal target anode by an accelerating voltage of typically 50 kV. The high-energy electrons interact with the atoms in the metal target. Sometimes the electron comes very close to a nucleus in the target and is deviated by the electromagnetic interaction. In this Bremsstrahlung process, the electron loses much energy and a photon (X-ray) is emitted. The energy of the emitted photon can take any value up to a maximum corresponding to the energy of the incident electron.

When X-rays are directed into an object, some of the photons interact with the particles of the matter and their energy can be absorbed or scattered. This absorption and scattering is called *attenuation*. Other photons travel completely through the object without interacting with any of the material's particles. The number of photons transmitted through a material depends on the thickness, density, and atomic number of the material and on the energy of the individual photons.

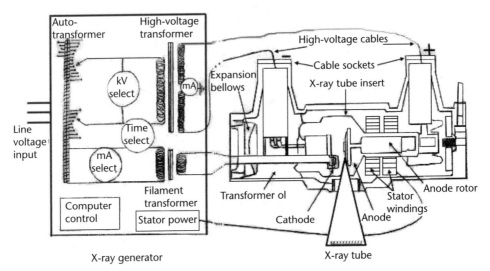

Figure 3.2   X-ray generator and X-ray tube components.

Even when they have the same energy, photons travel different distances within a material simply based on the probability of their encounter with one or more of the particles of the matter and the type of encounter that occurs. Because the probability of an encounter increases with the distance traveled, the number of photons reaching a specific point within the matter decreases exponentially with distance traveled. As shown in Figure 3.3 at the top, if 1,000 photons are aimed at ten 1-cm layers of a material and there is a 10% chance of a photon being attenuated in this layer, then 100 photons will be attenuated. This will allow 900 photons to travel into the next layer where 10% of these photons will be attenuated. By continuing this progression, the exponential shape of the curve becomes apparent. The formula that describes this curve is $I = I_0 e^{-\mu x}$, where

I = the intensity of photons transmitted across some distance x;

I_0 = the initial intensity of photons;

x = distance traveled;

μ = attenuation coefficient.

There are five different modes of interaction of X-rays with matter: coherent (Rayleigh) scattering, Compton scattering, photoelectric effect, photodisintegration, and pair production [6].

Coherent (Rayleigh) scattering represents a nonionizing interaction between X-rays and tissue. The X-ray energy is converted into harmonic motions of the electrons in the atoms of the tissue.

Compton scattering [6] refers to the interaction between an incident X-ray and a loosely bound electron in an outer shell of an atom in the tissue. A fraction of the

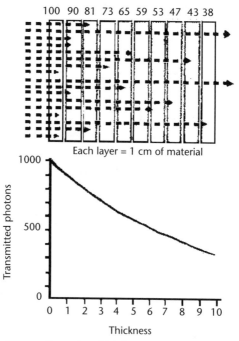

Figure 3.3 Penetration of X-rays through a thick layer.

X-ray energy is transferred to the electron, the electron gets ejected, and the X-ray is deflected from its original path as shown in Figure 3.4.

Photoelectric interactions [6] in the body involve the energy of an incident X-ray being absorbed by an atom in tissue, with a tightly bound electron being emitted from the K or L shell as a "photoelectron," as shown in Figure 3.5.

Photo disintegration [6] involves nuclear reactions. The photons interact with or are absorbed by the nucleus of the target atoms. In this modality, one or more nuclear particles are ejected. This results in one element becoming a different element.

Pair production [6] is characterized by a photon/nucleus interaction. In this process, high-energy photons are absorbed by a nucleus, and a positron (a positive electron—a form of antimatter) is emitted along with an electron.

The basic principle behind CT is that the two-dimensional internal structure of an object can be reconstructed from a series of one-dimensional "projections" of the object acquired at different angles. To obtain an image from a thin slice of tissue, the X-ray beam is collimated to give a thin beam. The detectors, which are situated opposite the X-ray source, record the total number of X-rays that are transmitted through the patient, producing a one-dimensional projection. The signal intensities

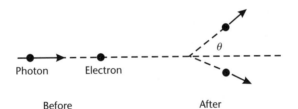

Figure 3.4   An elastic collision between the incident X-ray photon and a free electron.

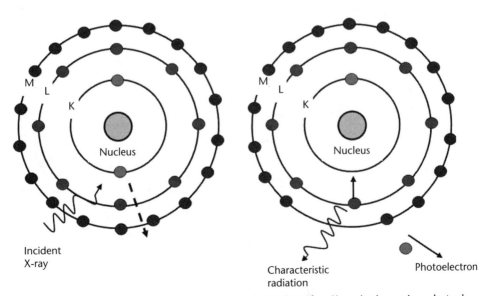

Figure 3.5   A schematic of the first two stages of absorption of an X-ray in tissue via a photoelectric interaction.

in this projection are dictated by the two-dimensional distribution of tissue attenuation coefficients within the slice. The X-ray source and the detectors are then rotated by a certain angle and the measurements repeated. This process continues until sufficient data have been collected to reconstruct an image with high spatial resolution. Figure 3.6 shows a fourth-generation CT scanner. Each generation shows improvements from its previous generation with an increased number of detectors and a wider fan beam.

3.1.2 Applications

Computed tomography has a wide range of medical applications. The original systems were dedicated to head imaging only, but "whole-body" systems with larger patient openings became available in 1976. However, since then, CT has been used for other purposes such as imaging of the eye. It is able to detect calcifications to a high degree of accuracy [7, 8], which is one determining factor of retinoblastoma. Studies have shown that it is more specific than magnetic resonance imaging (MRI) in reliably detecting elevated, very small calcified lesions. This is especially important because intraocular calcification in children, especially those younger than 3 years of age, is highly suggestive of retinoblastoma. Helical CT axial scanning with multiplanar reconstruction [9] is accurate at detecting and localizing intraocular and orbital metallic, glass, and stone foreign bodies. A CT scan is commonly used for evaluation of patients with trauma, fractures, tumors, and a multitude of other disorders where evaluation of tissues and structures deep within the body is necessary. Although a CT scan may not produce the resolution of an MRI, it is particularly useful for patients who have, or are suspected of having, metallic foreign bodies.

3.1.3 Advantages

One advantage of CT is that computed tomography is useful in showing foreign bodies and calcification in intraocular patients.

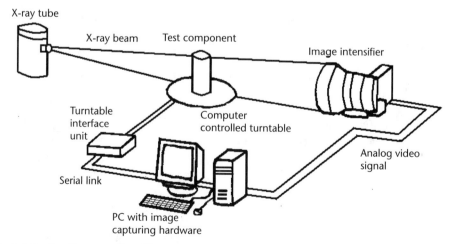

Figure 3.6 Fourth-generation CT scanner.

Another important advantage of computed tomography is its speed. It has a shorter scanning time compared to other scanning modalities such as MRIs. It takes only a few seconds to complete a scan based on the type of scan it is performing.

3.1.4   Limitations

One of the critical disadvantages of CT is that it utilizes ionizing radiation. Because ionizing radiation damages tissues, overexposure can cause irreparable damage to patients. Thus, there is a limit as to how much radiation a patient can be subjected to per year. Radiation dose is a matter of particular concern in pediatric and obstetric radiology [4].

Despite the inherent high resolution of CT, it is not very adept at distinguishing tissues with very similar density [7] because it lacks superior contrast resolution. It poorly differentiates subretinal fluid from tumor, reducing accuracy in measuring tumor size.

3.2   Confocal Laser Scanning Microscopy

Confocal microscopy has evolved greatly since it was pioneered by Marvin Minsky in 1955 [10]. But the principle of confocal imaging is still employed in all modern confocal microscopes [11].

A confocal microscope is capable of creating images that have better resolution than that of a conventional microscope. This is achieved by excluding most of the light from the specimen that is not from the microscope's focal plane, which results in an image with less haze and better contrast. Apart from that, a confocal microscope is able to build three-dimensional reconstructions of a volume of the specimen by assembling a series of thin slices taken along the vertical axis [11]. When there is a series of images, a three-dimensional representation of the object is produced by optical (as opposed to physical) sectioning. Although often thought of only as an instrument for creating three-dimensional images of live cells, the great versatility of the confocal microscope helps when examining structural details and also the dynamics of cellular processes that are being developed [10].

Two-dimensional images of an object will be formed on x and y planes in a conventional light microscope, generally parallel to the plane of sectioning. This is also true for confocal microscopy, but each image represents just a part of the sample's thickness. This is because the microscope's optics excludes features outside the plane of focus. Thus, we have the term *confocal*, in which two lenses are arranged to focus on the same point and are, therefore, sharing the same foci [10]. Also as in conventional microscopy, it is the objective lens of the microscope that forms the image in a confocal microscope, and the quality of the image, the magnification, and resolution are dependent on the quality of the objective lens used.

Three types of confocal microscopes are commercially available: confocal laser scanning microscopes, spinning-disk (Nipkow disk) confocal microscopes, and programmable array microscopes (PAMs). The confocal laser scanning microscope and Nipkow disk confocal microscope use an acousto-optic deflector (AOD) for steering the excitation light. Two deflectors can be used in series and at right angles, to provide full two-dimensional (x,y) scanning. The AOD is used to scan a laser beam over a range of angles or to control the output angle of a laser beam with great accuracy. A Bragg configuration produces a single first-order output beam, whose intensity is directly linked to the power of the radio-frequency (RF) control signal and whose angle is directly linked to the RF. By varying the frequency of the RF control signal, the angle of the output laser beam is adjusted.

3.2.1 Principle

Confocal laser scanning microscopy (CLSM) was an important invention of the 1980s, since it offers observation of thin optical sections in thick, intact specimens. CLSM uses a tiny laser spot, focused in a defined image plane, to excite fluorescence (Figure 3.7). This spot is scanned in lines across the field of view, a process that resembles that of image formation by an electron beam in a TV screen. Detection of fluorescence is by a photomultiplier with high sensitivity.

The light source is a laser that produces high-intensity, coherent light of a defined wavelength. Between the laser and beam splitter, the light is mirrored into an excitation aperture (pinhole) that produces a sufficiently thin laser beam.

The scanning unit (raster scan) is responsible for illuminating and recording the entire field. This unit "writes" the laser spot along columns and lines into the image plane. Usually two galvanometers are used to move a mirror that shifts the laser spot in the x-axis up to the end of a line, then one line down (one step in the y-axis) and again along the x-axis, and so forth.

In CLSM, images are not directly visible. The different intensities recorded by the photomultiplier (PMT) over time have to be converted back into x/y information by the computer controlling the CLSM. During scanning, the PMT readings are depicted on the CLSM monitor. The user can thus control the intensity by increasing or decreasing the voltage of the photomultiplier. If fluorescence is excited here, it forms one image point (pixel) out of a raster of usually 512×512 pixels. After trans-

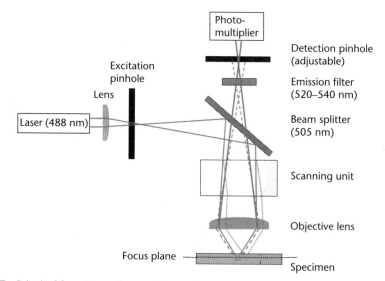

Figure 3.7 Principal function and ray pathway of a CLSM. Laser, beam splitter, and excitation filter wavelengths are enlarged for fluorescein detection.

mission through the beam splitter, only this pixel is detected by the photomultiplier, because the emission pinhole is in a confocal position (conjugate) to the excitation pinhole.

These changes smudge or sharpen the contrast in a CLSM image. The aim of voltage and offset changes is to produce an image with a black background and sufficiently outstanding fluorescence [13].

The image is displayed on the computer screen as a shaded gray image with 256 levels of gray, which can be suitably colored later for presentation [14]. For a 512×512-pixel image, this is typically done at a frame rate of 0.1 to 30 Hz.

Out-of-plane, unfocused light has been rejected, resulting in a sharper, better resolved images [12].

3.2.2   Applications

A confocal laser scanning microscope has a wide range of applications. One main application of CLSM is the two- and three-dimensional study of dynamics in living cells. Modern CLSMs are capable of producing an image of 512×512 pixels, a standard image format for CLSM, within 0.1 to 1 second. If the z-axis is unchanged, a series of CLSM scans represents a movie showing changes in a cell over time [15].

When used in conjunction with various antibodies conjugated with fluorescent markers, the relative distributions of epitopes of interest can also be studied. CLSM is able to perform a quantitative analysis and visualization of the spatial arrangement of the keratocyte network in living human corneal tissue [16]. This helps to study the alterations of the normal cellular arrangements in corneal disease.

In addition, CLSM of excised choroidal neovascularization (CNV) simulates fluorescein angiography and topographic localization of vascular channels, macrophages, retinal pigment epithelium, and other components of CNV [17]. This provides insight into the pathogenesis of CNV associated with age-related maculopathy.

3.2.3   Advantages

One of the hallmark features of this process is the ability to obtain serial optical sections with a confocal microscope. Using digital image processing techniques, these serial images can be reassembled to form three-dimensional representations of the structures [18, 19].

When compared with conventional epifluorescence microscopy, CLSM has two significant advantages. It has the ability to eliminate out-of-focus noise and obtain better resolution with high sensitivity of the machine. A CLSM has greatly increased sensitivity compared to a conventional epifluorescence microscope. In addition to having highly sensitive light detectors, the confocal microscope has the advantage of being able to accumulate or average images over time. This method of collection greatly increases the sensitivity of the machine, although with the disadvantage that cellular movement may distort the image [14].

Figure 3.8 shows the confocal microscopy images of the different conditions of the cornea [20].

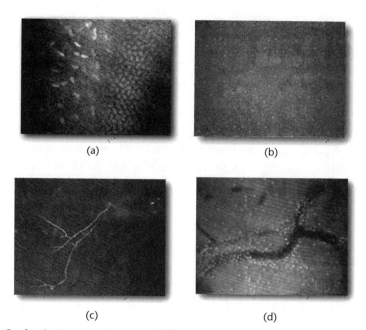

Figure 3.8 Confocal microscopy images of different corneal conditions. (a) This oblique view of the cornea shows the stromal keratocyte nuclei (left) adjacent to the endothelial cell layer (right). (b) A confocal image of the cornea used for research. (c) This patient presented with a corneal ulcer after trauma. Confocal microscopy revealed the fine filaments of *Aspergillus fumigatus*. (d) Thirty minutes after an applied stimulus, leukocytes are seen marginating within and surrounding the featured corneal limbal vessel. (*From:* [20]. © Reproduced with permission from the Louisana State University Eye Center of Excellence.)

Other advantages of confocal microscopy are the ability to use optical sectioning for the analysis, imaging of thick samples (up to a few hundred microns in thickness), and elimination of the out-of-focus information. The reduced risk of specimen or sample bleaching is another advantage. Bleaching is a major problem with fluorophores because they fade irreversibly when exposed to excitation light (Figure 3.9). The rectangular region near the center was faded after about 30 seconds of exposure to the excitation light [21, 22]. The scanning system of the confocal microscope reduces the exposure of each point to the strong light used for exciting fluorochromes, thus decreasing the risk of bleaching the sample [14].

3.2.4 Limitations

Despite the number of advantages, CLSM also has its limitations. The first limitation of confocal microscopy is that the resolution is limited by the wavelength of light. Although a confocal microscope pushes the limit of resolution to the theoretical limit of light microscopy, it does not resolve better than about 0.1 μm even under ideal conditions. This limitation in resolution greatly restricts imaging of biology structures (subcellular structures). The CLSM also lacks in millimeter penetration depth ability [14].

Photodamage is another limitation in the use of confocal microscopes. Although the damage is less than that of a conventional microscope, the high intensity of the

Figure 3.9 Dyed suspension of densely packed polymethylmethacrylate beads with significant photobleaching.

laser, when focused onto a fine spot within the sample, can still result in photodamage to both the dye being used and to other cellular components. It is possible to reduce this damage through the use of suitable antifade reagents that can control photobleaching of the dye. However, antifade reagents work well in fixed cell preparations, but are not as effective or may be toxic to live cells [10].

3.3   Magnetic Resonance Imaging

Magnetic resonance imaging is based on the physics of nuclear magnetic resonance (NMR), a property of atomic nuclei that was discovered in the 1940s [23].

An MRI scan is a radiology technique that uses magnetism, radio waves, and a computer. The MRI scanner (Figure 3.10) is a tube surrounded by a giant circular magnet. The patient is placed on a movable bed that is rolled into the magnet. The magnet creates a strong magnetic field that aligns the protons of hydrogen atoms. The subject is then exposed to a series of radio waves. This spins the various protons of the body, and a faint signal is generated that is detected by the receiver portion of the MRI scanner. The receiver information is processed by a computer, and an image is then produced. The image and resolution produced by MRI is fairly detailed and can detect tiny changes of structures within the body. For some procedures, dyes such as gadolinium are used to increase the accuracy of the images [24]. With the information collected by the computer, two-dimensional images or three-dimensional models can be created by integrating all of the information together.

The MRI technique is very valuable for the diagnosis of a broad range of pathological conditions in all parts of the body including cancer, heart and vascular disease, stroke, and joint and musculoskeletal disorders. MRI requires specialized equipment and expertise and allows evaluation of some body structures that may not be as visible with other imaging methods [25]. Several types of specialized MRI

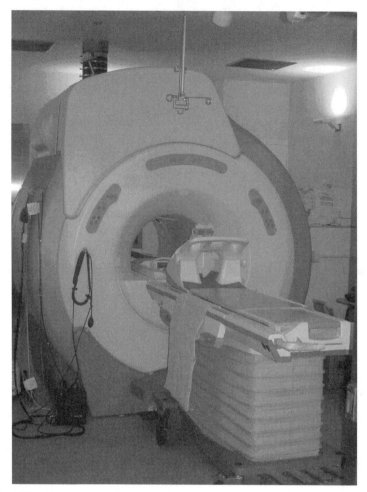

Figure 3.10   MRI scanner.

are available. For example, diffusion MRI measures the diffusion of water molecules in biological tissues, magnetic resonance angiography (MRA) generates pictures of the arteries in order to evaluate them for stenosis (abnormal narrowing) or aneurysms (vessel wall dilations at risk of rupture), and functional MRI (fMRI) measures signal changes in the brain that result from changing neural activity.

3.3.1   Principle

MRI typically measures the response of hydrogen molecules to a perturbation while in a magnetic field. The principles of MRI can be classified into three steps: (1) the alignment of atomic nuclei, (2) excitation by application of an RF pulse, and (3) T_1 and T_2 relaxation.

The alignment of atomic nuclei with the magnetic field is provided by an MRI scanner. The atomic nuclei are electrically charged and spin around their axes (Figure 3.11). Out of many nuclei within tissues, the hydrogen nucleus [25] is most commonly measured in MRI. Hydrogen nuclei are positively charged particles that

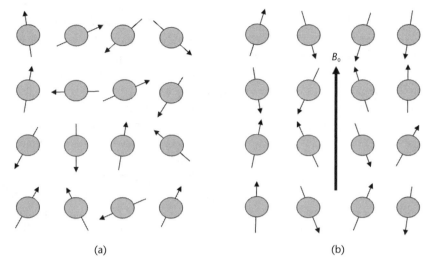

Figure 3.11   (a) A collection of hydrogen nuclei (spinning protons) in the absence of an externally applied magnetic field. (b) An external magnetic field B_0 is applied.

spin around their axis. When an electrically charged particle moves, it produces a magnetic field.

Next is the excitation of hydrogen nuclei through application of an RF pulse. The RF pulse is typically an electromagnetic wave resulting from the brief application of an alternating current perpendicular to the direction of the main magnetic field. The direction of the main magnetic field is referred as the *z-axis*. Before the excitation by RF pulse, the amplitude in the z-axis is zero and the amplitude in x-y plane is at maximal. During excitation, the amplitude in the z-axis slowly increases, whereas the amplitude in x-y plane slowly decreases. So there will be two forms of relaxation. First, the decay of the amplitude in the z-axis is known as T_1 (spin-lattice relaxation time) relaxation. T_1 is, by definition, the component of relaxation that occurs in the direction of the ambient magnetic field. This generally comes about by interactions between the nucleus of interest and unexcited nuclei in the environment, as well as electric fields in the environment (collectively known as the *lattice*). Therefore, T_1 is known as "spin-lattice" relaxation and is measured as the time required for the magnetization vector (M) to be restored to 63% of its original magnitude. It varies with magnetic field B_0.

Second, the regrowth of the amplitude in x-y plane is known as T_2 (spin-spin relaxation time) relaxation [26, 27]. T_2 is defined as the time required for the transverse magnetization vector to drop to 37% of its original magnitude after its initial excitation. Unlike T_1, T_2 is much less susceptible to variations of field strength B_0.

Different tissues have different T_1 and T_2 relaxation rates. When the MRI signal is measured at a point, the ratio of the amplitudes of different tissues in the z-axis is maximized and this signal is a T_1-weighted signal. Alternatively, when the MRI signal is measured at a point, the ratio of the amplitudes of different tissues in the x-y plane is maximized and this signal is known as a T_2-weighted signal. When the time from RF pulse to measurement of the signal (T_E) is kept short, while at the same time the time between two successive RF pulses (T_R) is also kept short, the difference in T_1 for the different tissues is maximized and the acquired scan is called a

T_1-weighted scan (Figure 3.12) [28]. T_1-weighted scans are also known as anatomi-cal scans, because they particularly show good contrast between gray and white matter. On the other hand, when the T_E and the T_R are both long simultaneously, the difference in T_2 for the different tissues is maximized and the acquired scan is called a T_2-weighted scan (Figure 3.13) [28]. T_2 weighted scans are also known as patho-logical scans, because lesions appear very bright in a T_2 weighted scan [26, 27].

To summarize, an MRI system uses magnetism to cause alignment of the hydro-gen nuclei in tissues and apply a brief RF pulse to energize or excite the hydrogen nuclei to a high-energy state. After the RF pulse, the hydrogen nuclei will start to "de-energize" or give out energy and this process is known as T_1 and T_2 relaxation. The system can be configured by changing certain scanning parameters to acquire either a T_1-weighted signal or T_2-weighted signal [29]. The signal will be processed by the computer and two-dimensional images or three-dimensional models will be generated and displayed.

3.3.2 Applications

MRI is widely used as biomedical scanning modality. Pre- and postcontrast mag-netic resonance studies help in differentiation of solid intraocular tumors [30], such as retinoblastoma, medulloepithelioma, retinal capillary hemangioma, leiomyoma, and choroidal melanoma, from intraocular lesions with primary retinal detachment such as Coats' disease, primary hyperplastic persistent vitreous, massive retinal

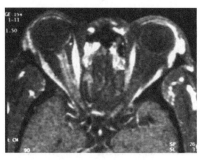

Figure 3.12   Axial, contrast-enhanced, T_1-weighted image showing optic nerve and ethmoid paranasal sinus. (*From:* [28]. © 2002 Pharmacotherapy. Reprinted with permission.)

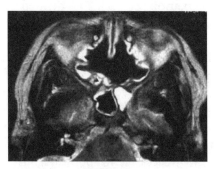

Figure 3.13   Axial T_2-weighted magnetic resonance image showing the maxillary sinuses and sphenoid sinuses. (*From:* [28]. © 2002 Pharmacotherapy. Reprinted with permission.)

gliosis (phthisis bulbi), premature retinopathy with total retinal detachment, and associated subretinal fluid or hemorrhage. However, various solid intraocular tumors were not reliably differentiated from one another based on magnetic resonance features.

Pre- and postcontrast-enhanced, T_1-weighted images with a fat-suppression technique [31] are also used to detect small intraocular tumors with a thickness of more than 1.8 mm and to evaluate intraocular neoplasms and simulating lesions, particularly when T_2-weighted images are not available.

Ocular dimensions can be calculated along the three cardinal axes with multislice magnetic resonance images of the eye [32]. This is used for eye diseases that can be diagnosed via ocular dimensions.

Cine MRI [33] is a technique in which multiple sequential static orbital MRI films are taken while the patient fixates a series of targets across the visual field. The excellent soft tissue differentiation of MRI, combined with the dynamic imaging, allows rapid visualization and functional assessment of the extraocular muscles. But this technique does not allow for the study of saccadic or pursuit eye movements. This is good for detection of ocular motility disorders, including thyroid-related ophthalmopathy, blowout fracture, postoperative lost or slipped muscle, and Duane's syndrome.

High-resolution magnetic resonance images of the eye can directly measure the relationship between ciliary muscle contraction and lens response [34]. This is used for diagnosis of presbyopia.

3.3.3 Advantages

One of the many noticeable advantages of MRI is that it is a noninvasive cross-sectional imaging modality. Patients just need to lie on a bed and all will be taken care of by the machine. MRI uses nonionizing radiation, which is comfortable for patients because they do not have to worry about possible side effects and, moreover, the contrast materials used for MRI scanning also have a very low incidence of side effects.

One major advantage of MRI is that it is capable of imaging in any plane. For CT, imaging is limited to the axial plane, whereas for MRI, imaging can be done at the axial, sagittal, and coronal planes and at any other degree in between without moving the patient.

3.3.4 Limitations

One critical drawback of MRI is its long acquisition time. Patients have to stay in the scanner for an extended period of time ranging from 20 to 90 minutes and they have to hold very still during the scanning period because even the slightest movement of the part being scanned can cause heavily distorted images and the process will have to be repeated. MRI subjects patients to a very strong magnetic field, so patients with pacemakers, cochlear implants, and certain other metallic foreign bodies are not eligible for MRI [35]. The only other disadvantage of the MRI system is that it is expensive.

3.4 Optical Coherence Tomography

Optical coherence tomography (OCT) is a noninvasive technique used for cross-sectional subsurface tissue imaging (micron-scale imaging) based on the principles of low-coherence interferometry [36]. OCT has developed rapidly since its potential for applications in clinical medicine was first demonstrated in 1991 by Huang et al. [37].

The technique is analogous to ultrasound imaging, with the exception that it measures the reflected intensity of near-infrared (NIR) light, rather than sound waves [38]. The much shorter wavelength of NIR light when compared to ultrasound allows spatial resolutions of around 10 μm, approximately 10 times higher than conventional ultrasound, and sensitivity may be greater than 110 dB [39]. Furthermore, unlike ultrasound, optical coherence tomography does not require direct contact with the tissue being imaged. Due to the fact that light is strongly scattered in most tissue, OCT's imaging depth is limited to a few millimeters into the body. So it would not be possible for OCT to perform direct imaging of deep structures within the body.

Despite this limitation, OCT has many applications in fields such as ophthalmology, dermatology, cardiology, and detection of cancers. In the detection of cancers, OCT may replace excision biopsy with "optical biopsy," negating the requirement for physical samples to be removed from the body for analysis and instead imaging the sample in situ and in real time.

3.4.1 Principles

Optical coherence tomography, as the name suggests, uses coherence gating in low coherence interferometry (LCI) to measure the echo time delay from reflective boundaries and backscattering sites within the sample. These reflective boundaries are due to changes of refractive index within the sample.

The main principle behind OCT is low coherence interferometry, which makes use of a Michelson interferometer (or a variant) to perform OCT. Figure 3.14 illustrates an optical configuration of a typical LCI system based on fiber optics. A single 2×2 optical fiber coupler is used to implement a Michelson interferometer. The power of a low-coherence source, typically a superluminescent diode, is divided into the sample arm and the reference arm of the interferometer through the fiber coupler. The sample arm fiber will collect the light reflected by the reflectors in the sample, and the light reflected by the reference mirror is re-collected by the reference arm fiber. The light from both the sample and reference arms is again coupled back in the fiber coupler and part of it is redirected toward the detector. Due to the finite coherence length of the source, optical interference is observed only when the optical path lengths of the beams reflected by the sample reflector and the reference mirror differ less than the coherence length [40].

Broad-bandwidth light sources are the requirement for high axial resolution, because the axial resolution is inversely proportional to the bandwidth of the light source. The preceding discussion focuses only on the axial resolution of the OCT image; in conventional optical microscopy, the axial and transverse resolutions are independent.

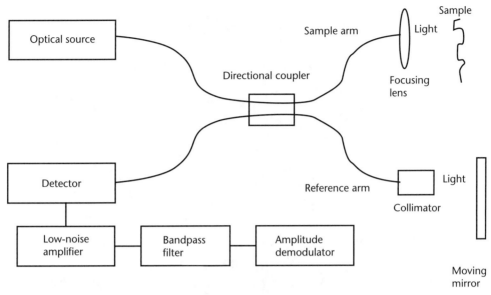

Figure 3.14   Typical configuration of a fiber-based system for low coherence interferometry.

3.4.2   Applications

The first OCT system was introduced by Huang et al. [37] and used a superluminescent diode (SLD) that operates at 830 nm. At this wavelength, the OCT is only useful for ophthalmic imaging because most tissues are nearly transparent, and the eye does not require high penetration power or wavelength to image any abnormalities within it. One application of ophthalmic OCT imaging is retinal OCT imaging (Figure 3.15). Retinal OCT imaging is now performed clinically, with commercially available instruments [41], to detect and diagnose early stages of ocular disease before physical symptoms and irreversible loss of vision occur. Another application of ophthalmic OCT imaging is the measurement of the retinal nerve fiber layer thickness in glaucoma and other diseases of the optic nerve [42, 43].

With advancements in technology, imaging in nontransparent tissue to depths of around 3 mm can be performed by means of longer wavelength sources (1,300 and 1,550 nm).

OCT is potentially a powerful tool for detecting and monitoring a variety of macular diseases, including macular edema, macular holes, and detachments of the neurosensory retina and pigment epithelium [44, 45].

Hence, OCT can be used for high-resolution, cross-sectional imaging of structures in the anterior segment of the human eye in vivo [46]. It has potential as a diagnostic tool for applications in noncontact biometry, anterior chamber angle assessment, identification and monitoring of intraocular masses and tumors, and elucidation of abnormalities of the cornea, iris, and crystalline lens.

3.4.3   Advantages

There are a few considerable advantages of OCT that make it the future of high-resolution biomedical imaging. One of the most prominent advantages would be that OCT is nonionizing over many wavelengths. Patients need not undertake the

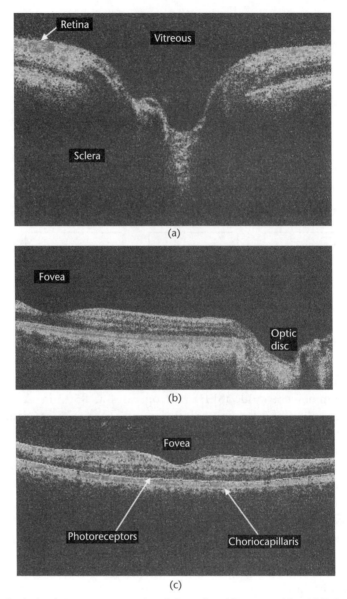

Figure 3.15 Optical coherence tomography of the retina. (Courtesy of Dr. Ali Tafreshi and Dr. Madhusudhan Balasubramanian, Hamilton Glaucoma Center, University of California, San Diego.)

risk of overexposure even when they are scanned multiple times over a short period of time. OCT utilization of NIR light permits relatively easy penetration of structures such as the skull, brain, and breast, which in turn aids the imaging of abnormalities within those structures. OCT is also more suitable for morphological tissue imaging than medical ultrasonography, MRI, or confocal microscopy [36, 37].

To summarize, NIR optical coherence tomography has been recognized as an ideal noninvasive diagnostic technique due to its potential low cost, and it has very few side effects.

3.4.4 Limitations

Like any other imaging system, OCT has its disadvantages. One of its main disadvantages would be the result caused by the scattering of light. It results in small strength and, thus, relatively large changes in the parameters of tissue tend to result in relatively small changes in the measurements. This leads to an issue of sensitivity and computational power because the system must utilize more computational power to compute the changes and display.

3.5 Ultrasound Imaging

Ultrasound utilizes high-frequency sound waves, which are reflected in specific ways by different tissues, normal or pathological, in the body. Ultrasound is mechanical, high-frequency, longitudinal vibration of molecules, and it differs from usual sound only by its frequency. It is does not emit ionizing radiation and is not harmful when used for diagnostic purposes [47].

In the last century, medical imaging was dominated by x-radiography, with a small contribution from radionuclide studies. At that time, ultrasound imaging was generally considered to be a laboratory curiosity with exceptions perhaps for applications in obstetrics, gynecology, and cardiology [48]. With the remarkable advances that have taken place in the physics and engineering of ultrasonic imaging, more than one out of every four medical diagnostic imaging studies in the world is now estimated to be an ultrasound study, and the proportion continues to increase [49].

Some major characteristic of ultrasound imaging is that it is a relatively inexpensive technology, spatial resolution of 1 mm can be obtained in abdominal scanning, and it exhibits good tissue contrast, which can be enhanced by means of contrast agents. In addition, it is a real-time method, convenient to use, very acceptable to patients, and apparently quite safe. However, images taken by ultrasonic imaging can be spoiled by the presence of bone or gas, and the operator needs a high level of skill, both in image acquisition and interpretation [48].

3.5.1 Principle

Basically, all ultrasound imaging is performed by emitting a pulse, which is partly reflected from a boundary between two tissue structures, and partially transmitted (Figure 3.16). The reflection depends on the difference in impedance of the two tissues. Figure 3.16 shows that part of the pulse energy is transmitted from the scatterer a, some from b, and the last of it from c. When the pulse returns to P, the reflected pulse provides information about two measurements: the amplitude of the reflected signal, and the time it takes to return, which is dependent on the distance from the probe (twice the time the sound uses to travel the distance between the transmitter and the reflector, as the sound travels back and forth). The amount of energy being reflected from each point is given in Figure 3.16 as the amplitude. When this is measured, the scatterer is displayed with amplitude and position. Thus, incoming pulse a is the full amplitude of P. At b, the incoming (incident) pulse is the pulse transmitted through a. At c, the incident pulse is the transmitted pulse from b [50].

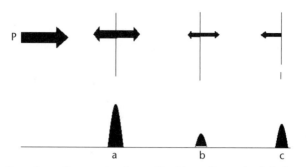

Figure 3.16 Reflection of an ultrasound pulse emitted from the probe P, being reflected at a, b, and c. Arrows pointing left represent reflected pulse. Arrows pointing right represent transmitted pulse. Part of the pulse energy is transmitted from the scatterer a, the rest is transmitted partially from b and the rest from c.

When an ultrasonic wave is scattered by a target that has a component of velocity along the direction of beam propagation, the frequency of the scattered ultrasound is shifted by the Doppler effect [48].

The basic properties of biological materials that are of importance in ultrasonic imaging are attenuation, speed, and reflectivity. Typical values for various tissues are given in Table 3.1; frequency dependences are within the range of 1 to 10 MHz (except for bone, for which the range is 1 to 2 MHz). The attenuation of an ultrasound wave in a medium is due partly to scattering and partly to absorption. In soft tissues, absorption is mainly due to the spectra of relaxation processes [51], which accounts for the nearly linear frequency dependence.

Ultrasonic imaging depends on the interactions between the structures of the human body and ultrasonic radiation. As a starting point, the ultrasonic wavelength can be considered to determine the spatial resolution. Structures of interest in abdominal cavity imaging are likely to be located at distances of up to about 150 mm beyond the skin surface and spatial resolution on the order of 1 mm is needed.

Table 3.1 Properties of Some Materials Relevant to Ultrasonic Imaging

Material	Propagation Speed, c ($m * s^{-1}$)	Characteristic Impedance, Z ($10^6 kg * m^{-2} * s^{-1}$)	Attention Coefficient, α at 1 MHz ($dB\ cm^{-1}$)	Frequency Dependence of α	Nonlinear Parameter, B/A
Air	330	0.0004	1.2	f^2	
Blood	1,570	1.61	0.2	$f^{1,3}$	6.1
Brain	1,540	1.58	0.9	f	6.6
Fat	1,450	1.38	0.6	f	10
Liver	1,550	1.65	0.9	f	6.8
Muscle	1,590	1.70	1.5–3.5	f	7.4
Skull bone	4,000	7.80	13	f^2	
Soft tissue (mean value)	1,540	1.63	0.6	f	
Water	1,480	1.48	0.002	f^2	5.2

Data collected from [50, 52]. Gaps in the table indicate that published data are not readily available [48].

To obtain this resolution, the wavelength of the ultrasound must not be greater than 1 mm (frequency of 1.5 MHz). The problem is that the attenuation increases with the frequency, so that the distance over which useful levels of energy can be propagated is reduced as the frequency is increased. An ultrasonic beam will also be distorted as it travels through tissue as a result of tissue nonhomogeneity [48].

The pulse-echo method depends on the measurement of the time that elapses between the transmission of a pulse of ultrasound and the reception of its echo from a reflecting or scattering target (from which the distance to the source of the echo can be calculated, if the propagation speed is known). The amplitude of the echo (which is related to the ultrasonic properties of the target) is measured. The spatial resolution is determined, in elevation and in azimuth, by the cross-sectional dimensions of the ultrasonic beam, and, in depth, by the duration of the ultrasonic pulse [48].

3.5.2   Color Doppler Imaging

Doppler ultrasound is a technique for making noninvasive velocity measurements of blood flow. Christian Doppler was the first to describe the frequency shift that occurs when sound or light is emitted from a moving source and the effect now bears his name. For the velocity measurement of blood, ultrasound is transmitted into a vessel and the sound that is reflected from the blood is detected. Because the blood is moving, the sound undergoes a frequency (Doppler) shift.

Three main techniques are used to make Doppler ultrasound velocity measurements of blood flow: continuous wave Doppler, pulsed Doppler, and color Doppler.

Color Doppler ultrasound (also referred to as color flow ultrasound) is a technique for visualizing the velocity of blood within an image plane. A color Doppler instrument measures the Doppler shifts in a few thousand sample volumes located in an image plane. For each sample volume, the average Doppler shift is encoded as a color and displayed on top of the B-mode image. The way in which the frequency shifts are encoded is defined by the color bar located to the left of the image. Positive Doppler shifts, caused by blood moving toward the transducer, are encoded as red, and negative shifts are encoded as blue. Color Doppler images are updated several times per second, thus allowing the flowing blood to be easily visualized [51]. However, color Doppler requires more electronics circuits and computational power from the Doppler instrument and, therefore, is relatively expensive.

Color Doppler ultrasound obviates many of these problems and can be used to examine orbital blood vessels not amenable to other methods of investigation. It also has the advantage of not requiring additional pharmacologic agents or abnormal physiologic circumstances to perform an examination. Figure 3.17 shows the grayscale ultrasound image of the retinal detachment [53].

3.5.3   Applications

Ultrasound is a popular imaging technique that has found applications in a very wide spectrum, both medically and nonmedically.

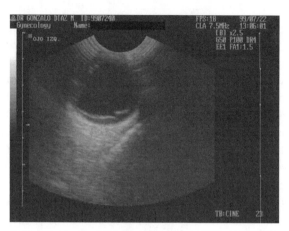

Figure 3.17 Ultrasonic image of retinal detachment. (*From:* [53]. © 2007 Dr. Gonzalo E. Diaz. Reprinted with permission.)

In the eye, it is most frequently used to measure axial eye length for estimation of intraocular lens (IOL) power (biometry). Other ocular applications include screening of eyes with opaque media, diagnosis of complex vitreoretinal conditions, and differentiation of intraocular masses. Ultrasound biomicroscopy (UBM) has allowed us to investigate subtypes of glaucoma (Figure 3.18), lesions in the iris, ciliary body, sclera, and anterior surface lens [54].

Color Doppler imaging [51] is a noninvasive ultrasound procedure that permits simultaneous grayscale imaging of structure and color-coded imaging of blood velocity. This improved technique allows the user to identify even very small blood vessels, such as those supplying the eye, from which measures of blood velocity and vascular resistance can be obtained

Color Doppler imaging has successfully demonstrated changes in orbital hemodynamics associated with a variety of pathologic conditions, including central retinal artery and vein occlusions, cranial arteritis, nonarteritic ischemic optic neuropathy, and carotid disease. In addition, the method has been used to detect the vascularization of orbital and ocular tumors, as well as to investigate altered hemodynamics associated with diseases such as glaucoma and diabetic retinopathy.

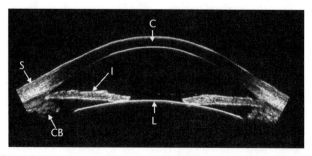

Figure 3.18 High-resolution ultrasound image of the anterior segment of the eye obtained with arc-scan geometry. Visualized structures include the cornea (C), sclera (S), iris (I), anterior lens surface (L), and ciliary body (CB). (*From:* [55]. © 2004 The Acoustical Society of America. Reprinted with permission.)

3.5.4 Advantages

Some known advantages of ultrasound are that ultrasound examination is noninvasive, and the technology used is relatively inexpensive, captures images quickly, and is convenient to use. It is completely harmless. Studies have been done, however, on the thermal and nonthermal effects of ultrasound in which tissues absorb the thermal energy and cause tissue damage. Ultrasound is also particularly suited for imaging of soft tissues such as the heart, eye, and unborn babies.

3.5.5 Limitations

One of the major disadvantages is that ultrasound is only efficient under the hand of a skilled clinician for image acquisition and interpretation of the image. Ultrasound does not pass well through solid objects such as bones. Good scans of the brain have been made, but greater detail can be obtained through an MRI. Ultrasound also does not pass well through a gaseous medium, so it is not useful when areas such as the lung need to be imaged.

3.6 Discussion

CT is an increasingly rapid imaging modality with excellent image resolution, enabling faster and more accurate diagnostic evaluation of patients over a wide spectrum of clinical indications. The data acquired in one scan can subsequently be manipulated to provide multiplanar and three-dimensional reconstructions.

Radiation can contribute to more than 20% of the radiation dose to the population by "medical X-rays." An *artifact* is a feature or appearance that is seen on an image that does not actually exist. Artifacts occur in all imaging modalities and are often unavoidable. Recognizing the presence of artifacts is important in order to avoid confusion with pathology.

The MRI does not use any ionizing radiation and provides good spatial resolution. In MRI three tissue parameters (M_0, T_1, and T_2) influence the resonance process and thereby the resulting images, whereas CT images are a function of the absorption coefficient. In the eye tissue being observed is diseased, the values of T_1 and T_2 change, thereby changing the images. So in order to diagnose a disease, MRI offers more information than CT. The MRI image, which represents a sectional view, can be created in any required direction in the human body.

The MRI scanner cannot be used by people with pacemakers or other implanted and movable metal parts. The price of an MRI scanner is still more expensive than the price for CT equipment. Bones are not visualized (or only indirectly and not precisely) in MRI images.

Ultrasound methods are relatively inexpensive, quick, harmless, and convenient when compared to techniques such as X-rays or MRI scans. This contrasts with methods based on X-rays or on radioactive isotopes, which have known risks associated with them, such that ultrasound methods are preferred whenever possible. Ultrasound is particularly suited to imaging soft tissues such as the eye in situations where opacities are present (corneal scarring, cataract, hyphema or vitreous hemorrhage, pupillary constriction). The major disadvantage is that the resolution of images is often limited.

Confocal microscopy yields noninvasive serial optical sectioning deep within intact, or even living, three-dimensional specimens. It rejects out-of-focus information, yielding raw images that are essentially blur-free. In confocal microscopy, axial resolution is significantly improved over conventional optical microscopy.

Optical coherence tomography is a powerful, noninvasive diagnostic imaging technology that provides high-resolution, cross-sectional images of the eye. Numerous qualitative and quantitative studies on eye diseases ranging from glaucoma to age-related macular degeneration have been conducted using OCT. Recently there have been dramatic advances in OCT technology using spectral/Fourier domain detection that enables imaging speeds of about 25,000 axial scans per second, about 50 times faster than time-domain detection. Because of the increased speed of image acquisition, motion artifacts are minimized, and images provide a more accurate representation of the retinal topography.

OCT is sensitive to the distorting effects associated with a light wave propagating through a turbid, scattering sample. The signal is strongly attenuated, and multiple scattering effects tend to corrupt the detected signal. OCT images are subject to the corrupting effects of speckle—the coherent interference of multiple light waves—which can limit image fidelity, resolution, and depth range.

3.7 Conclusions

CT detects choroidal melanomas and determines its size. Ultrasound imaging, unlike CT, does not use any ionizing radiation and can provide quantitative measurement and imaging of blood flow. Like CT, it can also detect and measure choroidal melanomas. However, its accuracy in comparison with CT remains much debated. OCT's penetration of high scattering tissue is less than that of ultrasound, CT, and MRI. It is good for imaging transparent tissue of the human eye to detect small changes that are overlooked by equipment with better tissue penetration. CLSM is mostly used for detecting glaucoma because it is able to image the optic disc and cup—the factors used to determine glaucoma. MRI allows for rapid visualization of ocular muscles, and its hydrogen nuclei excitation allows lesions to be seen clearly.

References

[1] http://www.aafp.org/afp/20060315/1039.html.

[2] Davis, L. M., and L. Davis, "CT Scan," *eMedicine*, 2002, http://www.emedicine.com/aaem/topic560.htm.

[3] Seibert, J. A., "X-Ray Imaging Physics for Nuclear Medicine Technologists. Part 1: Basic Principles of X-Ray Production," *J. Nucl. Med. Technol.*, Vol. 32, No. 3, 2004, 139–147.

[4] X-Ray Imaging and Computed Tomography, http://www.imt.liu.se/edu/courses/TBMT02/pdf/ct/X-Ray.pdf.

[5] Bushberg, J. T., et al., *Essential Physics of Medical Imaging,* 2nd ed., Philadelphia, PA: Lippincott Williams & Wilkins, 2002.

[6]   Dove, E. L., *Physics of Medical Imaging—An Introduction,* Lecture Notes, http://www.engineering.viowa.edu/ubme/060/Lecture/F03/xrayF03.pdf 2003.

[7]   Mafee, M. F., et al., "Magnetic Resonance Imaging Versus Computed Tomography of Leukocoric Eyes and Use of *In Vitro* Proton Magnetic Resonance Spectroscopy of Retinoblastoma," *Ophthalmology,* Vol. 96, No. 7, 1989, pp. 965–975.

[8]   Beets-Tan, R. G., et al., "Retinoblastoma: CT and MRI," *Neuroradiology,* Vol. 36, No.1, 1994, pp 59–62.

[9]   Lakits, A., et al., "Orbital Helical Computed Tomography in the Diagnosis and Management of Eye Trauma," *Ophthalmology,* Vol. 106, No. 12, 1999, pp. 2330–2335.

[10] Hibbs, A., "Confocal Microscopy for Biologists: An Intensive Introductory Course," *BIOCON,* 2000, pp. 2–9.

[11] Minsky, M., "Memoir on Inventing the Confocal Microscope," *Scanning,* Vol. 10, 1988, pp. 128–138.

[12] Semwogerere, D., and E. R. Weeks, "Confocal Microscopy," in *Encyclopedia of Biomaterials and Biomedical Engineering,* London: Taylor & Francis, 2005, pp.1–10.

[13] Sheppard, C. J. R., and D. M. Shotton, "Introduction," in *Confocal Laser Scanning Microscopy,* New York: Springer-Verlag, 1997, pp. 1–13.

[14] Rezai, N., "Taking the Confusion Out of Confocal Microscopy," *BioTeach Journal,* Vol. 1, 2003, pp. 75–80.

[15] Confocal Laser Scanning Microscopy, http://www.staff.kvl.dk/~als/confocal.htm.

[16] Hahnel, C., et al., "The Keratocyte Network of Human Cornea: A Three-Dimensional Study Using Confocal Laser Scanning Fluorescence Microscopy," *Cornea,* Vol. 19, No. 2, 2000, pp. 185–193.

[17] Grossniklaus, H. E., et al., "Correlation of Histologic Two-Dimensional Reconstruction and Confocal Scanning Laser Microscopic Imaging of Choroidal Neovascularization in Eyes with Age-Related Maculopathy," *Arch. Ophthalmol.,* Vol. 118, 2000, pp. 625–629.

[18] http://www.confocal-microscopy.org.

[19] Sheppard, C., *Confocal Laser Scanning Microscopy,* New York: Springer-Verlag, 1997.

[20] The Louisiana State University Eye Center of Excellence, http://www.lsu-eye.lsuhsc.edu/Research/confocalmicroscopy.htm.

[21] Becker, P. L., "Quantitative Fluorescence Measurements," in *Fluorescence Imaging Spectroscopy and Microscopy,* X. F. Wang and B. Herman, (eds.), New York: John Wiley & Sons, 1996, pp. 1–29.

[22] Chen, H., et al., "The Collection, Processing, and Display of Digital Three-Dimensional Images of Biological Specimens," in *Handbook of Biological Confocal Microscopy,* 2nd ed., J. B. Pawley, (ed.), New York: Plenum Press, 1995, pp. 197–210.

[23] Noll, C. D., "A Primer on MRI and Functional MRI," http://psy.jhu.edu/~fall200-312/MRIprimer2.pdf, June 2001.

[24] Magnetic Resonance Imaging (MRI Scan), http://www.medicinenet.com/mri_scan/article.htm, 2007.

[25] MR Imaging—Body, http://www.radiologyinfo.org/en/info.cfm?pg=bodymr&bhcp=1.

[26] Horowitz, A. L., *MRI Physics for Radiologists,* New York: Springer-Verlag, 1995.

[27] Jezzard, P., and S. Clare, "Principles of Nuclear Magnetic Resonance and MRI," in *Functional MRI: An Introduction to Methods,* P. Jezzard, P. M. Matthews, and S. M. Smith, (eds.), New York: Oxford University Press, 2001, pp. 67–92.

[28] Mondy, K. E., et al., "Rhinocerebral Mucormycosis in the Era of Lipid-Based Amphotericin B: Case Report and Literature Review," *Pharmacotherapy,* Vol. 22, No. 4, 2002, pp. 519–526, http://www.medscape.com/viewarticle/432400_2.

[29] de Haan, B., and c. Rorden, "An Introduction to Functional MRI," http://www.sphi.sc.edu/comd/rorden/fmri.guide/index.html

[30] Potter, P. D., et al., " The Role of Magnetic Resonance Imaging in Children with Intraocular Tumors and Simulating Lesions," *Ophthalmology*, Vol. 103, No. 11, 1996, pp. 1774–1783.

[31] De Potter, P., et al., "The Role of Fat-Suppression Technique and Gadopentetate Dimeglumine in Magnetic Resonance Imaging Evaluation of Intraocular Tumors and Simulating Lesions," *Arch. Ophthalmology*, Vol. 112, No. 3, 1994, pp. 340–348.

[32] Cheng, H. M., et al., " Shape of the Myopic Eye as Seen with High-Resolution Magnetic Resonance Imaging," *Optom. Vis. Sci.*, Vol. 69, No. 9, 1992, pp. 698–701.

[33] Bailey, C. C., et al., "Cine Magnetic Resonance Imaging of Eye Movements," *Eye*, Vol. 7, No. 5, 1993, pp. 691–693.

[34] Strenk, S. A., et al., " Age-Related Changes in Human Ciliary Muscle and Lens: A Magnetic Resonance Imaging Study," *Investigative Ophthalmology & Visual Science*, Vol. 40, pp. 1162–1169.

[35] Nguyen, M. M., et al., "Computed Tomography and Magnetic Resonance Imaging in Paediatric Urology," *British Journal of Urology*, Vol. 98, No. 2, 2006, pp. 273–277.

[36] Izatt, J. A., et al., "Optical Coherence Tomography for Biodiagnostics," *Opt. Photon. News*, Vol. 65, May 1997, pp. 41–47, 65.

[37] Huang, D., et al., "Optical Coherence Tomography," *Science,* Vol. 254, No. 5035, 1991, pp. 1178–1181.

[38] Fujimoto, J. G., et al., "Optical Biopsy and Imaging Using Optical Coherence Tomography," *Nature Med.*, Vol. 1, No. 9, 1995, pp. 970–972.

[39] Rollins, A. M., et al., "*In Vivo* Video Rate Optical Coherence Tomography," *Opt. Express*, Vol. 3, No. 6, 1998, pp. 219–229.

[40] Dufour, M. L., et al., "Low-Coherence Interferometry, An Advanced Technique for Optical Metrology in Industry," Boucherville Quebec, Canada: Industrial Materials Institute, National Research Council, 2004.

[41] Zeiss, C., "Retinal Imaging System,"pending FDA approval, http://www.fda.gov/cdrh/pdf6/k063343.pdf.

[42] New England Eye Center, http://www.neec.com/pages/Glaucoma_OCT.htm.

[43] Schuman, J. S., et al., " Quantification of Nerve Fiber Layer Thickness in Normal and Glaucomatous Eyes Using Optical Coherence Tomography," *Arch. Ophthalmology*, Vol. 113, No. 5, 1995.

[44] Puliafito, C. A., et al., "Imaging of Macular Diseases with Optical Coherence Tomography," *Ophthalmology,* Vol. 102, No. 2, 1995, pp. 217–229.

[45] Gaudric, A., et al., " Macular Hole Formation New Data Provided by Optical Coherence Tomography," *Arch. Ophthalmology*, Vol. 117, 1999, pp. 744–751.

[46] Izatt, J. A., et al., " Micrometer-Scale Resolution Imaging of the Anterior Eye *In Vivo* with Optical Coherence Tomography," *Arch. Ophthalmology*, Vol. 112, No. 12, 1994.

[47] GE Healthcare Medical Diagnostics, http://193.71.11.91/public/medical/ultrasound.shtml.

[48] Wells, P. N. T., "Ultrasonic Imaging of the Human Body," Rep. Prog. Phys., Vol. 62, 1999, pp. 671–722.

[49] Støylen, A., "Basic Ultrasound for Clinicians," 2006, http://folk.ntnu.no/stoylen/strainrate/Ultrasound.

[50] Duck, F. A., *Physical Properties of Tissues*, London, U.K.: Academic, 1990.

[51] Jaffe, R., and S. L. Warshof, *Color Doppler Imaging in Obstetrics and Gynecology*, New York: McGraw-Hill, 1992.

[52] Wells et al. 1997 [AU: Please supply missing reference.]

[53] http://www.drgdiaz.com/images/eye-rbn.shtml, 2007.

[54] Ultrasound Maps the Eye, http://www.acoustics.org/press/147th/Silverman.htm.

[55] Silverman, R. H., "Ultrasound Maps the Eye," *147th Acoustical Society of America Meeting*, New York, May 24, 2004, http://www.acoustics.org/press/147th/Silverman.htm.

Automatic Identification of Anterior Segment Eye Abnormalities in Optical Images

Rajendra Acharya, Eddie Y. K. Ng, Lim Choo Min, Caroline Chee, Manjunath Gupta, and Jasjit S. Suri

The eyes are complex sensory organs that are designed to optimize vision under conditions of varying light. A number of eye disorders can influence vision. Eye disorders among the elderly are a major health problem. With advancing age, the normal function of eye tissues decreases and there is an increased incidence of ocular pathology.

The most common symptoms elicited from ocular diseases are few in number and nonspecific in nature: blurred vision, pain, and redness. *Cataracts* occur most frequently in older people and have significant impact on an individual's quality of life. Effective therapies and visual aids are available for these potentially vision-limiting conditions. *Corneal haze* is a complication of refractive surgery characterized by the cloudiness of the normally clear cornea. *Iridocyclitis* is the inflammation of the iris and ciliary body.

These eye disorders need to be diagnosed. Automatic identification of these anterior segment eye diseases is very useful in diagnostics and it is carried out using an artificial neural network (ANN) classifier. This chapter presents an automatic classification of four kinds of eye data sets (three different kinds of eye diseases and a *normal* class). Features are extracted from these raw images, which are then fed to the classifier. Our protocol used 195 subjects consisting of four different kinds of eye disease conditions. We demonstrate an accuracy of 94%, sensitivity of more than 87%, and specificity of 100%. Our systems are ready clinically to run on large numbers of data sets.

Keywords: neural networks, iridocyclitis, cataract, normal eye, corneal haze, histogram equalization, ANOVA test, sensitivity, specificity, training, accuracy, homogeneity, positive predictive value, neuron.

4.1 Introduction

Most visual impairments are caused by disease and malnutrition. The World Health Organization (WHO) estimated that more than 42 million people in the world are currently blind [1]. According to 2002 WHO estimates, the most common causes of blindness around the world are cataracts (47.8%), glaucoma (12.3%), age-related macular degeneration (AMD) (8.7%), trachoma (3.6%), corneal opacity (5.1%), and diabetic retinopathy (4.8%). Almost half of all blindness is caused by cataracts. Currently approximately 1 million free cataract operations are performed annually. But there are 1.5 million new cases of cataract blindness occurring in the world every year. In other words, the world is slowly becoming blinder.

As the population ages, more people are experiencing the impaired vision caused by cataracts and glaucoma. These eye diseases develop slowly, and the patient normally may not be aware of the gradual loss of sight until his or her vision is seriously affected. The earlier the diseases are diagnosed and treated, the greater the chance of success in preventing visual loss and the higher the possibility of a cure. The most typical of the three diseases are cataract, iridocyclitis, and corneal haze.

Edwards et al. have classified *normal* and three types of *cataract* optical eye images based on their distance from the average profiles in Euclidean space [2]. Their system was able to classify the unknown class correctly with an accuracy of 98%. Findings suggest that brimonidine causes anterior uveitis as a late side effect [3]. The inflammation settles rapidly on stopping the medication and on using topical corticosteroids. Idiopathic recurrent acute anterior uveitis (RAAU) was a common reason for attendance at ophthalmic casualty departments [4]. It has been proven that stress was not a factor in triggering the recurrence of idiopathic acute anterior uveitis. Uveitis in patients with psoriasis had distinguishing clinical features [5, 6]. Further epidemiologic studies are required to determine the strength of association between psoriasis without arthritis but with uveitis.

The optical aberrations caused by cataract were detected and quantified objectively using a new system called the *focus detection system* (FDS) [7]. Improvement in FDS outcome measures were correlated with cataract severity and improvement in visual acuity. This approach may be useful in long-term studies of cataract progression. Monestam and Wachmeister have compared the functional outcome of cataract surgery in terms of visual ability between patients younger than 84 years, 85 to 89 years, and 90+ years [8]. The patients ages 85 years and older had improved visual ability, acuity, and satisfaction after cataract surgery. Ferraro et al. have assessed the validity of a digital nonmydriatic fundus camera in detecting cataract as a cause of visual impairment [5]. They showed that the nonmydriatic fundus camera may be an alternative method for screening of visually significant cataract in the community. Zangwill et al. have studied the effect of pupil size and cataract on the quality of the image obtained with confocal scanning laser ophthalmoscopy [9]. It was found that pupillary dilation improved the image quality in many subjects by a small amount. Subjects with small undilated pupils and/or cataracts were able to benefit from pupillary dilation.

The effect of intact corneal epithelium on stromal haze and myofibroblast cell formation after excimer laser surgery was studied [10]. Kunihiko et al. proposed that the intact corneal epithelium plays an important part in curbing subepithelial haze and differentiation of myofibroblasts in corneal wound healing. It was found that minimally invasive radial keratotomy (mini-RK) enhancement after photorefractive keratectomy (PRK) induces central corneal haze and reduces corneal integrity [11]. Deep lamellar keratoplasty for refractory corneal haze after refractive surgery was useful in this eye. Song et al. have proved that the topical tranilast reduces the corneal haze by suppressing transforming growth factor-beta$_1$ expression in keratocytes after PRK [12].

The performance of five classifiers—linear discriminant analysis, support vector machine, recursive partitioning with regression tree, generalized linear model, and generalized additive model—was compared for detecting glaucomatous abnormalities using optical coherence tomography (OCT) data [13]. Their study show that the support vector machine (SVM) classifier was found to be very effective, with a high percentage of correct classifications.

Dynamic corneal imaging induced a reproducible and reversible change in corneal topography corresponding to the different indentation depths [14]. The results indicate that several clinical parameters were correlated with corneal elastic behavior in vivo and that the technology increased the predictability of refractive corneal surgery. This helped in the early diagnosis of corneal diseases. Twa et al. have described a new decision tree induction learning classification method to discriminate between normal and keratoconic corneal shapes [15]. They showed that Zernike polynomials can be used for automated decision tree classification of corneal shapes.

Recently, Acharya et al. have classified normal and abnormal eye classes using the K-means algorithm and radial basis function classifier with an accuracy of 90% [16]. In this chapter, we illustrate the extraction of features using image processing and identification of the eye classes using a feedforward neural network.

4.2   Data Acquisition

For the purpose of the present work, 195 subjects—patients suffering from cataract, iridocyclitis, corneal haze, as well as those in normal health—were studied. These data were taken from the Kasturba Medical Hospital Eye Centre in Manipal, India. The number and details of subjects in each group is shown in Table 4.1. Images were stored in 24-bit TIFF format with an image size of 128×128 pixels. The camera interfaced to the computer is shown in Figure 4.1(a). Figure 4.1(b) shows the typical optical eye images of different subjects.

Table 4.1   Range of Age, Gender, and Number of Subjects in Each Group

Types	Normal	Cataract	Iridocyclitis	Corneal Haze
Age	42 ± 8	60 ± 13	56 ± 12	62 ± 10
Gender	35 males and 25 females	37 males and 25 females	22 males and 15 females	24 males and 12 females

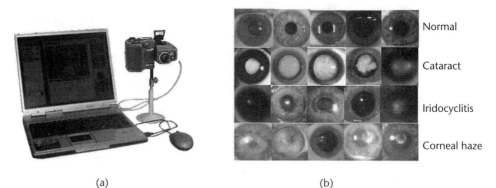

(a)	(b)

Figure 4.1   (a) Camera interfaced to the computer. (b) Typical normal, cataract, iridocyclitis, and corneal haze images.

4.3   Preprocessing

Feature extraction is most important in the classification process, and extractions are made from the preprocessed images. It is thus necessary to improve the contrast of the image, which will aid in the feature extraction process. In this study, the preprocessing steps consist of image contrast improvement based on histogram equalization, followed by binarization. These techniques are briefly explained in the following subsections.

4.3.1   Histogram Equalization

The dynamic range of the histogram of an image can be increased using histogram equalization [17]. This process assigns the intensity values of pixels in the input image such that the output image contains a uniform distribution of intensities. As a result, the contrast of the image is improved. This technique was used for all eye images in our study.

Histogram equalization redistributes intensity distributions. If the histogram of any image has many peaks and valleys, it will still have peaks and valleys after equalization. But the peaks and valleys will be shifted. This "spreading" of the histogram concept is used in histogram equalization. Each pixel is assigned a new intensity value based on the previous intensity level. Histogram equalization operates on an image in three steps: histogram formation, calculation of new intensity values for each intensity level, and replacement of the previous intensity values with the new intensity values. Figure 4.2 shows the result of histogram equalization.

4.3.2   Binarization

When binarizing an image, a suitable threshold has to be carefully chosen. Too small of a threshold may yield an image with edges linked together, and too big of a threshold may result in an image with edge segments containing curves that are not closed. In our study, we obtained good results by setting the threshold at 25% of the gray intensities (25% of the lower gray intensities are discarded) and this was set empirically. Figure 4.3 shows the result of histogram binarization.

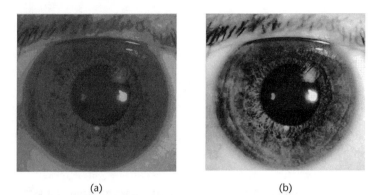

(a)   (b)

Figure 4.2 Result of histogram equalization: (a) original image and (b) image after histogram equalization.

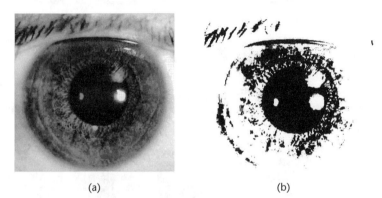

(a)   (b)

Figure 4.3 Result of binarization: (a) original image and (b) image after binarization.

4.4 Features Used for the Classification

Four features are extracted from the preprocessed eye images. They are big ring area, small ring area, homogeneity, and BW morph. Brief descriptions of these features are given next.

4.4.1 Big Ring Area

The color at the outer surface of the cornea is not the same in all four image classes. In the case of iridocyclitis and the majority of the cataract images, the outer surfaces of the corneas are bright in color compared to the normal and corneal haze images. Hence, we have chosen big ring area (BRA) as one of the parameters for the classification. The abstract representation of the BRA is in the form of a template as shown in Figure 4.4. It consists of two circles: an inner circle C_i and an outer circle C_o. The region between the C_i and C_o is the BRA with C_i ($r_i = 55$) and C_o ($r_o = 59$) in this work. Figures 4.5(a–d) show, respectively, the schematic diagram of the BRA detection for normal, cataract, iridocyclitis, and corneal haze images of different subjects respectively.

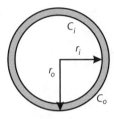

Figure 4.4 Template for big ring area detection.

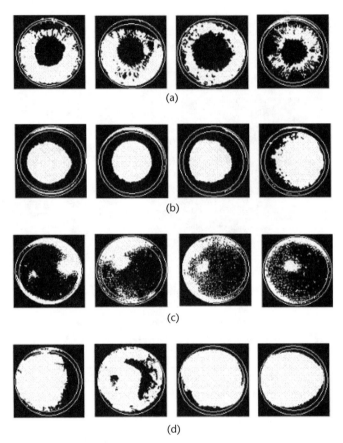

Figure 4.5 Big ring area detection in (a) normal, (b) cataract, (c) iridocyclitis, and (d) corneal haze for different subjects.

4.4.2 Small Ring Area

The color at the inner surface of the cornea is not the same in all four image classes. The inner surfaces of the corneas are more whitish in cataract and iridocyclitis eyes compared to the normal and corneal haze. The abstract representation of the small ring area (SRA) is in the form of a template as shown in Figure 4.6. It is the region between C_i (5) and C_o (15) in this work. Figures 4.7(a–d) show, respectively, the schematic diagram of the SRA detection for normal, cataract, iridocyclitis and corneal haze images of different subjects.

Figure 4.6 Template for small ring area detection.

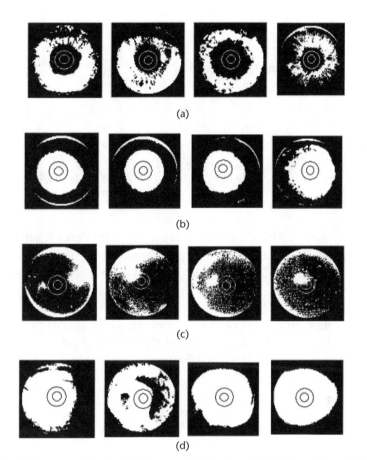

Figure 4.7 Small ring area detection in (a) normal, (b) cataract, (c) iridocyclitis, and (d) corneal haze for different subjects.

4.4.3 Homogeneity

Texture has been described in many ways. Intuitively, texture descriptors provide measures of properties such as smoothness, coarseness, and regularity [18–20]. One way to describe texture is to consider it as being composed of elements of texture primitives. Texture can also be defined as a mutual relationship among intensity values of neighboring pixels repeated over an area larger than the size of the relationship.

A good texture analysis can lead to a successful recognition system. A conventional texture recognition system can be grouped into three classes: *structural, statistical,* and *spectral*. Statistical approaches yield characterizations of textures as smooth, coarse, grainy, and so on. Statistical algorithms are based on the relationship between the intensity values of pixels; measures include entropy, contrast, and correlation based on the gray level co-occurrence matrix (GLCM) [18–20].

In statistical methods, we describe features using a spatial gray level dependency (SGLD) matrix. For a two-dimensional image $f(x,y)$ with N discrete gray levels, we define the spatial gray level dependency matrix $P(d, \Phi)$ for each d and Φ, which is given by

$$P(d,\Phi) = \begin{vmatrix} p_{00} & p_{01} & \cdot & \cdot & p_{0,N-1} \\ p_{10} & p_{11} & \cdot & \cdot & p_{1,N-1} \\ \cdot & & \cdot & \cdot & \cdot \\ \cdot & & \cdot & \cdot & \cdot \\ p_{N-1,0} & p_{N-1,1} & \cdot & \cdot & p_{N-1,N-1} \end{vmatrix} \qquad (4.1)$$

where

$$P_{i,j} = \frac{\text{number of pixel pair with intensity } (i,j)}{\text{total number of pairs considered}}$$

The term p_{ij} is defined as the relative number of times gray level pair (i, j) occurs when pixels separated by the distance along the angle Φ are compared. Each element is finally normalized by the total number of occurrences giving co-occurrence matrix P. A spatial gray level dependency matrix is also called a co-occurrence matrix and is shown in (4.1).

Homogeneity (H) is obtained from the co-occurrence matrix and is defined as follows: Homogeneity is a measure of closeness of the distribution of elements in the GLCM to the GLCM diagonal (Figure 4.8). (Range = [0 1] in MATLAB.) Homogeneity is 1 for a diagonal GLCM.

$$\phi_5 = \sum_{i=0}^{N-1} \sum_{j=0}^{N-1} \frac{p_{ij}}{1+|i-j|} \qquad (4.2)$$

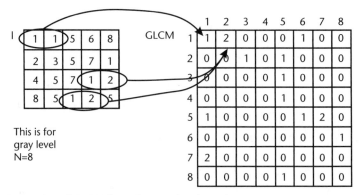

Figure 4.8 Generation of GLCM from the gray level matrix.

Energy is also known as *uniformity, uniformity of energy,* and *angular second moment. Contrast* is also known as *variance* and *inertia.*

4.4.4 BW Morph

The normal, cataract, and iridocyclitis images have too many sudden changes in the gray levels. Hence, there will be many edges in their images as compared to the corneal haze images. After binarizing, the image becomes black and white. The BW Morph (BWM) function can be used to find the area between the edges. BWM is a function in MATLAB that does the particular morphological operation on binary images. Many different types of morphological operations are available. However, in our study, the best result was obtained by using an operation called *erode.* It performs erosion using the structuring element of size 3×3 with all 1s. It can be used to obtain the boundaries/edges of a binary image in order to distinguish between the different types of eye conditions. Results of the BW Morph for the normal, cataract, iridocyclitis and corneal haze eye diseases are shown in Figures 4.9(a–d), respectively.

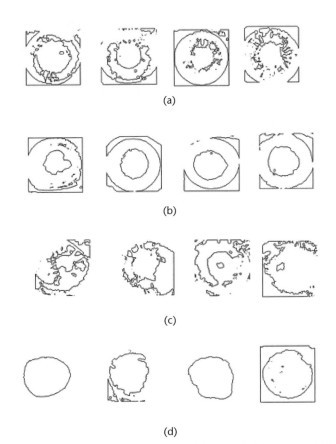

(a)

(b)

(c)

(d)

Figure 4.9 Result of inversion (negative) of BWM for (a) normal, (b) cataract, (c) iridocyclitis, and (d) corneal haze on different subjects.

4.5   Artificial Neural Network–Based Classifier

Artificial neural networks are biologically inspired networks—inspired by the human brain and the way it organizes its neurons and decision-making process. ANNs are useful in application areas such as pattern recognition and classification [21–23]. An ANN's decision-making process is more holistic than that of a conventional computer, because it is based on the aggregate of entire input patterns, whereas a conventional computer has to wade through the processing of individual data elements to arrive at a conclusion.

Neural networks derive their power from their massively parallel structure and an ability to learn from experience. They can be used for fairly accurate classification of input data into categories, provided they have been previously trained to do so. The accuracy of the classification depends on the efficacy of training, which in turn depends on the rigor and depth of the training. The knowledge gained by the learning experience is stored in the form of connection weights, which are used to make decisions about fresh input.

Three issues need to be settled when designing an ANN for a specific application: (1) the topology of the network, (2) the training algorithm, and (3) the neuron activation function. A network may have several "layers" of neurons, and the overall architecture may be either a feedback or feedforward structure. If the task is merely to distinguish linearly separable classes, a single-layer perceptron classifier is quite adequate [Figure 4.10(a)]. If the class separation boundaries can be piecewise linear approximated, then a two-layer perceptron classifier needs to be used. If the class boundaries are more complex, a three-layer *feedforward* neural network, with a sigmoid activation function [Figure 4.10(b)] is more suitable [24]. The most important reason in favor of such a network is that the sigmoid function $f(x)$ is differentiable for all values of x, which allows the use of the powerful *backpropagation* learning algorithm [24]. In the present case, the nature of class boundaries is not

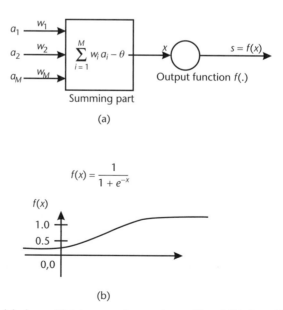

Figure 4.10   (a) Model of an artificial neuron (processing unit) and (b) sigmoid activation function.

clearly known; therefore, the three-layer network with sigmoid activation function is being used as classifier.

Of the three distinct layers of the network, the input layer consists of nodes to accept data (inputs are not weighted); the subsequent layers process the data. During the training phase, the connection weights of the last two layers are modified according to the *delta rule* of the backpropagation algorithm [25].

4.5.1   Backpropagation Learning Algorithm

Learning algorithms are of two kinds: supervised and unsupervised. In the former, the system weights are randomly assigned at the beginning and then progressively modified in light of *desired* outputs for a set of training inputs. The difference between the desired output and the actual output is calculated for every input, and the weights are altered in proportion to the error factor. The process is continued until the system error is reduced to an acceptable limit.

The modified weights correspond to the boundary between various classes, and to draw this boundary accurately, the ANN requires a large training data set that is evenly spread throughout the class domain. For quick and effective training, it is desirable to feed the ANN the data from each class in a routine sequence, so that the right message about the class boundaries is communicated to the ANN.

The *error backpropagation algorithm* (BPA), aims at reducing the overall system error to a minimum. The weight increment is directed toward the minimum system error; therefore, it is termed a *gradient descent* algorithm [22]. There is no definite rule for selecting the step size for the weight increment, but the step length certainly has a bearing on the speed of convergence. It has been observed that for good speed, the step size should be "moderate," that is, neither "too large" nor "too small." In the present case, an near optimum learning constant of $\eta = 0.9$ (which controls the step size) was chosen by trial and error. Because weight incrementing is accomplished in small steps, the algorithm also bears the name *delta rule*.

The whole process of updating the weight matrix is a slow (in small incremental steps) movement toward a global minima of system error function. Sometimes, the possibility may arise of the system entering a local minima and being unable to exit it. To remedy such a possibility, the algorithm incorporates a *momentum term* into its update increment. The momentum term is a fraction of the increment of its previous step; this term tends to push the present increment in the same direction as that of the *previous* step. This term is no guarantee against the algorithm getting stuck at the local minima, but helps to get out of "small" dips in the path.

Figure 4.11 shows the block diagram for the neural network classifier used for the classification of the eye disorders. Figure 4.12 shows the structure of the feedforward neural network classifier used for this work. The output layer has four neurons, giving rise to an output domain of 16 possible classes. However, the network is trained to identify only four classes given by decoded binary outputs [0001, 0010, 0100, 1000]. The weights associated with the hidden layers get adjusted, which enables the ANN to learn. For this study, we tried one to four hidden layers, and performance was found to be better with one hidden layer and 12 nodes (neurons).

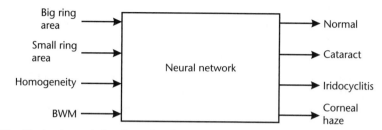

Figure 4.11   Block schematic for ANN classification model.

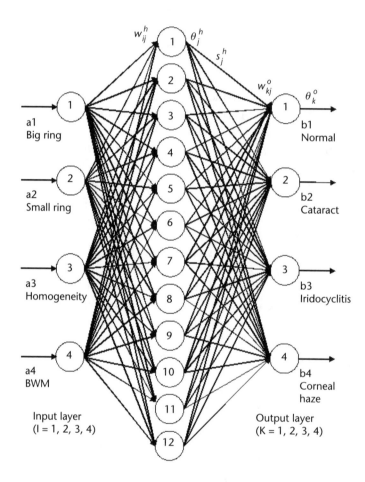

Hidden layer (j = 1, 2, 3, 4, 5, 6, 7, 8, 9, 10, 11, 12)

Figure 4.12   Three-layer feedforward neural network classifier.

The outputs of the hidden layer s_j^h and output layer (b_k) are evaluated using (4.3) and (4.4):

$$s_j^h = f\left(\sum_{i=1}^{4} w_{ji}^h s_i - \theta_j^h\right) \tag{4.3}$$

$$b_k = f\left(\sum_{j=1}^{4} w_{kj}^o s_j^h - \theta_k^o\right) \tag{4.4}$$

where w_{ji}^h and w_{kj}^o are the connection weights and θ_j^h and θ_k^o are the bias terms, respectively.

The error vectors of hidden layer (e_j) and output layer (e_k) are calculated using (4.5) and (4.6), respectively:

$$e_k = b_k(1-b_k)(d_k - b_k) \tag{4.5}$$

$$e_j = s_j^h(1-s_j^h)\sum_{k=1}^{4} w_{kj}e_k \tag{4.6}$$

where d_k is the desired output.

The weight update equations of the output and hidden layers are as follows:

$$w_{kj}(\text{new}) = w_{kj} + \eta s_j^h e_k \tag{4.7}$$

$$w_{ji}(\text{new}) = w_{ji} + \eta s_i e_j \tag{4.8}$$

$$\theta_k^o(\text{new}) = \theta_k^o + \eta e_k \tag{4.9}$$

$$\theta_j^h(\text{new}) = \theta_j^h + \eta e_j \tag{4.10}$$

4.6   Results

Table 4.2 shows the ranges of the three features used as input to the ANN. For the purpose of training and testing the classifier, a database of 195 patient samples was divided into two sets: a training set of 131 arbitrarily chosen samples and a test set of 64 samples (Table 4.3). The training consisted of 1,200 iterations. Table 4.4 shows the result of sensitivity, specificity, and positive predictive values for the four classes of eye images using the neural network classifier.

During the *training* phase, each output of the ANN is an analog value in the range of $0 \rightarrow 1.0$, whereas the "desired" output is either 0 or 1.0. During the recall phase, the output signal is approximated to binary levels by comparing it with a threshold value of 0.5. Figure 4.13 shows the convergence of weights during the training period. The mean square error (MSE) of the ANN was set to 0.001 and is shown in the straight line. The dotted curve shows the variation of MSE with

Table 4.2   Range of Input Features to ANN Classification Model

Parameter	Normal	Cataract	Iridocyclitis	Corneal Haze	p-Test
BRA (pixels)	826.08 ± 164.41	706.89 ± 212.51	937.00 ± 126.56	569.17 ± 225.70	<0.0001
SRA (pixels)	26.90 ± 60.08	593.37 ± 65.60	172.11 ± 181.38	381.00 ± 278.32	<0.0001
Homogeneity	0.056 ± 0.0027	0.067 ± 0.0045	0.058 ± 0.0049	0.063 ± 0.0072	<0.0001
BWM (pixels)	1.23E + 03 ± 203.39	1.0085E + 03 ± 170.5905354	1.004E + 03 ± 183.18	8.64E + 02 ± 335.94	<0.0001

Table 4.3   Training and Testing Data Set

Classes	Number of Data Sets Used for Training	Number of Data Sets Used for Testing	Percentage (%) of Correct Classifications
Normal	40	20	100.00
Cataract	42	20	95.0
Iridocyclitis	25	12	90.00
Corneal Haze	24	12	91.66
Average	94.16		

Table 4.4   Results of Sensitivity, Specificity, and Positive Predictive Value for Complete Eye Classes Using a Neural Network

Classifier	True Negative	True Positive	False Positive	False Negative	Sensitivity	Specificity	Positive Predictive Accuracy
ANN	25	34	0	5	87.17	100	100

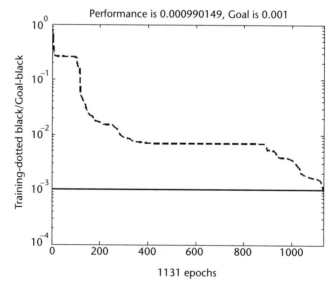

Figure 4.13   Convergence of the weights.

respect to the training iterations. During the early part of the training, this MSE will be high and remains constant during the middle part of the training. It gradually falls in the later part of the training.

The p-value can be obtained using the ANOVA (ANalysis Of VAriance between groups) test. ANOVA uses variances to decide whether the *means* are different. This test uses the variation (variance) *within* the groups and translates into variation (i.e., differences) *between* the groups, taking into account how many subjects there are in the groups. If the observed differences are high then it is considered to be statistically significant.

As mentioned earlier, in this study, the eye disorders are classified into four categories: normal, cataract, iridocyclitis, and corneal haze. The outer surface of the cornea is more white as compared to the rest in iridocyclitis images; hence, the BRA is higher (937.00 ± 126.56). And it is lower (569.17 ± 225.70) in the case of corneal haze, due to the black surface in the outer layer of the cornea. The SRA value is high (593.37 ± 65.60) for the cataract subjects and the lowest (26.90 ± 60.08) for the normal subjects. In cataract, there is a white ring (cataract) present on the cornea that does not exist in the normal eye images. This contributes to the white pixels in the cataract eye images. The number of dark pixels within the two boundaries (BWM) is higher ($1.23E + 03 \pm 203.39$) for the normal eye and the lowest ($8.64E + 02 \pm 335.94$) for the corneal haze images. In a normal eye, the cornea is darker, compared to the other images. Hence, this value is bigger for the normal images. In corneal haze, the two boundaries are not clearly visible (see Figure 4.9). Homogeneity measures the closeness of the distribution of elements in the GLCM to the GLCM diagonal values. This value is high (0.067 ± 0.0045) for the cataract due to the presence of the white ring in the cornea, and this value is very low (0.056 ± 0.0027) for the normal eye, because the whole cornea is dark (i.e., there is no change in the color).

4.7 Discussion

A diagnostic test can be used to investigate the statistical relationship between test results and the presence of disease. For all diagnostic tests, two critical components determine its accuracy: sensitivity and specificity.

The sensitivity of a test is the proportion of people with the disease who have a positive test result. The higher the sensitivity, the greater the detection rate and the lower the false negative (FN) rate. The specificity of the test is the proportion of people without the disease who have a negative test. The higher the specificity, the lower the false positive (FP) rate and the lower the proportion of people having the disease who will be unnecessarily worried or exposed to unnecessary treatment. The positive predictive value (PPV) of a test is the probability of a patient with a positive test actually having a disease. In our study, we found a sensitivity of 87%; specificity and PPV were 100%. This indicates that these results are clinically significant.

Cataracts are lens opacities that account for approximately 10% of blindness in children. An early transvaginal anomaly scan at 14 to 16 gestational weeks has been proposed to diagnose fetal eye anomalies (especially cataract) using ultrasound [26]. Specifically, using the ultrasonography technique, cataract was detected in the fetus in more than 80% of cases.

The eyes were photographed with the Topcon SL-45 Scheimpflug camera and the images scanned and processed to obtain one-dimensional profiles through a 40 × 440-μm axial window [2]. The system was found to be very sensitive (98%) in detecting the presence of cataracts and 100% in identifying normal eyes. The system was able to classify pure cataracts into the various classes with an accuracy of 98% cases. Recently, the normal and abnormal optical eye images have been classified using neural network classifiers with an accuracy of more than 90% [16]. In this study, the abnormal class consisted of cataract, iridocyclitis, corneal haze, and arcus eye images. In our study, we were able to identify four types of optical images with an accuracy of 94%, sensitivity of more than 87%, and specificity of 100% using the feedforward neural network classifier.

The accuracy of the system can further be increased by increasing the size and quality of the training data. The classification results can be improved by extracting the proper features from the optical images. The environmental conditions like the reflection of the light influences the quality of the optical images and hence the percentage of classification efficiency. The software for feature extraction and classification of eye images are written in MATLAB 7.0.4.

4.8   Conclusions

Eye disorders such as iridocyclitis, corneal haze, and cataract can influence a person's vision. These diseases are the largest cause of blindness due to their late diagnosis. Factors such as diet, lifestyle, and aging also influence the normal function of eye tissues.

In this chapter, we discussed the performance of a neural network classifier as a diagnostic tool to aid the physician in the early detection of these eye abnormalities. However, these tools generally do not yield results with 100% accuracy. The accuracy of the tools depends on several factors, such as the size and quality of the training set, the rigor of the training imparted, and also parameters chosen to represent the input. It is evident from the results that the classifier is effective to the tune of more than 94% accuracy. In this work, we have demonstrated a sensitivity of 87% and specificity of 100% for the classifier. Our classification system produces encouraging results. However, the robustness of diagnostic systems can be further improved by choosing better features.

References

[1]   http://www.owsp.org/problem.htm.

[2]   Edwards, P. A., et al., "Computerized Cataract Detection and Classification," *Current Eye Research*, Vol. 9, No. 6, 1990, pp. 517–524.

[3]   Burgansky-Eliash, Z., et al., "Optical Coherence Tomography Machine Learning Classifiers for Glaucoma Detection: A Preliminary Study," *Invest. Ophthalmol. Vis. Sci.*, Vol. 46, No. 11, 2005, pp. 4147–4152.

[4]   Mulholland, B., M. Marks, and S. L. Lightman, "Anterior Uveitis and Its Relation to Stress," *Br. J. Ophthalmol.*, Vol. 84, No. 10, 2000, pp. 1121–1124.

[5]   Durrani, K., and C. S. Foster, "Psoriatic Uveitis: A Distinct Clinical Entity?" *Am. J. Ophthalmol.*, Vol. 139, No. 1, 2005, pp. 106–111.

[6]   Nusz, K. J., et al., "Rapid, Objective Detection of Cataract-Induced Blur Using a Bull's Eye Photodetector," *J. Cataract Refract. Surg.*, Vol. 31, No. 4, 2005, pp. 763–770.

[7]   Monestam, E., and L. Wachmeister, "Impact of Cataract Surgery on the Visual Ability of the Very Old," *Am. J. Ophthalmol.*, Vol. 137, No. 1, 2004, pp. 145–155.

[8]   Ferraro, J. G., et al., "Detecting Cataract Causing Visual Impairment Using a Nonmydriatic Fundus Camera," *Am. J. Ophthalmol.*, Vol. 139, No. 4, 2005, pp. 725–726.

[9]   Zangwill, L., et al., "Effect of Cataract and Pupil Size on Image Quality with Confocal Scanning Laser Ophthalmoscopy," *Arch. Ophthalmol.*, Vol. 115, No. 8, 1997, pp. 983–990.

[10] Kunihiko, N., et al., "Intact Corneal Epithelium Is Essential for the Prevention of Stromal Haze After Laser Assisted *In Situ* Keratomileusis," *Br. J. Ophthalmol.,* Vol. 85, 2001, pp. 209–213.

[11] Shoji, N., et al., "Central Corneal Haze Increased by Radial Keratotomy Following Photorefractive Keratectomy," *J. Refract. Surg.,* Vol. 19, No. 5, 2003, pp. 560–565.

[12] Song, J. S., H. R. Jung, and H. M. Kim, "Effects of Topical Tranilast on Corneal Haze After Photorefractive Keratectomy," *J. Cataract Refract. Surg.,* Vol. 31, No. 5, 2005, pp. 1065–1073.

[13] Byles, D. B., P. Frith, and J. F. Salmon, "Anterior Uveitis as a Side Effect of Topical Brimonidine," *Am. J. Ophthalmol.,* Vol. 130, No. 3, 2000, pp. 287–291.

[14] Grabner, G., et al., "Dynamic Corneal Imaging," *J. Cataract Refract. Surg.,* Vol. 31, No. 1, 2005, pp. 163–174.

[15] Twa, M. D., et al., "Automated Decision Tree Classification of Corneal Shape," *Optom. Vis. Sci.,* Vol. 82, No. 12, 2005, pp. 1038–1046.

[16] Acharya, U. R., et al., "Automatic Identification of Anterior Segment Eye Abnormality," *Innovations Technol. Biol. Med. (ITBM-RBM),* Vol. 28, No. 1, 2007, pp. 35–41.

[17] Gonzalez, R. C., and P. Wintz, *Digital Image Processing,* 2nd ed., Reading, MA: Addison-Wesley, 1987.

[18] Parker, J. R., *Algorithms for Image Processing and Computer Vision,* New York: John Wiley & Sons, 1997.

[19] Haralick, R. M., K. Shanmugam, and I. Dinstein, "Textural Features for Image Classification," *IEEE Trans. on Systems, Man and Cybernetics,* Vol. 3, No. 6, 1973, pp. 610–621.

[20] Kulkarni, A. D., *Artificial Neural Networks for Image Understanding,* New York: Van Nostrand Reinhold, 1994.

[21] Bart, K., *Neural Networks and Fuzzy Systems,* New Delhi: Prentice-Hall India, 1992.

[22] Lippman, R. P., "Pattern Classification Using Neural Network," *IEEE Communication Mag.,* Vol. 27, No. 11, 1989, pp. 47–64.

[23] Haykin, S., *Neural Networks: A Comprehensive Foundation,* New York: Macmillan, 1995.

[24] Patterson, P., and S. Draper, "A Neural Net Representation of Experienced and Non-Experienced Users During Manual Wheelchair Propulsion," *J. Rehab. Res. Dev.,* Vol. 35, No. 1, 1998, pp. 43–51.

[25] Yegnanarayana, B., *Artificial Neural Networks,* New Delhi: Prentice-Hall India, 1999.

[26] Mashiach, R., et al., "Early Sonographic Detection of Recurrent Fetal Eye Anomalies," *Ultrasound Obstet. Gynecol.,* Vol. 24, No. 6, 2004, pp. 640–643.

CHAPTER 5

Identification of Different Stages of Diabetic Retinopathy Using Retinal Optical Images

Wong Li Yun, Rajendra Acharya, Caroline Chee, Eddie Y. K. Ng, Lim Choo Min, and Jagadish Nayak

Diabetes is a disease that occurs when the pancreas does not secrete enough insulin or when the body is unable to break down glucose in the blood. Over time, diabetes affects the circulatory system, including that of the retina. As diabetes progresses in a patient, vision may eventually be affected, causing diabetic retinopathy. In our study, 200 retinal photographs were analyzed. Three groups were identified: *normal*, *nonproliferative diabetic retinopathy* (mild, moderate, severe diabetes retinopathy), and *proliferative diabetic retinopathy*. This work presents classification of three eye classes using a Gaussian mixture model (GMM) classifier. The desired features were extracted from the raw images using image processing techniques and fed to the GMM classifier for classification.

5.1 Introduction

Fundus imaging in ophthalmology plays an important role in a variety of medical diagnoses such as hypertension, diabetes, and cardiovascular disease. Moreover, computerized image processing of the fundus image is highly desirable in the case of diabetic retinopathy, the primary cause of blindness in the world, which requires the screening of a large number of patients chosen from specialized groups.

Of the main features of fundus retinal images, the blood vessels are the most common parameter. Several morphological features of retinal veins and arteries have diagnostic relevance such as diameter, length, branching angle, and tortuosity. Accurate vessel detection is a difficult task for several reasons: the presence of noise, the low contrast between vessels and background, and the variability of vessel width, brightness and shape. Moreover, the presence of lesions, exudates, and other pathological effects may dominate the image.

The World Health Organization estimates that 135 million people have diabetes mellitus worldwide and that the number of people with diabetes will increase to 300 million by 2025 [1]. More than 18 million Americans currently have diabetes and the number of adults with the disease is projected to be more than double by 2050 [2]. An additional 16 million adults between the ages of 40 and 74 have prediabetes, the state that occurs when a person's blood glucose levels are higher than normal but not high enough for the diagnosis of diabetes; these people are at high risk of developing diabetes. Visual disability and blindness have a profound socioeconomic impact on the diabetic population and diabetic retinopathy (DR) is the leading cause of new blindness in working-age adults in the industrialized world [2]. The prevalence rates for DR and vision-threatening DR in adults with diabetes over age 40 is 40.3% and 8.2%, respectively [3].

Globally, diabetic retinopathy is the primary cause of blindness not because it has the highest incidence, but because it often remains undetected until severe vision loss occurs. Diabetic retinopathy is characterized by changes in the retina that include blood vessel diameter changes, microaneurysms, lipid and protein deposits referred to as hard exudates or cotton wool spots depending on the appearance, hemorrhages, and new vessel growth [4, 5]. These pathological changes are known risk factors for severe vision loss, hypertension, and cardiovascular disease. If the disease is detected early, treatment is effective at reducing eyesight loss [6]. Advances in shape analysis and the development of strategies for the detection and quantitative characterization of blood vessel changes in the retina are therefore of great clinical importance.

We are not aware of any completely automated system capable of locating the optic disc, detecting the blood vessels, and making useful measurements such as the arteriolar-venule diameter ratios in the vicinity of the optic disc. Li et al. have proposed a semiautomatic system to locate the optic disc and blood vessels [7].

The optic disc is the brightest object in the healthy retinal fundus and several algorithms have been described to locate its center and boundary [8–11]. These authors have reported the use of Canny edge detection, template matching, and Haar transforms. Optic disc detection, however, remains a problematic task due to the discontinuities along the boundary where blood vessels cross, as well as dramatic hue changes within the optic disc boundary, with accompanying intradisc hemorrhage.

It is also possible to locate the optic disc by tracking vessels back to their origin, because all vessels emerge into the retina via the optic disc [12, 13]. In this method, however, one had to detect a good portion of the blood vessel network first. This area of automated classification of retinal blood vessels has not received much attention in the literature, although methods of delineating blood vessels into arteries and veins are currently used by various groups [7, 12]. Most often color (red, green, blue [RGB] mean and standard deviation) and hue (HSV mean and standard deviation) information was used for classifying retinal blood vessels into arteries and veins. The availability of toolboxes for performing a variety of statistical and machine learning methods makes this an accessible approach.

In DR, the microvascular changes cause detectable changes in the appearance of retinal blood vessels. Dua et al. have proposed a new blood vessel detection technique in retinal images that is based on the regional recursive hierarchical decomposition using quadtrees and postfiltration of edges [14]. The technique provides information on retinal blood vessel morphology that can be calibrated to normal blood vessel diameters and also helps to detect the blood vessel pathology and hence aid early detection of diabetic retinopathy.

Exudates were detected using their high gray level variation, and their contours were determined by means of morphological reconstruction techniques. The optic disc was detected by means of morphological filtering techniques and watershed transformation [15]. Morphologic classifications of diabetic retinopathy were performed based on the number, location, and type of discrete microvascular lesions in the fundus of the eye [16]. The performance evaluation was conducted for an automated fundus photographic image analysis algorithm in high-sensitivity and/or high-specificity segregation of patients with diabetes with untreated DR from those without retinopathy [17]. The automated lesion detection method correctly identified 90.1% of patients with retinopathy and 81.3% of patients without retinopathy.

Cree et al. have proposed use of an integrated automated analyzer of the retinal blood vessels in the vicinity of the optic disc using retinal images, and they showed a mean accuracy of 70% using Bayes' rule [18]. The features of the red lesions were extracted and classified using a k-nearest neighbor classifier. The automated system showed a sensitivity of 100% and specificity of 87% for all the test images [19]. Hoover et al. have located and outlined blood vessels in images by means of a novel method to segment blood vessels that complements local vessel attributes with region-based attributes of the network structure [13]. Blood vessels were detected using two-dimensional matched filters [20]. Gray level profiles of cross sections of blood vessels were approximated by a Gaussian-shaped curve. The concept of matched filter detection of signals was used to detect piecewise linear segments of blood vessels after the vessel approximation.

The work discussed so far focuses mainly on the automatic identification of diabetic retinopathy and normal retinal images. However, in this chapter, we discuss the automatic classification of normal, nonproliferative, and proliferative stages of diabetic retinopathy.

5.2   Computer Methods and Theory

In this work, 200 retinal photographs of mild nonproliferative diabetic retinopathy (NPDR), moderate NPDR, severe NPDR, proliferative diabetic retinopathy (PDR), and normal cases were studied. These patient data were provided by the National University Hospital in Singapore. The number and details of photographs in each group are shown in Table 5.1. Images were stored in 24-bit JPEG format with an image size of 256×256 pixels. The Zeiss Visucam^{lite} fundus camera was used to acquire the fundus image. Figure 5.1 shows the normal, mild NPDR, moderate NPDR, severe NPDR, and PDR fundus images.

Table 5.1   Range of Age and Number of Subjects in Each Group

Types	Normal	Mild NPDR	Moderate NPDR	Severe NPDR	PDR
Age (years)	32 ± 8	53 ± 21	65 ± 16	60 ± 22	70 ± 10
Number of subjects	31	40	40	42	47

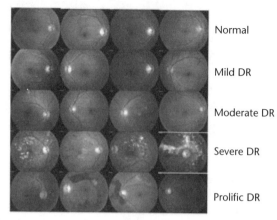

Normal

Mild DR

Moderate DR

Severe DR

Prolific DR

Figure 5.1   Normal, mild NPDR, moderate NPDR, severe NPDR, and PDR fundus images.

5.2.1   Imaging Techniques

Feature extraction plays the most important role toward the eventual DR classification. Preprocessed images have their features extracted to improve the contrast of the image. The preprocessing step consists primarily of image shade correction improvement based on median filtering and top-hat transform operators, followed by use of binarization.

5.2.1.1   Image Shade Correction

Fundus images often contain intensity variations in the background across the image (called *vignetting*). This may interfere with the performance of the proposed system as the features of the image are being selected. Therefore, the slow variation in the background of the three channels in the fundus color image was removed using shade correction. This was performed by estimating the background image using median filtering. Then this image was subtracted from the original RGB component of the fundus image. The size of the median filtering used in this system was 40×40; it depends on the size of the image and was chosen appropriately. The typical normal fundus image and shade corrected red channel is shown in Figure 5.2.

5.2.1.2   Top-Hat Transform

The grayscale top-hat transform is a grayscale morphological algorithm [21, 22]. It is beneficial in finding the pixel clusters that are light on a surrounding relatively

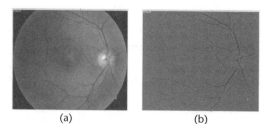

(a)   (b)

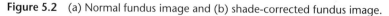

Figure 5.2   (a) Normal fundus image and (b) shade-corrected fundus image.

dark background. The first step in the top-hat transform was the operation to open an image; then the opened image was subtracted from the original image, resulting in a final image with highlighted blood vessels. The structuring element is a linear structuring element. Various top-hat transforms are performed with different orientations of the structuring element. A total of 12 rotated structuring elements were used with a resolution of 15°. The length of the structuring element used was selected appropriately depending on the size of the image. By taking the maximum pixel value at each location in all 12 images of the top-hat transform, an image showing only vasculature was obtained.

5.2.1.3   Binarization

To binarize the image, a threshold should be carefully chosen. Too small of a threshold will produce an image that has edges linked together. Too big of a threshold will produce edge segments that contain curves that are not closed. We obtained good results by setting the threshold at 12% of the gray intensities contained into the image (12% of the lower gray intensities are discarded) and this was set empirically.

5.2.2   Features

Six features—red layer of perimeter (RLP), green layer of perimeter (GLP), blue layer of perimeter (BLP), red layer of area (RLA), green layer of area (GLA), and blue layer of area (BLA)—are extracted from the images after the preprocessing by means of morphological operations. Figures 5.3 and 5.4 are illustrations of the layers extracted for the perimeter and area, respectively, of a normal retina.

5.2.2.1   Perimeter of Veins, Hemorrhages, and Microaneurysms

The perimeter is determined by the number of pixels that outline the veins. Figures 5.5(a–e) are illustrations of these veins for normal retina, mild, moderate, and severe NPDR, and PDR retina images.

5.2.2.2   Area of Veins, Hemorrhages, and Microaneurysms

The area is the number of black pixels within the veins and blood vessels. Figure 5.6 illustrates the extracted vein and blood vessels area for normal retina, mild, moderate, and severe NPDR, and PDR retina images.

Figure 5.3   Perimeter of red, green, and blue layers, respectively, for a normal retina.

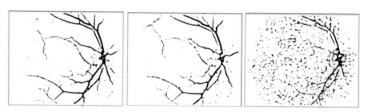

Figure 5.4   Area of red, green, and blue layers, respectively, for a normal retina.

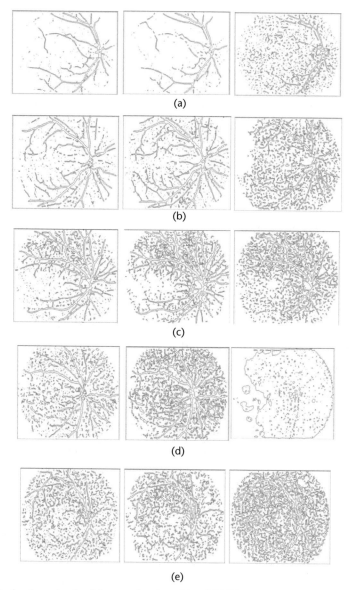

Figure 5.5   Perimeter veins for (a) normal retina, (b) mild NPDR retina, (c) moderate NPDR retina, (d) severe NPDR retina, and (e) PDR retina.

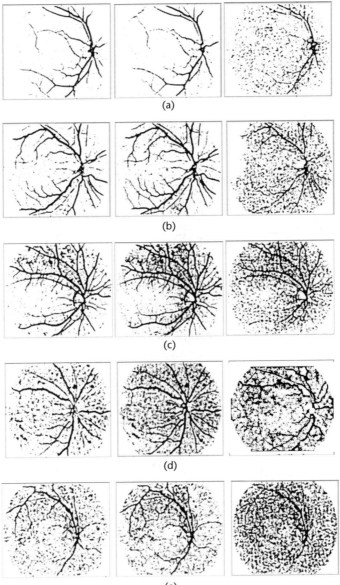

Figure 5.6   Area of veins and blood vessels for (a) normal retina, (b) mild NPDR retina, (c) moderate NPDR retina, (d) severe NPDR retina, and (e) PDR retina.

5.3   System Description

In this work, we have used a Gaussian mixture model for classification of the three eye classes.

5.3.1   Gaussian Mixture Model for Classification

A Gaussian mixture model is a parametric model used to estimate a continuous probability density function from a set of multidimensional feature observations. A

GMM probability density is described by the additive contribution of N multidimensional Gaussian components. The Gaussian mixture model is described by mixture component weights w_i, means μ_i, and covariances $\sum_i$. For a single observation, x, the probability density of given a GMM described by λ is given by:

$$p(x|\lambda) = \sum_{i=1}^{N} w_i g\left(x|\mu_i, \sum_i\right)$$ (5.1)

The probability density of a single Gaussian component of D dimensions is given by

$$g\left(x|\mu_i, \sum_i\right) = \frac{1}{\sqrt{(2\pi)^D|\sum_i|}} \exp\left(-\frac{1}{2}(x-\mu_i)\sum_i^{-1}(x-\mu_i)\right)$$ (5.2)

The vector or matrix transpose is represented by a prime ('). The solution for determining the parameters of the GMM is arrived at by using the maximum likelihood (ML) parameter estimation criterion. The joint likelihood of T independent and identically distributed feature vector observations, $X = \{x_1, x_2, x_3, \dots, x_T\}$ may be specified according to the following equation:

$$p(X|\lambda) = \prod_{i=1}^{T} p(x_i|\lambda)$$ (5.3)

This may conveniently be represented in log form:

$$L(\lambda) = \log p(x|\lambda) = \sum_i \log p(x_i|\lambda)$$ (5.4)

In terms of the mixture component densities, the log-likelihood function to be maximized is given here:

$$L(\lambda) = \sum_{i=1}^{T} \log\left(\sum_{i=1}^{N} w_i g\left(x_t|\mu_i, \sum_i\right)\right)$$ (5.5)

The model parameters are estimated such that they maximize the likelihood of the observations. A method for maximizing the log-likelihood of the observations is by the general form of the expectation-maximization (E-M) algorithm. The E-M algorithm, given the parameters of an initial estimate, $\hat{\lambda} = \left\{\{\hat{w}_1, \dots, \hat{w}_N\}, \{\hat{\mu}_1, \dots, \hat{\mu}_N\}, \left\{\hat{\sum}_i, \dots, \hat{\sum}_N\right\}\right\}$, will determine new estimates of the parameters, $\lambda_i = \left\{\{w_i, \dots, w_N\}, \{\mu_1, \dots, \mu_N\}, \{\sum_1, \dots, \sum_N\}\right\}$, such that $p(x|\lambda) \geq p(x|\hat{\lambda})$.

One of the important attributes of the GMM is its ability to form smooth approximations for any arbitrarily shaped densities. Because real-world data has multimodal distributions, the GMM is a great tool for modeling the characteristics of the data. Another extremely useful property of GMMs is the possibility of employing a diagonal covariance matrix instead of a full covariance matrix [23]. Thus, the amount of computational time and complexity can be reduced significantly. GMMs have been widely used in many areas of pattern recognition and classification, with great success in the area of speaker identification and verification [24].

A Gaussian mixture model must be trained with input data. Usually, the initial estimates of the parameters are obtained from a sample of the training data using a simpler procedure such as *K*-means. The *K*-means procedure starts with randomly chosen initial means and assumed unit variances for the diagonal covariance matrix. This method has been adopted in this work. Sometimes, a background model can be obtained by pooling together all available data for all classes (this provides a large set and consequently more reliable estimates of the parameters), and then individual models for each class are obtained by updating only the mean vectors using E-M on training data specific to the class.

The six features extracted are fed into GMM inputs to distinguish them into the three classes.

5.4 Statistics of System

The range of perimeter and area values for each stage of DR with the different RGB layers are shown in Tables 5.2 and 5.3, respectively. The *p*-values (significance level) shown were obtained using the ANOVA (ANalysis Of VAriance between groups) test. ANOVA uses variances to decide whether the *means* are different. This test uses the variation (variance) *within* the groups and translates it into variation (i.e., differences) *between* the groups, taking into account the number of subjects in the groups. If the observed differences are high, then it is statistically significant

Table 5.2 Mean and Variance for Perimeter Ranges

Type	Perimeter of Red Layer	Perimeter of Green Layer	Perimeter of Blue Layer
Normal	13,476 ± 1,923	13,267 ± 1,953	32,698 ± 10,570
NPDR	21,076 ± 11,010	18,981 ± 10,530	36,476 ± 11,230
PDR	32,426 ± 12,440	28,883 ± 12,720	31,819 ± 11,410
p-Value	<0.0001	<0.0001	0.0015

Table 5.3 Mean and Variance for Area Ranges

Type	Area of Red Layer	Area of Green Layer	Area of Blue Layer
Normal	61,906 ± 11,670	59,871 ± 11,880	184,497 ± 69,770
NPDR	11,4595 ± 68910	106,600 ± 73370	262,541 ± 82,180
PDR	19,3408 ± 83,380	192,570 ± 101,500	276,842 ± 74,560
p-Value	<0.0001	<0.0001	<0.0001

(lower *p*-value). Then the classification can be used as a tool for the automatic classification of the different stages of diabetic retinopathy.

The graphical plots of the ANOVA test of the perimeter and area ranges of the extracted features for different kinds of images are shown in Figures 5.7 and 5.8. The box plot function in MATLAB enables clear comparisons between different classes. This function displays a box and whisker plot. The box has lines at the lower quartile, median, and upper quartile values. The whiskers are lines extending from each end of the box to show the extent of the rest of the data. Outliers are data with values beyond the ends of the whiskers. In a notched box plot, the notches represent a robust estimate of the uncertainty about the medians for box-to-box comparison. Boxes whose notches do not overlap indicate that the medians of the two groups differ at the 5% significance level.

The red-green (R-G) box plots (or ANOVA) show only linear relationships. However, GMMs are able to relate complex, nonlinear relationships. When the blue (B) layer was removed in the training stage, a decrease in the accuracy of classification resulted.

An analysis on the relationship of the confidence intervals of the different classes in the same layer provides information on how to group the different classes for input into the GMM analysis. This improves the accuracy of distinguishing between the classes. Figures 5.7 and 5.8 indicate the *means* of the normal, NPDR, and PDR classes, and they are significantly different from the means of one another. These data can be put into three groups; normal, NPDR, and PDR. The six features extracted are clinically very significant because the *p*-value is very low (< 0.005) for the three groups. Figure 5.9 shows the graphical representation of the training and testing data set, and Table 5.4 shows the result of the GMM classifier and the amount of data used for training and testing.

During the training phase, we used 133 pieces of data for the normal, NPDR, and PDR cases. During the testing phase, 67 DR images, different from those used for the training phase, were used to test for the classification efficiency of the system. Our results show that the classifier was able to identify the normal retinal images correctly in up to 90% of these cases; the NPDR and PDR classes, however, were classified with a lower accuracy of about 86%. Overall, our GMM classifier is able to classify an average of 88% of the fundus images correctly.

Table 5.5 indicates that the system shows sensitivity and specificity percentages of 89% and 81%, respectively. These percentages indicate that these results are clinically significant. System performance can be increased by using a diverse, extremely large training set.

5.5 Discussion

Sinthanayothin et al. [11] have differentiated diabetic retinopathy from a normal retina using image processing algorithms. Retinal images were preprocessed using adaptive, local, contrast enhancement. The optic disc was detected by means of the highest variation in the intensity of adjacent pixels. Blood vessels were identified by

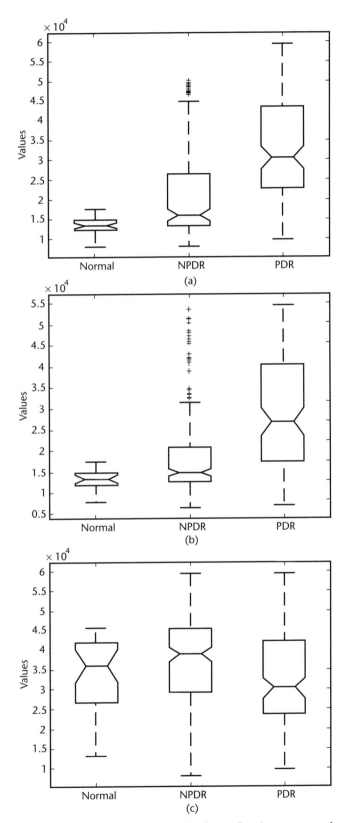

Figure 5.7 (a) Red layer, (b) green layer, and (c) blue layer of perimeter ranges from an ANOVA analysis.

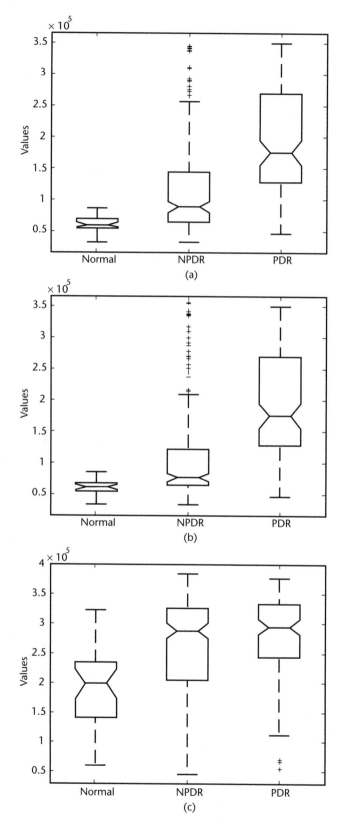

Figure 5.8 (a) Red layer, (b) green layer, and (c) blue layer of perimeter ranges from an ANOVA analysis.

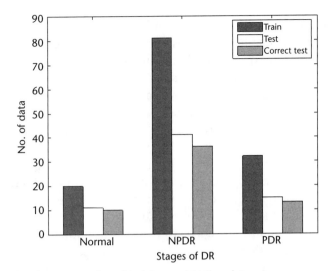

Figure 5.9   Graphical representation of training and testing data set.

Table 5.4   Results of the Classification

Type of Image	Training Data	Test Data	Correctly Classified Test Data	Percentage Correctly Classified (%)
Normal	20	11	10	90.1
NPDR	81	41	36	87.8
PDR	32	15	13	86.6
Average				88.0

Table 5.5   Results of the Sensitivity and Specificity Classifications

Classifier	True Positive	True Negative	False Positive	False Negative	Sensitivity	Specificity
GMM	50	9	2	6	89.29%	81.82%

using a multilayer perceptron neural network. A recursive region growing segmentation algorithm was applied to detect hard exudates. The automatic classification of the system performed well with sensitivity and specificity percentages of 80.21% and 70.66%, respectively.

Kahai et al. [25] have proposed a decision support framework for automated screening of diabetic retinopathy for the univariate case based on the Bayesian framework. This system distinguishes normal from DR images. The sensitivity of the decision is 100%, while its specificity is 67%. Samuel et al. have developed a computer-based assessment of NPDR based on three lesions: hemorrhages and microaneurysms, hard exudates, and cotton wool spots [26]. The software was able to identify the correct eye disease to the tune of 81.7%.

The vessel tracker algorithm was developed to determine the retinal vascular network captured using the digital camera [27]. These tracker algorithms were developed to detect the optic disc, bright lesions such as cotton wools spots, and dark lesions such as hemorrhages. This algorithm correctly identifies arteries and veins with an accuracy of 78.4 % and 66.5%, respectively. The fundus images were subjected to segmentation to extract the lesions and later subjected to the neural network for classification [28]. The system showed a sensitivity of 95.1% and specificity of 46.3%. An automatic method to detect hard exudates, a lesion associated with diabetic retinopathy, was proposed [29]. These features were fed into the statistical classifier and yielded a sensitivity of 79.62%.

Wong et al. have classified normal, moderate NPDR, severe NPDR, and PDR stages using the image processing and neural network techniques [30].

In this present work, we have also focused on the early detection of mild NPDR, which will prevent the loss of vision if detected at an early stage. We have identified normal, NPDR, and PDR correctly with an accuracy of 88% and sensitivity and specificity of more than 80%. However, we can improve the efficiency of the correct classification by extracting better features such as microaneurysms, exudates, and hemorrhages, and by increasing the amount of data in each class.

5.6   Conclusions

Diabetic retinopathy is a complication of diabetes and a leading cause of blindness. It occurs when diabetes damages the tiny blood vessels inside the retina, the light-sensitive tissue at the back of the eye.

Three kinds of retinal conditions—normal retina, NPDR, and PDR—were considered for classification using a GMM classifier. The features are extracted from the raw images using image processing techniques and fed to the GMM classifier for classification. We demonstrated an accuracy of more than 88% of correct classification, sensitivity of 89%, and specificity of 81% for the classifier. The accuracy of the system can be increased further by using proper input features, by using a diverse quality of images for training, and by increasing the size of the training data set.

References

[1]   Amos, A. F., D. J. McCarty, and P. Zimmet, "The Rising Global Burden of Diabetes and Its Complications: Estimates and Projections to the Year 2010," *Diabetic Med.*, Vol. 14, 1997, pp. S57–S85.

[2]   "National Diabetes Fact Sheet," Centers for Disease Control and Prevention, 2003, http://www.cdc.gov.

[3]   Stellingwerf, C., P. Hardus, and J. Hooymans, "Two-Field Photography Can Identify Patients with Vision-Threatening Diabetic Retinopathy: A Screening Approach in the Primary Care Setting," *Diabetes Care*, Vol. 24, 2001, pp. 2086–2090.

[4]   Klein, R., et al., "The Relation of Retinal Vessel Caliber to the Incidence and Progression of Diabetic Retinopathy: XIX: The Wisconsin Epidemiologic Study of Diabetic Retinopathy," *Arch. Ophthalmol.*, Vol. 122, 2004, pp. 76–83.

[5]   "Management of Diabetic Retinopathy: Clinical Practice Guidelines," Canberra: National Health and Medical Research Council, Australian Government Publishing Service, 1997.

[6]   Wong, T. Y., et al., "Retinal Microvascular Abnormalities and Their Relationship with Hypertension, Cardiovascular Disease, and Mortality," *Survey Ophthalmol.*, Vol. 46, 2001, pp. 59–80.

[7]   Li, H., et al., "Automated Grading of Retinal Vessel Caliber," *IEEE Trans. on Biomedical Engineering*, Vol. 52, 2005, pp. 1352–1355.

[8]   Abdel-Ghafara, R. A., et al., "Detection and Characterisation of the Optic Disk in Glaucoma and Diabetic Retinopathy," *Medical Image Understanding and Analysis*, September 23–24, 2004.

[9]   Ege, B. M., et al., "Screening for Diabetic Retinopathy Using Computer Based Image Analysis and Statistical Classification," *Computer Methods and Programs in Biomedicine*, Vol. 62, 2000, pp. 165–175.

[10] Osareh, A., et al., "Classification and Localisation of Diabetic-Related Eye Disease," *ECCV2002*, 2002.

[11] Sinthanayothin, C., et al., "Automated Localisation of the Optic Disc, Fovea and Retinal Blood Vessels from Digital Color Fundus Images," *Br. J. Ophthalmol.*, Vol. 83, 1999, pp. 902–912.

[12] Forracchia, M., M. E. Grisan, and A. Ruggeri, "Extraction and Quantitative Description of Vessel Features in Hypertensive Retinopathy Fundus Images," *CAFIA2001*, 2001.

[13] Hoover, A., V. Kouznetsova, and M. Goldbaum, "Locating Blood Vessels in Retinal Images by Piecewise Threshold Probing of a Matched Filter Response," *Medical Imaging*, Vol. 19, 2000, pp. 203–210.

[14] Dua, S., N. Kandiraju, and H. W. Thompson, "Design and Implementation of a Unique Blood-Vessel Detection Algorithm Towards Early Diagnosis of Diabetic Retinopathy," *Int. Conf. on Information Technology: Coding and Computing*, Vol. 1, No. 4–6, 2005, pp. 26–31.

[15] Walter, T., et al., "A Contribution of Image Processing to the Diagnosis of Diabetic Retinopathy—Detection of Exudates in Color Fundus Images of the Human Retina," *IEEE Trans. on Medical Imaging*, Vol. 21, No. 10, 2002, pp. 1236–1243.

[16] Early Treatment Diabetic Retinopathy Study Research Group, "Grading Diabetic Retinopathy from Stereoscopic Color Fundus Photographs: An Extension of the Modified Airlie House Classification" (ETDRS Report No. 10), *Ophthalmology*, Vol. 98, 1991, pp. 786–806.

[17] Nicolai, L., et al., "Automated Detection of Diabetic Retinopathy in a Fundus Photographic Screening Population," *Investigative Ophthalmol. Vis. Sci.*, Vol. 44, 2003, pp. 767–771.

[18] Cree, M. J., et al., "Comparison of Various Methods to Delineate Blood Vessels in Retinal Images," *Proc. 16th Australian Institute of Physics Congress*, Canberra, 2005.

[19] Niemeijer, M., et al., "Automatic Detection of Red Lesions in Digital Color Fundus Photographs," *IEEE Trans. on Medical Imaging*, Vol. 24, No. 5, pp. 584–592.

[20] Chaudhuri, S., et al., "Detection of Blood Vessels in Retinal Images Using Two-Dimensional Matched Filters," *IEEE Trans. on Medical Imaging*, Vol. 8, No. 3, 1989, pp. 263–269.

[21] Gonzalez, R. C., and P. Wintz, *Digital Image Processing*, 2nd ed., Reading, MA: Addison-Wesley, 1987.

[22] Soille, P., *Morphological Image Analysis: Principles and Applications*, New York: Springer-Verlag, 1999.

[23] Reynolds, D. A., and R. C. Rose, "Robust Text-Independent Speaker Identification Using Gaussian Mixture Speaker Models," *IEEE Trans. on Speech and Audio Processing*, Vol. 3, 1995, pp. 72–83.

[24] Seo, C., K. Y. Lee, and J. Lee, "GMM Based on Local PCA for Speaker Identification," *Electronics Lett.*, Vol. 37, 2001, pp. 1486–1488.

[25] Kahai, P., K. R. Namuduri, and H. Thompson, "A Decision Support Framework for Automated Screening of Diabetic Retinopathy," *Int. J. of Biomedical Imaging*, 2006, pp. 1–8.

[26] Samuel, C. L., et al., "Computer Classification of a Nonproliferative Diabetic Retinopathy," *Arch. Ophthalmol.*, Vol. 123, 2005, pp. 759–764.

[27] Englmeier, K. H., et al., "Early Detection of Diabetes Retinopathy by New Algorithms for Automatic Recognition of Vascular Changes," *Eur. J. Med. Res.*, Vol. 9, No. 10, 2004, pp. 473–488.

[28] Usher, D., et al., "Automated Detection of Diabetic Retinopathy in Digital Retinal Images: A Tool for Diabetic Retinopathy Screening," *Diabet. Med.*, Vol. 21, No. 1, pp. 84–90.

[29] Sanchez, C., et al., "Retinal Image Analysis to Detect and Quantify Lesions Associated with Diabetic Retinopathy," *Conf. Proc. IEEE Eng. Med. Biol. Soc.*, Vol. 3, 2004, pp. 1624–1627.

[30] Wong, L. Y., et al., "Identification of Different Stages of Diabetic Retinopathy Using Retinal Optical Images," *Information Sciences*, Vol. 178, No. 1, 2008, pp. 106–121.

CHAPTER 6

Computer-Based Detection of Diabetes Maculopathy Stages Using Higher-Order Spectra

Chua Kuang Chua, Vinod Chandran, Rajendra Acharya, Eddie Y. K. Ng, Caroline Chee, Manjunath Gupta, Lim Choo Min, Melissa Tan Yan Jun, Gracielynne Flores, and Jasjit S. Suri

Diabetic maculopathy is a common sight-threatening condition among diabetics. Screening to detect maculopathy can lead to successful early treatments in preventing visual loss. In diabetic maculopathy, fluid rich in fat and cholesterol leaks out of damaged vessels. The fluid accumulates near the center of the retina (the macula) and leads to the distortion of central vision and can ultimately cause blindness. Screening to detect maculopathy can lead to successful early treatments and avert loss of vision.

This chapter presents a computer-based intelligent system for the identification of clinically significant and nonclinically significant maculopathy fundus eye images (Figure 6.1). Our protocol uses the Sugeno fuzzy model–based classifier for classification of these two stages. Features are extracted from these raw fundus images using higher-order spectra (HOS), which are then fed to the fuzzy classifier. Our protocol uses 80 subjects, each presenting one of the two kinds of maculopathy conditions. The results are very promising and reveal a sensitivity of 95% and specificity of 90% for the classifier.

6.1 Introduction

Biomedical image processing is one of the research areas that has currently attracted intense interest among scientists and physicians. It consists of the study of digital images with the objective of providing computational tools that assist quantification and visualization of interesting pathology and anatomical structures. The progress that has been achieved in this area during recent years has significantly improved the type of health care that is available to patients. Physicians can now examine the inside of the human body and diagnose, treat and monitor conditions, and plan appropriate treatments more successfully than before. The physician can obtain decision support, be reassured of repetitive tasks, and consistently receive

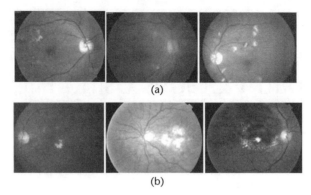

(a)

(b)

Figure 6.1   Fundus images: (a) clinically nonsignificant maculopathy and (b) clinically significant maculopathy.

valuable data. However, the implementation of medical image analysis systems is a multidisciplinary task and requires comprehensive knowledge of many disciplines, such as image processing, computer vision, pattern recognition, and artificial intelligence.

Fundus imaging in ophthalmology plays an important role in the treatment of medical conditions such as hypertension, diabetes, and cardiovascular disease. Moreover, computerized image processing of the fundus image is highly desirable in the diagnosis of diabetic maculopathy, a leading cause of blindness in the world, which requires the screening of a large number of patients by specialized medical personnel.

The blood vessels are the most common feature used in fundus retinal imaging. Several morphologic features of retinal veins and arteries have diagnostic relevance such as diameter, length, branching angle, and tortuosity. Accurate vessel detection is a difficult task for several reasons: the presence of noise, the low contrast between vessels and background, and the variability of vessel width, brightness, and shape. Moreover, due to the presence of lesions, exudates, and other pathological effects, the image may have large abnormal regions.

Diabetic retinopathy is the most common microvascular complication in diabetes, and it can cause severe vision losses [1–3]. The pathogenetic mechanisms involved in the onset and progression of retinopathy are poorly understood [4, 5]. Independent of diabetic retinopathy, severe visual impairment among diabetic patients may also be caused by diabetic maculopathy. Diabetic maculopathy, resulting from diabetic retinopathy, is defined as the presence of retinal thickening within one disc diameter from the fovea [6–10]. Macular edema results from the accumulation of fluid at the posterior pole of the retina, and visual acuity will be threatened due to the thickening of the center of the macula [11]. Factors associated with the development of maculopathy are unknown for the most part [8, 11, 12]. Since diabetic maculopathy is characterized by an increase in capillary leakage in the main retinal vessels and by alterations in the microcirculation of the macula, several previous reports have suggested that poor metabolic control might be involved in hemodynamic changes of retinal circulation, and thereby lead to maculopathy [13, 14]. It is conceivable that an increase in the retinal blood flow could play a part in hemodynamic changes of increased intracapillary retinal pressure and shear stress, thereby leading to diabetic maculopathy [4, 15, 16].

In diabetic maculopathy, fluid rich in fat and cholesterol leaks out of damaged vessels. If the fluid accumulates near the center of the retina (the macula), distortion of central vision will occur. If excess fluid and cholesterol accumulate in the macula, permanent loss of central vision can result. Clinically significant macular edema (CSME) is the term given to describe water logging of the macular area. Most patients with CSME need laser treatment [5]. Early diagnosis of maculopathy can help to save the loss of vision [1].

Most studies about diabetic maculopathy and associated risk factors were hospital-based cohort studies [17–20]. Only one population-based cohort study has been extended to a longitudinal study [12, 21]. Some data have shown that better glycemic and blood pressure control were beneficial in reducing the incidence of macular edema [21].

As shown in Figure 6.2, the macula, or central area, has the following components from center to periphery: foveola, fovea, parafovea, and perifovea. The central part is the foveola, which measures 350 μm. This avascular area consists of densely packed cones that are elongated and connected by the external membrane. In pathological conditions, loss of the normal foveolar reflex may indicate a glial disturbance or it may indicate traction or edema of glial cells. The fovea consists of the thin bottom and measures about 1,500 μm. The parafovea is a belt that measures 0.5 mm in width and surrounds the foveal margin. The perifovea surrounds the parafovea as a belt that measures 1.5 mm wide. The region is characterized by several layers of ganglion cells and six layers of bipolar cells. The foveola, fovea, parafovea, and perifovea together constitute the macula, or central area. The central area can be differentiated from the extra-areal periphery by the ganglion cell layer.

6.1.1   Clinically Nonsignificant Maculopathy

The presence of exudates and swelling in the macula region indicates the maculopathy. Depending on the severity, it may interfere with vision, particularly for reading and seeing fine details. The smaller blood vessels (capillaries) become narrowed or obstructed, whereas others form balloon-like sacs. These changes cause the vessels to leak blood and fluid called *exudates*. Figure 6.1(a) shows the

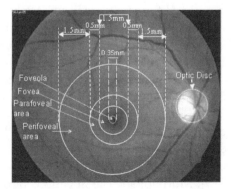

Figure 6.2   Normal fundus image showing major vascular arcade.

fundus image of a nonclinically significant macular edema (non-CSME) subject. Mild maculopathy is the first stage of maculopathy, and usually the patient will not realize whether she is affected, because there are no visible symptoms. The patient's vision may not be seriously affected because the locations of the exudates are a distance away from the fovea. Treatment is not required, but the patient's condition must be monitored to prevent it from worsening.

6.1.2 Clinically Significant Maculopathy

When maculopathy reaches the severe stage, other complications of diabetic retinopathy usually arise, such as cotton wool spots and hemorrhages. But the location of the exudates will be the main concern. Most retinal blood vessels are damaged and the leakage area becomes bigger. The exudates leak out and are deposited very close or on the fovea, which greatly affects visibility because the image cannot focus on the macula properly. Figure 6.1(b) shows a CSME fundus image.

The quantification of diabetic maculopathy and detection of exudates on fundus images have been studied [22, 23]. Global and local threshold values were used to segment exudate lesions from the red-free images. Before applying thresholds, the digitized color photographs were preprocessed to eliminate photographic nonuniformities (shade correction), and the contrast of the exudates was enhanced subsequently. The lesion-based sensitivity of the exudate identification technique was reported to be between 61% and 100% (mean: 87%) [23].

Irrespective of the type of diabetes, patients with long-standing diabetes have a high risk for the development of diabetic maculopathy. Diabetic maculopathy is closely associated with diabetic nephropathy and neuropathy and with several atherosclerotic risk factors, which suggests that these factors might have an important role in the pathogenesis of maculopathy [24].

Kandiraju et al. [25] have used a blood vessel detection algorithm in retinal images that is based on regional recursive hierarchical decomposition using quadtrees and postfiltration of edges to extract blood vessels. The algorithm was able to decrease false dismissals of predominantly significant edges and do so faster when compared to the existing approach with reduced storage requirements for the edge map. Robust extraction of the retinal vessel structure was performed by using seed point extraction to identify a set of points as starting positions and bubble analysis to identify underlying vessels and a direction for each point [26]. Vessel points in a cross section were found by means of a fuzzy C-means classifier. Hoover et al. [27] have located and outlined blood vessels in images by the use of a novel method to segment blood vessels that complements local vessel attributes with region-based attributes of the network structure. Blood vessels were detected using two-dimensional matched filters [28]. Gray level profiles of the cross sections of blood vessels was approximated by a Gaussian-shaped curve. The concept of matched filter detection of signals was used to detect piecewise linear segments of blood vessels after the vessel approximation.

Exudates were detected using their high gray level variation, and their contours were determined by means of morphological reconstruction techniques. The optic disc was detected by means of morphological filtering techniques and the watershed transformation [23]. Morphologic classifications of diabetic retinopathy were performed based on the number, location, and type of discrete microvascular lesions in the fundus of the eye [29].

The presence of microaneurysms in retinal fluorescent angiograms was identified by first locating the fovea by subsampling the image by a factor of 4 in each dimension [30]. Then the image was subjected to median filtering with a 5×5 mask to reduce high-frequency components. The image was then correlated with a two-dimensional circularly symmetric triangular function with modeled gross shading of the macula.

Most of the work carried out so far in this area consider either fluorescent angiogram images [31] or gray level images [27]. The former is time consuming for physicians, inconvenient for patients, costly, and causes nonuniform illumination across the image due to varying amounts of background fluorescence. In the latter, monochrome images of the retina do not always capture all of the available information for more accurate segmentation. Other semiautomated methods for measuring exudates have been developed that need human intervention for defining a threshold, thus reducing the objectivity of the technique [23]. Gardner et al. have used artificial neural networks for identification of exudates by classifying whole regions 20×20 pixels in size [32].

A wide range of applications have developed, exploiting unique properties of higher-order spectra in image reconstruction, pattern recognition, image restoration, and edge detection. Higher-order spectra preserve both amplitude and phase information from the Fourier transform of a signal, unlike the power spectrum. The phase of the Fourier transform contains important shape information [33]. Feature extraction from two-dimensional images and the generation of similarity transformation invariant features using the bispectrum are described in [34].

Our present work deals with the detection of exudates in the macular region. The measurement of severity was carried out based on the four regions of the macular area. The distribution of the exudates in the macular region decides the severity of the maculopathy. The automatic identification of the severity of the diabetic maculopathy system was studied.

6.2   Data Acquisition and Processing

In this work, about 80 retinal photographs of clinically nonsignificant and clinically significant diabetic maculopathy were studied. These patient data were provided by the Kasturba Medical Hospital in Manipal, India. The number and details of the images in each group are shown in Table 6.1. Images were stored in 24-bit JPEG

Table 6.1 Range of Age, Gender, and Number of Subjects in Each Group

Types	Clinically Nonsignificant Maculopathy	Clinically Significant Maculopathy
Age	68 ± 13	72 ± 12
Gender	35 males and 25 females	38 males and 22 females

format with an image size of 576×720 pixels. A Zeiss Visucam^{lite} fundus camera was interfaced to the computer. During the image acquisition process, a patient's chin is positioned on the chin rest and the patient is asked to look into the camera. The photographer aligns the digital system with the center of the patient's papillary axis. Figure 6.1 shows the fundus images of clinically nonsignificant and clinically significant diabetic maculopathy.

The block diagram of the proposed system for classification of maculopathy is shown in Figure 6.3. Feature vectors for the classification of CSME and non-CSME are extracted using higher-order spectra (HOS). Fuzzy logic is used for the classification. Preprocessing, feature extraction, and classification methodologies are described in the following sections.

6.2.1   Preprocessing of Image Data

Before feature extraction it is necessary to preprocess the images and make them suitable. Preprocessing is used to improve the contrast of the image based on histogram equalization.

6.2.1.1   Histogram Equalization

Histogram equalization is the technique by which the dynamic range of the histogram of pixel intensities in an image is increased [35]. Histogram equalization assigns the intensity values of pixels in the input image such that the output image contains a uniform distribution of intensities. It improves contrast. This technique was applied to the whole image, that is, as a global operation.

6.2.2   Feature Extraction

The grayscale image is subjected to the Radon transform to convert the image into one-dimensional data, and then bispectral invariant features [34–37] are extracted from the one-dimensional projections. The steps involved are explained in the following sections.

6.2.2.1   Radon Transform

The Radon transform is used to detect features within a two-dimensional image. It transforms lines through an image to points in the Radon domain, or the Radon backprojection, where each point in the Radon domain is transformed to a straight line in the image (Figure 6.4).

Given a function $A(x, y)$, the Radon transform is defined as:

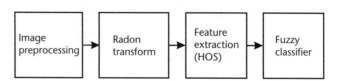

Figure 6.3   Proposed system for identification of diabetic maculopathy.

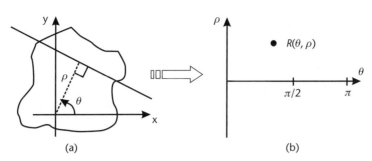

Figure 6.4   (a) Image domain. (b) Radon domain.

$$R(\rho,\theta) = \int_{-\infty}^{\infty} A(\rho\cos\theta - s\sin\theta, \rho\sin\theta + s\cos\theta)ds \qquad (6.1)$$

This equation describes the integral along a line s through the image, where ρ is the distance of the line from the origin and θ is the angle from the horizontal. So, Radon transform converts a two-dimensional signal into the one-dimensional parallel beam projections, at various angles, θ and in this work, we have used $\theta = 20°$. Figure 6.4 shows the schematic diagram of the working of Radon transform.

6.2.3 Higher-Order Spectra (HOS)

Higher-order spectra consist of moment and cumulant spectra and can be defined for both deterministic signals and random processes [36]. In particular, this work uses features derived from the third order statistics of the signal and the corresponding spectrum, namely, the bispectrum. The bispectrum is the Fourier transform of the third order correlation of the data and is given by

$$B(f_1,f_2) = E\left[X(f_1)X(f_2)X^*(f_1+f_2)\right] \qquad (6.2)$$

where $X(f)$ is the Fourier transform of the signal $x(nT)$ and $E[.]$ stands for the expectation operation. In practice, the expectation operation is replaced by an estimate that is an average over an ensemble of realizations of a random signal. For deterministic signals, the relationship holds without an expectation operation with the third order correlation being a time-average. For deterministic sampled signals, $X(f)$ is the discrete-time Fourier transform and in practice is computed as the discrete Fourier transform (DFT) at frequency samples using the FFT algorithm. The frequency f may be normalized by the Nyquist frequency to be between 0 and 1.

The bispectrum may be normalized (by power spectra at component frequencies) such that it has a value between 0 and 1, and indicates the degree of phase coupling between frequency components [36]. The normalized bispectrum or bicoherence is given by

$$B_{co}(f_1,f_2) = \frac{E\left(x(f_1)X(f_2)X^*(f_1+f_2)\right)}{P(f_1)P(f_2)P(f_1+f_2)} \qquad (6.3)$$

where $P(f)$ is the power spectrum.

6.2.3.1 Higher-Order Spectral Features

One set of features in our experiments is based on the phases of the integrated bispectrum [37] and is described briefly next.

Assuming that there is no bispectral aliasing, the bispectrum of a real signal is uniquely defined with the triangle $0 = f_2 = f_1 = f_1 + f_2 = 1$. Parameters are obtained by integrating along the straight lines passing through the origin in bifrequency space. The region of computation and the line of integration are depicted in Figure 6.5. The bispectral invariant, $P(a)$, is the phase of the integrated bispectrum along the radial line with the slope equal to a. This is defined by

$$P(a) = \arctan\left(\frac{I_i(a)}{I_r(a)}\right) \tag{6.4}$$

where

$$I(a) = I_r(a) + jI_i(a)$$
$$= \int_{f_L=0^+}^{\frac{1}{1+a}} B(f_1, af_1)df_1 \tag{6.5}$$

for $0 < a = 1$, and $j = \sqrt{-1}$. The variables I_r and I_i refer to the real and imaginary part of the integrated bispectrum, respectively.

These bispectral invariants contain information about the shape of the waveform within the window and are invariant to shift and amplification and robust to time-scale changes. They are particularly sensitive to changes in the left-right asymmetry of the waveform. For windowed segments of a white Gaussian random process, these features will tend to be distributed symmetrically and uniformly about zero in the interval $[-\pi, +\pi]$. If the process is chaotic and exhibits a colored spectrum with third-order time correlations or phase coupling between Fourier components, the mean value and the distribution of the invariant feature may be used to identify

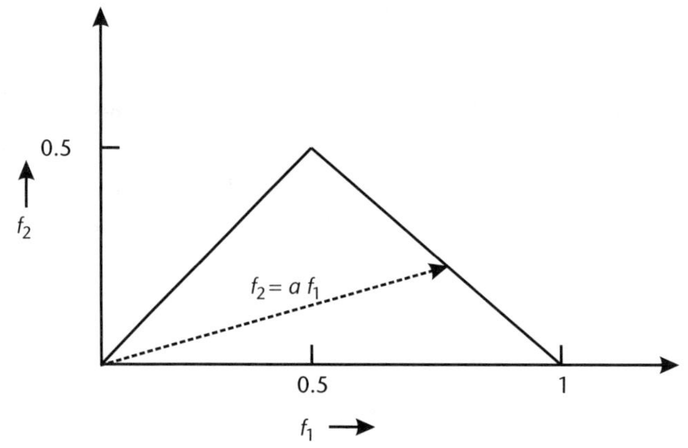

Figure 6.5   Nonredundant region of computation of the bispectrum for real signals. Features are calculated by integrating the bispectrum along the dashed line with slope = a. Frequencies are shown normalized by the Nyquist frequency.

the process. In this work, we have used these bispectral invariant features for every 20°.

6.2.4   Fuzzy Classifier

In a fuzzy classification system, pattern space is divided into multiple subspaces. For each subspace, the relationships between the target patterns and their classes are described by if-then type fuzzy rules [38]. The advantage of this system is that a non-linear classification boundary can be easily implemented. Unknown patterns are classified by fuzzy inference, and patterns that belong to an unknown class, which was not considered by learning, can be easily rejected. Ishibuchi et al. [39, 40] proposed methods to acquire a fuzzy classification system automatically by a simple learning procedure and a genetic algorithm. With these methods, however, a pattern space is divided in to lattice-like structure. Therefore, many fuzzy rules corresponding to fine subspaces are required to implement a complicated classification boundary.

A fuzzy classifier [41] using subtractive clustering and a Sugeno fuzzy inference system is implemented as a classifier as shown in Figure 6.6. The algorithm for implementation is as follows:

Step 1
Fuzzify inputs: The input is fuzzified using a symmetric Gaussian membership function given by

$$f(x;\sigma,\mu) = \frac{e^{-(x-\mu)^2}}{2\sigma^2} \tag{6.6}$$

where σ and μ are the variance and mean of data, respectively.

Step 2
Fuzzy inference: Fuzzy inference is the process of formulating the mapping from a given input to an output using fuzzy logic for making decisions. From the fuzzified inputs, the cluster centers are determined using a subtractive clustering method. In this method:
- The data point with the highest potential to be the first cluster center is selected.
- All data points in the vicinity of the first cluster center (as determined by radii) are removed in order to determine the next data cluster and its center location.
- This process is iterated until all of the data are within the radii of a cluster center.

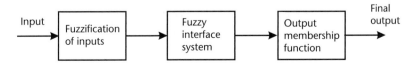

Figure 6.6   Fuzzy classification system.

Step 3

Obtaining the membership computation: Final output is obtained using the Sugeno fuzzy model. The output membership function is linear and is given by

$$r = ax + by + cz + d \tag{6.7}$$

where a, b, c, and d are the adaptive parameters, x, y, z are the inputs and r is the output membership function.

Output level r_i of each rule is weighted by the firing strength w_i of the rule. The final output of the system is the weighted average of all rule outputs, computed as

$$\text{Final output } (F) = \frac{\sum_{i=1}^{N} w_i r_i}{\sum_{i=1}^{N} w_i} \tag{6.8}$$

where N is the total number of fuzzy rules.

6.3   Results

Table 6.2 shows the ranges of the three parameters used to feed as input to the fuzzy classifier. We extracted 18 bispectrum invariants. From them we chose three features, which are clinically significant. These features are subjected to the ANOVA (ANalysis Of VAriance between groups) test to obtain the p-value. The ANOVA test uses variances to decide whether the means are different. This test uses the variation (variance) *within* the groups and translates it into variation (i.e., differences) *between* the groups, taking into account how many subjects are in each group. If the observed differences are high, then it is considered to be statistically significant. In

Table 6.2   ANOVA Applied to Bispectral Invariant Features (Mean Values and Standard Deviations Are Indicated for Each Class in Columns 2 and 3, P-Values in Column 4)

Parameter	Clinically Nonsignificant Maculopathy	Clinically Significant Maculopathy	P-Test
P(1/18)	0.0780 ± 0.0501	0.0523 ± 0.0376	<0.0001
P(5/18)	0.0779 ± 0.0146	0.0975 ± 0.0241	<0.0001
P(13/18)	0.0777 ± 0.0129	0.0940 ± 0.0195	<0.0001

Table 6.3   Classification Accuract

Cases	Total Samples	Number of Data Samples Used for Training	Number of Data Samples Used for Testing	Number of Correctly Classifications	Percentage Classification (%)
Non-CSME	60	40	20	18	90
CSME	60	40	20	19	95
Average					92.5

Table 6.4   Results of Sensitivity, Specificity, and Positive Predictive Value, for Complete Eye Classes Using Fuzzy Classifier

Classifier	True Positive	True Negative	False Positive	False Negative	Sensitivity	Specificity	Positive Predictive Accuracy
Fuzzy	19	18	2	1	95%	90%	90.48%

our work, we have obtained a p-value of less than 0.005, indicating that it is statistically significant. For the purpose of training and testing of the classifier, a database of 120 patient samples is divided into two sets: a training set of 80 arbitrarily chosen samples and a test set of 40 samples (Table 6.3).

Classification accuracy results are shown in Table 6.3 and are above 90%. Table 6.2 shows the hypothesis test conducted using ANOVA, which gives p-value of less than 0.001, indicating a high level of confidence.

We have evaluated the performance of the system using performance measures such as sensitivity, specificity, and positive predictive accuracy. These results are given in Table 6.4. From Tables 6.3 and 6.4, we can see that the proposed system yeilds a promising 92.5% diagnostic accuracy. The system is evaluated to have a sensitivity of 95% and specificity of 90%.

The accuracy of the system can be tested further by increasing the size and quality of the training set. The classification results can be verified by extracting other features from the optical images. Environmental conditions such as the reflection of light influence the quality of the optical images and hence the percentage of classification efficiency. The software for feature extraction and classification of eye images is written in MATLAB 7.0.4.

The proposed system is intended to aid physicians when making clinical diagnoses. This system can give preliminary diagnostics for evaluating the stage of maculopathy. This is helpful in diagnosing clinically nonsignificant maculopathy such that one can detect the maculopathy at an early stage and hence can prevent the loss of vision. The results are promising allowing human observation and automatic detection of the two classes denoted as *non-CSME* and *CSME*.

Figure 6.7 shows a snapshot of the graphical user interface (GUI) of the system. There is an *Upload* pushbutton provided to the left side of the GUI unknown input image. There is an *Analysis* pushbutton provided to extract the relevant features using HOS and a *Results* button to display the classification result in the text box named *Output*. In the case shown, the class of the input image is clinically significant case and is displayed in the left top side of the GUI. Typical CSME and non-CSME images are displayed on the right-hand side of the GUI and after pressing the *Result* button the output is displayed in the bottom of the GUI as clinically significant.

6.4   Discussion

Retinal exudates are typically manifested as spatially random yellow/white patches of varying sizes and shapes. They are a characteristic feature of retinal diseases such

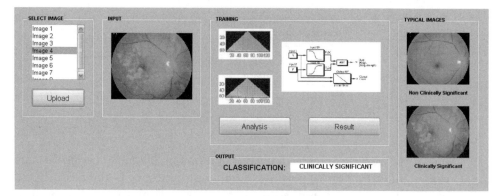

Figure 6.7   Snapshot of the graphical user interface of the system.

as diabetic maculopathy. An automatic method for the detection of exudate regions was introduced comprising image color normalization, enhancing the contrast between the objects and background, segmenting the color retinal image into homogenous regions using fuzzy C-means clustering, and classifying regions into exudate and nonexudate patches using a neural network [16]. Experimental results indicate that we are able to achieve 95% sensitivity and 90% specificity.

Gardner et al. have detected blood vessels, exudates, and hemorrhages using a backpropagation neural network and achieved success rates of 91.7%, 93.1%, and 73.8% respectively [20, 32]. When compared with the results of ophthalmologists, the network achieved a sensitivity of 88.4% and a specificity of 83.5% for the detection of diabetic retinopathy.

The study was designed to compare the severity of age-related maculopathy as graded from photographs taken using three different techniques [42]. Two methods of nonstereoscopic 45° retinal photography of the macula (through a nonpharmacologically dilated pupil and through a pharmacologically dilated pupil) were compared with results from standard 30° stereoscopic photographs in 112 subjects. The results of the study suggest that 45° nonstereoscopic fundus photographs, when graded according to a standard classification scheme, should be considered for detection of age-related maculopathy in situations where the pupils cannot be pharmacologically dilated and retinal specialists are not available to examine the fundus.

Recently, Nayak et al. have identified clinically significant maculopathy using a backpropagation neural network with an accuracy of more than 95% using 350 images [43].

6.5   Conclusions

Diabetic maculopathy is a complication of diabetes and a leading cause of blindness. It occurs when diabetes damages the tiny blood vessels inside the retina, the light-sensitive tissue at the back of the eye. An automatic system for identification of non-CSME and CSME retinal fundus images was proposed. In this chapter we have proposed the use of bispectral invariant features to the fundus eye images. These

HOS features capture the variation in the shapes and contours in these images. The features are extracted from the fundus images using image processing techniques and fed to the fuzzy classifier for classification. We demonstrated an accuracy of correct classification of 92.5%, sensitivity of 95%, and specificity of 90% for the classifier. The accuracy of the system can be tested further by using proper input features, by using a diverse quality of images for training, and by increasing the size of the training data set. The system can be made more reliable by choosing other features such as number of microaneurysms and hemorrhagic areas.

Acknowledgments

The authors wish to thank Dr. Vijaya Pai, professor and head of the Department of Ophthalmology, Manipal Academy of Higher Education, Manipal, India, for providing the eye data and helping with diagnosis during this work.

References

[1] Kanski, J. J., *Clinical Ophthalmology,* 3rd ed., London: Butterworth, 1994.

[2] Klein, R., et al., "The Wisconsin Epidemiologic Study of Diabetic Retinopathy III. Prevalence and Risk of Diabetic Retinopathy When Age at Diagnosis Is 30 or More Years," *Arch. Ophthalmol.,* Vol. 102, No. 4, April 1984, pp. 527–532.

[3] Klein, R., B. E. K. Klein, and S. E. Moss, "Epidemiology of Proliferative Diabetic Retinopathy," *Diabetes Care,* Vol. 15, No. 12, 1992, pp. 1875–1891.

[4] Grunwald, J. E., J. Du Pont, and C. E. Riva, "Retinal Haemodynamics in Patients with Early Diabetes Mellitus," *Br. J. Ophthalmol.,* Vol. 80, 1996, pp. 327–331.

[5] Klein, R., S. E. Moss, and B. E. K. Klein, "New Management Concepts for Timely Diagnosis of Diabetic Retinopathy Treatable by Photocoagulation," *Diabetes Care,* Vol. 10, No. 5, 1987, pp. 633–638.

[6] Cunha-Vaz, J. G., et al., "Early Breakdown of the Blood-Retina Barrier in Diabetes," *Br. J. Ophthalmol.,* Vol. 59, 1975, pp. 649–656.

[7] Cunha-Vaz, J. G., and A. Travassos, "Breakdown of the Blood-Retina Barriers and Cystoid Macular Edema," *Surv. Ophthalmol.,* Vol. 28, 1984, pp. 485–492.

[8] Ferris, F. L., and A. Patz, "Macular Edema. A Complication of Diabetic Retinopathy," *Surv. Ophthalmol.,* Vol. 28, 1984, pp. 452–461.

[9] Ferris, F. L., M. D. Davis, and L. M. Aiello, "Treatment of Diabetic Retinopathy," *N. Engl. J. Med.,* Vol. 341, 1999, pp. 667–678.

[10] Patel, V., et al., "Retinal Blood Flow in Diabetic Retinopathy," *Br. Med. J.,* Vol. 305, 1992, pp. 678–683.

[11] Moss, S. E., R. Klein, and B. E. K. Klein, "The Incidence of Vision Loss in a Diabetic Population," *Ophthalmology,* Vol. 95, 1988, pp. 1340–1348.

[12] Klein, R., et al., "The Wisconsin Epidemiologic Study of Diabetic Retinopathy IV. Diabetic Macular Edema," *Ophthalmology,* Vol. 91, No. 12, 1984, pp. 1464–1474.

[13] Arend, O., et al., "Retinal Microcirculation in Patients with Diabetes Mellitus: Dynamic and Morphological Analysis of Perifoveal Capillary Network," *Br. J. Ophthalmol.,* Vol. 75, No. 9, September 1991, pp. 514–518.

[14] Grunwald, J. E., et al., "Total Retinal Volumetric Blood Flow Rate in Diabetic Patients with Poor Metabolic Control," *Invest. Ophthalmol. Vis. Sci.,* Vol. 33, 1992, pp. 356–363.

[15] Kohner, E. M., et al., "The Retinal Blood Flow in Diabetes," *Diabetologia,* Vol. 11, No. 1, 1975, pp. 27–33.

[16] Osareh, A., et al., "Automatic Recognition of Exudative Maculopathy Using Fuzzy C-Means Clustering and Neural Networks," *Proc. Medical Image Understanding Analysis Conf.,* 2001, pp. 49–52.

[17] Aiello, L. M., et al., "Nonocular Clinical Risk Factors in the Progression of Diabetic Retinopathy," in *Diabetic Retinopathy,* H. L. Little et al., (eds.), New York: Thieme-Stratton, 1983, pp. 21–31.

[18] British Multicenter Study Group, "Photocoagulation for Diabetic Maculopathy: A Randomized Controlled Clinical Trial Using the Xenon Arc," *Diabetes,* Vol. 32, 1983, pp. 1010–1016.

[19] Klein, B. E. K., et al., "The Wisconsin Epidemiologic Study of Diabetic Retinopathy XIII. Relationship of Serum Cholesterol to Retinopathy and Hard Exudates," *Ophthalmology,* Vol. 98, No. 8, 1991, pp. 1261–1265.

[20] Myers, F. L., M. D. Davis, and Y. L. Magli, "The Natural Course of Diabetic Retinopathy: A Clinical Study of 321 Eyes Followed One Year or More," *Symp. on Treatment of Diabetic Retinopathy,* M. F. Goldberg and S. L. Fine, (eds.), (PHS Pub. No. 1890), Washington, D.C.: U.S. Government Printing Office, 1969, pp. 81–85.

[21] Klein, R., et al., "The Wisconsin Epidemiologic Study of Diabetic Retinopathy XVII. The 14-Year Incidence and Progression of Diabetic Retinopathy and Associated Risk Factors in Type 1 Diabetes," *Ophthalmology,* Vol. 105, No. 10, 1998, pp. 1801–1815.

[22] Phillips, R., et al., "Quantification of Diabetic Maculopathy by Digital Imaging of the Fundus," *Eye,* Vol. 5, Pt. 1, 1991, pp. 130–137.

[23] Philips, R., J. Forrester, and P. Sharp, "Automated Detection and Quantification of Retinal Exudates," *Graefe's Arch. Clin. Exper. Ophthalmol.,* Vol. 231, 1993, pp. 90–94.

[24] Zander, E., et al., "Maculopathy in Patients with Diabetes Mellitus Type 1 and Type 2: Associations with Risk Factors," *Br. J. Ophthalmol.,* Vol. 84, 2000, pp. 871–876.

[25] Kandiraju, N., S. Dua, and H. W. Thompson, "Design and Implementation of a Unique Blood Vessel Detection Algorithm Towards Early Diagnosis of Diabetic Retinopathy," *Proc. Int. Conf. on Information Technology: Coding and Computing,* Los Alamitos, CA: IEEE Computer Society, 2005, pp. 26–31.

[26] Grisan, I. E., et al., "A New Tracking System for the Robust Extraction of Retinal Vessel Structure," *Proc. 26th Annual Int. Conf. of the IEEE EMBS,* San Francisco, CA, 2004, pp. 1620–1623.

[27] Hoover, A., V. Kouzanetsova, and M. Goldbaum, "Locating Blood Vessels in Retinal Images by Piecewise Threshold Probing of a Matched Filter Response," *IEEE Trans. on Medical Imaging,* Vol. 19, No. 3, 2000, pp. 203–210.

[28] Chaudhuri, S., et al., "Detection of Blood Vessels in Retinal Images Using Two-Dimensional Matched Filters," *IEEE Trans. on Medical Imaging,* Vol. 8, No. 3, 1989, pp. 263–269.

[29] Early Treatment Diabetic Retinopathy Study Research Group, "Grading Diabetic Retinopathy from Stereoscopic Color Fundus Photographs: An Extension of the Modified Airlie House Classification" (ETDRS Report No. 10), *Ophthalmology,* Vol. 98, 1991, pp. 786–806.

[30] Cree, M. J., et al., "Automated Microaneurysm Detection," *Proc. IEEE Int. Conf. on Image Processing,* Vol. 3, 1996, pp. 699–702.

[31] Frame, A. J., et al., "A Comparison of Computer Classification Methods Applied to Detection of Microaneurysms in Ophthalmic Fluorescein Angiograms," *Computers Biol. Med.,* Vol. 28, 1998, pp. 225–238.

[32] Gardner G., et al., "Automatic Detection of Diabetic Retinopathy Using an Artificial Neural Network: A Screening Tool," *Br. J. Ophthalmol.,* Vol. 80, 1996, pp. 940–944.

[33] Oppenheim, A., and J. Kim, "The Importance of Phase in Signals," *Proc. IEEE,* Vol. 69, 1981, pp. 529–541.

[34] Chandran V., et al., "Pattern Recognition Using Invariants Defined from Higher Order Spectra: 2-D Image Inputs," *IEEE Trans. on Image Processing*, Vol. 6, 1997, pp. 703–712.

[35] Gonzalez, R., and R. Woods, *Digital Image Processing*, Reading, MA: Addison-Wesley, 1993.

[36] Nikias, C. L., and A. P. Petropulu, *Higher-Order Spectra Analysis: A Nonlinear Signal Processing Framework*, Englewood Cliffs, NJ: Prentice-Hall, 1993.

[37] Chandran, V., and S. L. Elgar, "Pattern Recognition Using Invariants Defined from Higher Order Spectra One-Dimensional Inputs," *IEEE Trans. on Signal Processing*, Vol. 41, 1997, pp. 205–212.

[38] Guanrong, C., and T. T. Pham, *Introduction to Fuzzy Systems*, Boca Raton, FL: CRC Press, 2006.

[39] Nozaki, N., and T. Ishibuchi, "Selecting Fuzzy If-Then Rules with Forgetting in Fuzzy Classification Systems," *Journal of Japan Society for Fuzzy Theory and Systems*, Vol. 6, No. 3, 1994, pp. 585–602.

[40] Ishibuchi, H., T. Murata, and H. Tanaka, "Construction of Fuzzy Classification Systems Using Genetic Algorithms," *Journal of Japan Society for Fuzzy Theory and Systems*, Vol. 7, No. 5, 1995, pp. 1022–1040.

[41] George, K., and Y. Bo, *Fuzzy Sets and Fuzzy Logic: Theory and Applications*, Upper Saddle River, NJ: Prentice-Hall, 1995.

[42] Klein, R., et al., "Detection of Drusen and Early Signs of Age-Related Maculopathy Using a Nonmydriatic Camera and a Standard Fundus Camera," *Ophthalmology*, Vol. 99, No. 11, 1992, pp. 1686–1692.

[43] Nayak, J., P. S. Bhat, and U. R. Acharya, "Automatic Detection of Diabetic Maculopathy Stages Using Fundus Images," *Journal of Medical & Engineering Technology*, 2008.

CHAPTER 7

Algorithms for Detecting Glaucomatous Structural Changes in the Optic Nerve Head

Madhusudhanan Balasubramanian, Christopher Bowd, and Linda M. Zangwill

In this chapter, we present a brief overview of the anatomy of the eye and pathophysiology of glaucoma followed by a comprehensive review of the glaucoma literature in the context of detecting progression of glaucomatous structural defects in the optic nerve head (ONH) region of an eye. Confocal scanning laser ophthalmoscopes (CSLOs) are routinely used in clinical and laboratory research applications in ophthalmology to capture the three-dimensional architecture of the ONH of a human eye in vivo and to generate reproducible ONH topographs of an eye. Because glaucoma is a progressive optic neuropathy, the structural changes associated with a glaucomatous condition in an eye can be captured and analyzed using CSLOs. The CSLO-generated ONH topographs taken during baseline and follow-up visits can be used to detect and quantify the structural changes in the ONH region. We review the existing statistical and computational methods of detecting structural changes in the ONH region of an eye using global and regional ONH summary parameters (called stereometric parameters) and using pixel-level height measurements. Data sets of selected participants from the University of California San Diego (UCSD) Diagnostic Innovations in Glaucoma Study are used for demonstration.

Section 7.1 provides a brief overview of the anatomy of the eye and pathophysiology of glaucoma. In Section 7.2, we discuss the general appearance of the ONH of a healthy eye and observable characteristic changes in the ONH region in eyes with glaucoma. A brief description of the principle of CSLOs and their use in examining the ONH region of an eye by acquiring ONH topographs is described in Section 7.3. In addition, methods of automatically estimating various ONH stereometric parameters using the Heidelberg Retina Tomograph (Heidelberg Engineering, Heidelberg, Germany), including the existing methods of locating the lower extents of the retinal nerve fiber layer using a *reference plane*, are presented in detail. In Section 7.4, we discuss in detail the existing methods of detecting change over time in the CSLO ONH topographs using their ONH stereometric parameters and the pixel-level analysis methods of topographic change analysis and nonparametric permutation tests.

7.1 Anatomy of the Eye and Pathophysiology of Glaucoma

7.1.1 Anatomy of the Eye

The retina of the eye is a photosensitive layer approximately 0.25 mm thick and is an important part of the visual system. The light enters the eye through the cornea and passes through the pupil; the lens then focuses the light onto the retinal layers. The cornea of a healthy eye is a clear medium, and the corneal layers contribute to the majority of the refractive power of the eye. The iris controls the amount of light reaching the retinal layers. The retina is composed of 10 distinct layers as shown in Chapter 1. The photoreceptor layer consists of two types of photoreceptors called *rods* and *cones*. Rods are more sensitive to light than cones but insensitive to color and are used for low-light vision, whereas cones are responsible for daylight color vision. In opposition to an intuitive expectation, the photoreceptor layer lies beneath the neuronal layers close to the retinal pigment epithelium (RPE) in the human retina [1–3]; therefore, the light has to pass through the neuronal layers to reach the photoreceptors. Also, the RPE absorbs light and avoids backscatter on to the photoreceptors.

The spacing of the neural cells and the fibers in the neuronal layers allows the light to reach the photoreceptors. The *bipolar* and *horizontal cells* in the inner nuclear layer combine the chemical responses of the photoreceptors to the incident light and pass it on to the *ganglion cells* of the ganglion cell layer. The ganglion cells convert the chemical responses into electrical signals. The axons of the ganglion cells group together to form the *optic nerve* and carry the visual electrical impulses to the visual cortex of the brain for image formation. These nerve fibers run parallel to the retinal surface [4] and exit the eye at the *optic disc* (the blind spot), physiologically identified as the inner aperture of the scleral canal, through a collagenous matrix called *lamina cribrosa*.

The optic disc is also referred to as the optic nerve head (ONH) and exhibits a natural cup shape due to the arrangement of the nerve fibers exiting the eye. Figure 7.1 shows an ONH photograph of a human eye with the optic disc region demarcated using a manually drawn contour line guided by the inner aperture of the scleral canal. As the nerve fibers exit the eye through the scleral canal, they appear as a sloping region within the optic disc called the *neuroretinal rim,* and they exhibit a natural cup shape within the optic disc called the *optic cup* as shown in Figure 7.1. A healthy human eye has more than 1 million optic nerve fibers, and the nerve fiber count is estimated to increase by a factor of 175,000 fibers per additional millimeter of scleral canal area or optic disc area [5].

Maintaining the spherical shape of the eye and the convex shape of the cornea is crucial for the proper functioning of the eye. This is primarily accomplished by a pressure in the anterior chamber, called intraocular pressure (IOP), in the range of 10 to 21 mmHg above the atmospheric level maintained by the aqueous humor production in the posterior chamber of the eye and its outflow into Schlemm's canal through the *trabecular meshwork* in the anterior chamber of the eye. The *ciliary*

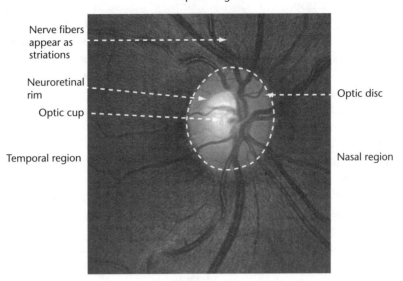

Figure 7.1 A photograph of the ONH region of the right eye of a participant in the UCSD Diagnostic Innovations in Glaucoma Study: The optic disc region is marked using a manually drawn contour line. The nerve fibers from the ganglion cells in the inferior, superior, temporal, and nasal regions of the optic disc follow an accurate pattern and converge into the optic disc.

body produces the aqueous humor in the posterior chamber. The aqueous humor is a clear liquid that distributes nutrients and immune responses in case of inflammation or infections to the lens, iris, and trabecular meshwork; it also removes metabolic wastes. The angle between the cornea and the iris needs to be clear of any obstruction to provide a path of least resistance for the aqueous humor outflow. An increased aqueous outflow resistance at the angle may cause elevated IOP levels.

7.1.2 Pathophysiology of Glaucoma

Glaucoma is a progressive optic neuropathy that, when left untreated, may result in progressive vision impairment and eventual blindness. It is the second leading cause of blindness in the world next only to cataract. Current estimates based on population-based prevalence surveys predict that 60 million people worldwide will be affected by glaucoma in 2010, increasing to 80 million by 2020, disproportionately affecting women and Asians [6]. Retrospective estimates of glaucoma-related expenses such as ophthalmologist visits, glaucoma-related surgical procedures, medications, and other indirect costs and services using multicenter resource chart reviews and Medicare claims indicate an annual cost of about $1.2 billion to $2.5 billion in the United States [7–9]. The mean annual cost of treating glaucoma increases with disease severity [8]. Near the end stage of glaucoma, the treatment costs go down with an increase in the indirect costs of vision care and rehabilitation charges.

The leading factors that influence development of glaucoma in an individual are age, race, family history, and IOP elevation levels. Risks of developing glaucoma are high for older populations, with a glaucoma incidence rate of 1% at 50 years that increases up to 8% at 80 years [10]. Individuals of African origin have a high prevalence of open-angle glaucoma [11], with the largest absolute number of people with open-angle glaucoma and angle-closure glaucoma in China, followed by Europe and India [6]. In many cases, elevation in IOP levels, typically above 21 mmHg (normal range), causes optic nerve and ganglion cell damage either directly: (1) due to pressure-induced stress in the retinal layers or (2) due to an increased pressure gradient across the lamina cribrosa resulting in its deformation, compressing the axons or indirectly through ischemia [12]. However, elevated IOP cannot be used as a defining characteristic of glaucoma, because eyes with normal IOP levels may undergo progressive structure and vision changes, a situation sometimes referred to as *normal-tension glaucoma* (an arbitrary distinction that has become controversial in recent years), yet eyes with IOP levels above the normal range may undergo no structural damages and vision changes (a condition called *ocular-hypertension*). Geometrically, the size and shape of the scleral canal (optic disc), inner radius of the eye, and the scleral thickness affect the biomechanics of the ONH and influence the damaging effects of the elevated IOP levels [13, 14].

Based on the cause of the elevated IOP levels, glaucomatous conditions are classified as *primary* and *secondary* glaucoma. In primary glaucoma, there are no associated pathological conditions in the eye that cause the IOP elevations, whereas in secondary glaucoma pathological conditions such as an injury, cataract, or tumor cause an elevation in IOP. Further, depending on the source of blockage in the anterior chamber angle, the primary and secondary glaucomatous conditions are subclassified into *open-angle* and *angle-closure* glaucoma.

In primary open-angle glaucoma (POAG) the resistance to the aqueous humor drainage increases without any clinically visible obstruction in the trabecular meshwork [15]. POAG is the most common type and a leading cause of blindness due to glaucoma. Structural defects in the optic disc region with confirmed visual defects consistent with glaucoma are used as a defining characteristic of POAG [16].

Currently, IOP is the only proven treatable risk factor associated with glaucoma [15]. IOP measurements are confounded to some degree by the central corneal thickness (CCT) of an eye. Ocular-hypertensive eyes with a thinner CCT are at greater risk of developing POAG [17]. Also, the IOP measurements are overestimated in eyes with thicker CCTs and underestimated in eyes with thinner CCTs. The IOP-related treatment for glaucoma involves reducing and maintaining the IOP at a target level using topical hypotensive medications, *laser trabeculoplasty* (a laser treatment aimed at reducing the resistance to the aqueous humor drainage), and surgical techniques such as *trabeculectomy*. IOP reduction and maintenance have been proven beneficial in lowering the rate of glaucoma progression and delaying the progression in various studies involving participants with early glaucoma [18], normal-tension glaucoma [19], and in animal models [20] and are beneficial in preventing the onset of POAG or delaying the onset of POAG in ocular-hypertensive participants [21]. A recent study suggests that an IOP reduction in early glaucoma participants not only reduces the rate of progression and delays glaucoma progression, but also revives the dysfunctional RGCs possibly preventing their eventual death and causing improvements in visual response [22]. Currently, there is conflicting evidence in favor of either the mean IOP level over time or the fluctuations in the IOP level over time influencing glaucoma progression [23].

7.2   Morphological Changes in the ONH Region in Glaucoma

Glaucoma is a chronic disease; therefore, it typically results in a gradual loss of nerve fibers and death of retinal ganglion cells. The loss of nerve fibers results in *characteristic changes* in the appearance of the retinal nerve fiber layer as localized or diffuse defects and as changes in the configuration of the optic disc.

7.2.1   Optic Disc Configuration

In a normal eye, there is large variability in the axonal nerve fiber count, appearance of the optic nerve, and size of the optic disc. There are approximately 800,000 to 1.5 million nerve fibers in a normal eye [5, 24–26]. The size of the optic disc varies from approximately 0.80 to 6 mm^2 [27]. As mentioned earlier, the number of nerve fibers increases proportionately with the increase in the optic disc size in the range of about 175,000 fibers per square millimeter increase in the optic disc area [5]. Despite this large variability in the appearance of the ONH, some general patterns emerge. The optic disc is usually vertically oval (vertical optic disc diameter $d_{v(D)} \approx$ 1.1 × the horizontal disc diameter $d_{h(D)}$), and the optic cup is horizontally oval (horizontal cup diameter $d_{h(C)} \approx 1.08$ × the vertical cup diameter $d_{v(C)}$) as shown in Figure 7.2, thus resulting in the characteristic shape of the neuroretinal rim [27]. The neuroretinal rim is usually thicker along the vertical poles of the optic disc with the absolute rim thickness measuring higher in the inferior region followed by the superior, nasal, and temporal regions, in that order [27, 28]. Any violation of this arrangement (Inferior–Superior–Nasal–Temporal: ISNT) may indicate a glaucomatous condition in the eye [28, 29]. Figure 7.3 shows the optic disc photographs of a participant in the UCSD Diagnostic Innovations in Glaucoma Study (DIGS) taken in 1998 and 2002. A localized rim thinning can be observed in the inferior region marked with an arrow in the photograph taken in 2002 with reference to the photograph taken in 1998, with an eventual increase in the optic cup size and a physiological change in the rim that violates the ISNT rule.

   Diffuse rim loss may be harder to detect and may not be obvious in the initial stages. However, it has been suggested that the neuroretinal rim area does not decrease with the axonal nerve fiber loss related to age and nonglaucomatous optic nerve damage [26]. Therefore, the sectorial rim thickness measurements have an important role in identifying optic disc abnormalities, including diffuse rim loss. Optic disc geometrical parameters such as the vertical and horizontal diameters of the optic disc ($d_{v(D)}$ and $d_{h(D)}$) and the vertical and horizontal diameters of the optic cup ($d_{v(C)}$ and $d_{h(C)}$) can be measured clinically using a handheld slit-lamp biomicroscope, optic disc stereo photographs, and modern imaging instruments such as CSLOs (discussed in detail later). Rim area and volume, cup area and volume, cup-to-disc area ratio, vertical CDR (VCDR), and horizontal CDR (HCDR) are some of the optic disc parameters useful in identifying glaucoma. The optic cup enlargement due to any rim loss results in higher cup-to-disc area and diameter

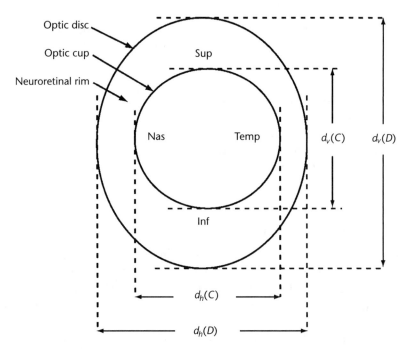

Figure 7.2 In a normal optic disc, the shape of the optic disc is typically vertically oval and the optic cup is horizontally oval, resulting in the characteristic shape of the neuroretinal rim. Using the vertical disc diameters of the disc and cup ($d_{v(D)}$ and $d_{v(C)}$, respectively), the vertical cup-to-disc ratio (CDR) can be estimated as $d_{v(C)}/d_{v(D)}$; similarly the horizontal CDR can be estimated as $d_{h(C)}/d_{h(D)}$ using the horizontal cup diameter $d_{h(C)}$ and horizontal disc diameter $d_{h(D)}$.

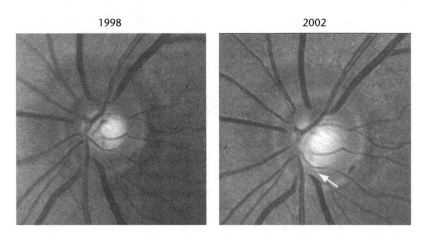

Figure 7.3 ONH of the left eye of a DIGS participant exhibiting changes in the inferior neural retinal rim (white arrow).

ratios. Because rim losses are more prominent along the vertical poles of the optic disc in the initial stages, the VCDR may be more sensitive in the initial stages. These cup-to-disc area and diameter ratios can be measured with good reproducibility [30,

31]. Previous glaucoma prevalence surveys have used a 97.5th percentile and 99.5th percentile of the CDRs observed in normal eyes as a cutoff to identify the presence of glaucoma in combination with the presence of associated visual function defects [16]. However, the absolute values of the rim thickness measures should be interpreted with caution because rim thickness is correlated with optic disc size [5]. For example, a smaller cup in a small optic disc may be glaucomatous, whereas a larger cup in a large disc may be normal. Also, population-based studies have observed a high variability of VCDR measurements in normal eyes between 0.07 and 0.85 [32, 33], resulting in a large overlap in these parameters among the normal and glaucoma population. Therefore, these parameters should be adjusted for the disc size before applying the percentile cutoffs of these parameters for classification [32, 33].

7.2.2 Retinal Nerve Fiber Layer Defects

In a healthy eye, the optic nerve axons appear as striations in the retinal surface. Also the regions with higher axonal fiber density appear relatively brighter in the optic disc photographs. Histological studies have highlighted the large variability in the number of retinal nerve fiber layer (RNFL) fibers and axons in the healthy eye [5, 24–26]. Upon axonal losses, the affected regions appear relatively smoother (i.e., less striated) and darker than in a normal eye. Usually the retinal vessels are embedded in the RNFL and may not appear with good clarity. After axonal fiber loss, the retinal vessels become more visible. The defects in the RNFL can be localized or diffuse. The localized changes in the RNFL appear as a wedge defect running toward the optic disc as shown in Figure 7.4(b) in the inferotemporal (7 o'clock) region. Figure 7.4(c) shows an eventual enlargement of the wedge defect after 5 years, a clear indication of progressing glaucoma [27, 28]. Also the nerve fiber visibility decreases with age and it correlates with an estimated 4,000 axonal losses per year.

7.2.3 Parapapillary Atrophy and Optic Disc Hemorrhages

The loss of RPE cells and marked reduction of photoreceptors near the optic disc region appear as a central β-zone. The frequency of β-zones is higher and the areas of the β-zones are larger in glaucomatous eyes than normal eyes. Also, enlargement

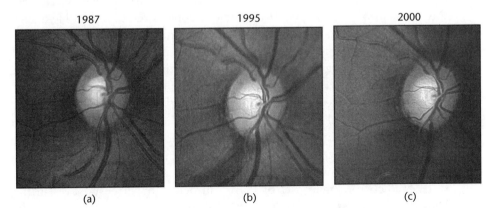

Figure 7.4 An RNFL wedge defect can be observed in part (b) in the 7 o'clock position. The defect has grown during a 5-year period, as shown in part (c).

of the β-zones is one of the earliest signs of glaucoma onset in eyes with ocular-hypertension [27]. Enlargement of the β-zone and its area, in combination with the optic disc parameters discussed earlier, is useful in identifying glaucoma, especially in eyes with small optic discs. When used alone, these may not be the most sensitive parameters indicative of a glaucomatous condition [34]. Irregular hyper/hypopigmentation and thinning of the chorioretinal tissue layer appear as α-zones. The α-zones can be observed both under normal and glaucomatous conditions. When both α- and β-zones are present, the α-zone appears peripheral to the β-zone.

Optic disc hemorrhages are more common in glaucomatous eyes than normal eyes. In glaucoma, the hemorrhages have a characteristic splinter shape and they appear near the optic disc border [27, 28, 34]. Hemorrhages are more common in normal-tension glaucoma than in POAG [35].

The structural changes discussed earlier that appear as changes in the optic disc configuration or as defects in the nerve fiber layer originate either directly from the defects in the underlying neuronal cell layers or indirectly influence the function of the neural cell layers and cause eventual changes in the visual function of the eye. However, the order of the temporal sequence of observing the structural and the visual function defects varies [36]. Although visual function defects are associated with structural defects such as dysfunction and death of retinal ganglion cells and loss of axonal nerve fibers, the associated structural defects may not always be visible in the retinal surface at the same time. For example, the optic nerve axons of the ganglion cells associated with the wedge defect may be located in the middle or deeper layers. Thus, the structural defect may not be visible in the retinal surface until a significant number of optic nerve fibers are lost [27], in which case the visual defects may be observed before the appearance of the structural defects. These findings are also reflected in the sensitivities of the tests used to monitor visual function and retinal topography.

7.3 Confocal Scanning Laser Ophthalmoscope for ONH Analysis

Confocal microscopes are used routinely nowadays in clinics and for laboratory research in ophthalmology. A class of confocal microscope, CSLOs utilize a rotating mirror arrangement to scan an imaging area using laser light beams and capture the three-dimensional architecture of a microscopic structure. The Heidelberg retina tomograph (HRT) is the CSLO most commonly clinically used for imaging the ONH architecture. HRT generates a reflectance image similar to the optic disc photographs and a topograph for analysis. Two of the clear advantages of using CSLO images compared to optic disc photographs are: (1) reduced requirements for pupil dilation and clear media, and (2) the immediate availability of images for analysis. Using CSLO, several optic disc parameters including disc and cup diameters, disc and cup area, and cup-to-disc area and diameter ratios can be estimated automatically after manually marking the optic disc margin. The CSLO estimated disc parameters were shown to be in good agreement with clinicians [30] and were found to be predictive of the onset of POAG in ocular-hypertensive eyes in the Ocular Hypertensive Treatment Study [37]. Also in another study in a set of 46 ocular-hypertensive participants, 46 POAG participants, and 46 normal participants, the

CSLO optic disc parameters were found to be able to discriminate ocular-hypertensive participants from POAG and normal participants [38]. Commercial CSLOs such as the HRT have been proven to be reliable and suitable for regular ophthalmic use in clinical and laboratory settings [39–46]. See Chapter 3 for more details about confocal microscopes.

7.3.1   Principle of Confocal Microscopy

The concept of confocal microscopy was invented by Marvin Minsky in 1955 [47, 48]. A confocal microscope allows imaging of a microscopic structure with suitable refractive properties at various depths along its optical axis without physically sectioning the structure. Thus, the initial depth images acquired using a confocal microscope are called *optical sections*. Figure 7.5 shows a schematic representation of a confocal microscope. An objective lens with focal length f focuses a thin ray of light passing through a light-source pinhole and illuminates a small volume of the structure. With an imaging plane at a distance i from the objective lens, a layer at a

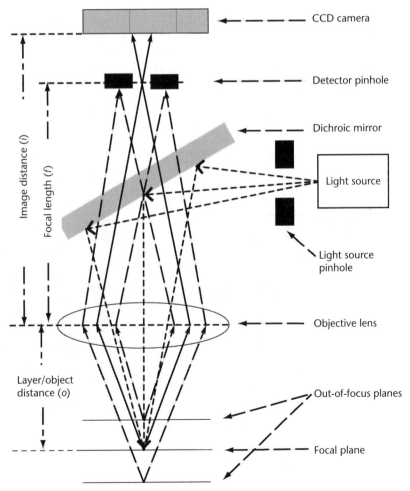

Figure 7.5   Principle of confocal microscopy.

depth o will be in focus. A finitely small detector pinhole placed at the focal plane f will effectively filter the aberrant light from the out-of-focus layers and allow the reflected light from the layer on focus to reach the detector as shown in Figure 7.5. To form an entire image, the whole on-focus layer needs to be imaged one point at a time. Confocal microscopes that utilize white-light sources for illumination and imaging use a Nipkow scanning disc for imaging multiple scan points at a time [49]. CSLOs utilize a laser light source and a rotating mirror arrangement to scan an entire imaging area.

The HRT was introduced into clinical practice in 1993, the HRT-II in 2002, and the HRT-3 in 2005. All instruments are theoretically backward compatible and utilize a laser diode light source of wavelength 670 nm. The HRT-3 acquires a total of 16 to 64 optical sections in about 1 to 6 seconds per ONH scan maintaining a constant z-axis resolution of 1/16 mm between optical sections each of size 384 × 384 pixels, thus generating three-dimensional volume data sets of size 384 × 384 ×16 voxels up to 384 × 384 × 64 voxels in a 15° field of view. Using the optical section images, HRT constructs *topograph* and *reflectance* images. The topograph is constructed by identifying the peak intensity along the z-axis from all the optical sections at each pixel location. Thus, a topograph represents a height matrix identifying the height profile of the ONH region. The reflectance images are constructed as a summation of intensities along the z-axis from all of the optical sections at each pixel location and thus represent the surface reflectance profile of the ONH surface. The topograph height matrices are useful in measuring the ONH parameters including the parameters of the optic disc.

Using mean topographs constructed from three or more individual scans per visit can improve the HRT measurement repeatability [39]. Acquiring more than three scans during each imaging session may not be clinically feasible. Therefore, HRT takes a set of three three-dimensional scans per exam and computes a mean topograph and reflectance images for further analysis. Figure 7.6 shows the mean

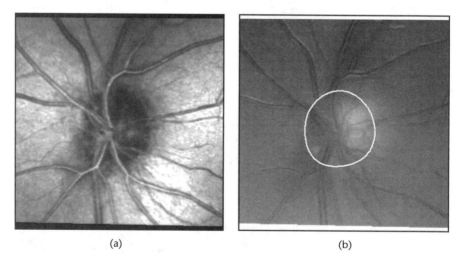

(a)   (b)

Figure 7.6   (a) The mean HRT reflectance image and (b) the corresponding mean topograph image of a subject in the UCSD DIGS database. The optic disc margin is marked using a manually drawn contour line on the mean topograph in part (b).

HRT reflectance image and topograph of a participant in the UCSD DIGS database. As with the optic disc photographs, a good-quality HRT is essential for obtaining reliable estimates of the optic disc parameters. HRT uses the pixel standard deviation of topographs as an image quality metric to assess the quality of the ONH scans automatically. Topograph scans with pixel standard deviation below 30 μm are recommended and scans with standard deviation above 50 μm are considered very poor. Though HRT does not require pupil dilation for obtaining ONH scans, pupil dilation is recommended for eyes with particularly small pupil diameters or with any media opacities such as cataract.

7.3.2 Measuring Optic Disc Stereometric Parameters from the CSLO Optic Disc Topographs

The first step in estimating the optic disc parameters is to identify the optic disc and manually outline its margin. Recently, an *active contour*–based automatic delineation of the optic disc, neuroretinal rim, and the region with parapapillary atrophy was described [50]. In brief, the method normalizes a topograph to correct for any uneven illumination. In the illumination-corrected topographs, only the ONH region and vessels appear as the dark region and the neuroretinal rim appears as a very dark ring. The vessels are removed using their geometric properties. From the remaining regions of the neuroretinal rim area, an approximate estimate of the optic disc margin is estimated using the *Hough transform*. Further, the contour is refined using an *active contour model*. They report good agreement between the automatic optic disc delineation method and the manually drawn contour lines in 71% of the study data sets. Despite the initial promising results, their reproducibility is very challenging due to the variability in the size and shape of the optic discs and neuroretinal rim among normal and glaucomatous populations. In optic discs with very thin neuroretinal rims, the proposed method may fail. Also, the presence of parapapillary atrophy may artificially result in a larger optic disc in this method. At present, manual outlining of the optic disc margin in the HRT topographs continues to be the most reliable method of generating reproducible optic disc stereometric parameters. Using the HRT software, a user manually outlines the optic disc margin following the inner margin of the scleral ring.

The lower extent of the nerve fiber layer needs to be identified to calculate the ONH stereometric parameters such as RNFL thickness, rim area and volume, cup area and volume, and cup-to-disc area and diameter ratios from the CSLO topographs. Typically a *reference plane* is placed parallel to the retinal surface in the acquired CSLO three-dimensional data sets corresponding to an approximate location of the lower extent of the RNFL. Once the lower extent of the RNFL is identified, topographic measurements above the reference plane can be used to derive nerve fiber–related measurements such as RNFL thickness and neuroretinal rim area and volume. The topographic measurements below the reference plane within the optic disc margin can be used to estimate optic cup–related parameters such as cup area and volume. Additionally, the topograph heights within the cup are measured relative to a *curved surface* for estimating the cup shape and cup depth measures. The curved surface separates the optic disc surface from the vitreous cavity and is constructed as a linear interpolation functional between the topograph measurements along the contour line and their mean height. Thus, it can be seen that the location of the reference plane influences the stereometric parameter estimates and should ideally remain in the same location during subsequent exams irrespective of the physiological changes in the ONH region.

Currently, three different methods are used to estimate the reference plane location: (1) a 320-μm reference plane, (2) a standard reference plane (50 μm) in reference to the optic disc margin, and (3) a new ONH-specific reference plane. A recent study comparing the rim area measurements using the 320-μm reference plane and the standard reference plane found that the 320-μm reference plane is superior with a reduced test–retest variability in the rim area measurements [51]. The construction details of these reference planes are given in the following sections.

7.3.2.1 The 320-μm Reference Plane

In this method, a reference plane is located at a distance of 320 μm from an *image plane* to delineate the neuroretinal rim and optic cup as shown in Figure 7.7 [52] and is not based on a manually drawn optic disc contour line. The image plane or a focal plane of a topograph identifies an approximate mean height of the peripheral retinal surface, which is assumed to be relatively stable across multiple scans. The location of the image plane in a topograph is calculated as a mean height of the pixels within a reference ring, with the outer diameter being 94% of the topograph size and a ring width that is 3% of the topograph size, located at the topograph center. The 320-μm offset for the reference plane was empirically derived as the location of least variability from a set of 300 topographs [52]. Because the location of the reference depends on the mean peripheral retinal height, any changes in the peripheral retinal region affect the location of the reference plane and the stereometric parameters.

7.3.2.2 The Standard Reference Plane

The current HRT software version (HRT 3.0) uses the standard reference plane for estimating the stereometric parameters [53]. As shown in the Figure 7.7, the standard reference plane is located at an offset of 50 μm below the mean height of the pixels between 350° and 356° along the optic disc contour line. The location along the optic disc margin between 350° and 356° physiologically corresponds to the papillomacular bundle, which is expected to be relatively stable until a very advanced stage of glaucoma is reached [54]. This sectorial location (between 350° and 356°) was experimentally determined as the average optic disc surface inclination angle from a set of 180 normal eyes and 99 glaucomatous eyes [52]. The height reproducibility along the contour line between 350° and 356° was 16.0 μm ± 13.6 μm in normal eyes and 23.4 μm ± 20.2 μm in glaucomatous eyes. Therefore, the 50-μm offset is chosen approximately three-standard deviation of height variability in normal eyes or approximately two-standard deviation of height variability in glaucomatous eyes. Note that the HRT topograph quality metric recommended cutoffs of 30 and 50 μm are directly linked to the reliability of the stereometric parameter estimates using the standard 50-μm reference plane.

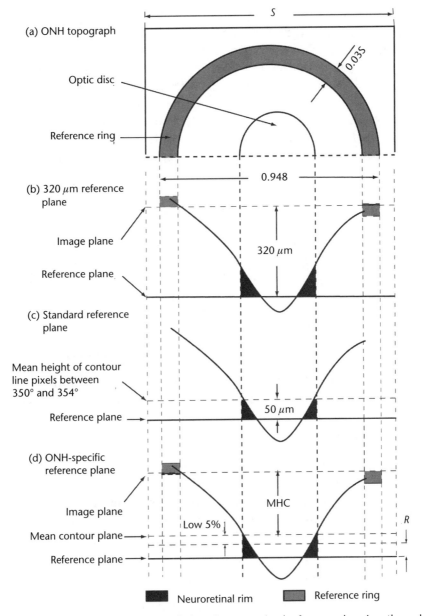

Figure 7.7   (a–d) Methods of automatically locating an optimal reference plane in a three-dimensional ONH HRT topograph to delineate the optic cup and neuroretinal rim. (MHC: mean height of the contour.)

7.3.2.3   The ONH Specific Reference Plane

The location of the 320-μm reference plane is relatively stable when there are no significant changes in the peripheral retinal surface. Also it was observed that the 320-μm offset may be too large to observe small changes in the nerve fiber layer; therefore, the stereometric parameters may be less sensitive in detecting glaucomatous changes [52]. The standard reference plane utilizes a regional (350° to 356°) mean contour height information specific to each eye, in defining the reference plane as discussed earlier; however, the reference plane placement may be affected by image tilt since the reference plane is located in reference to the contour line measurements within a small 6° segment. Recently, a new method of positioning the reference plane was introduced that combines the strength of the 320-μm reference plane and the standard reference plane [55]. The reference plane placement parameter has two components: (1) an eye-specific offset estimated from contour line information for each eye ($Low_{5\%}$) and (2) a constant offset ($R = 100$ μm). $Low_{5\%}$ is calculated as the mean height of the lower fifth percentile of the contour height measurements along the contour line at a 1° angular interval. From the $Low_{5\%}$, the reference plane is located at a constant distance R. The value of R was estimated to be an optimal offset of 100 μm from $Low_{5\%}$, calculated for each eye, in a set of normal eyes that resulted in the least variability in the stereometric parameter estimates. Once the $Low_{5\%}$ is estimated for an eye, it is used as a constant offset to locate the reference plane of the eye during all subsequent scans with respect to the current mean contour height as shown in Figure 7.7. Thus, the reference plane position is relatively stable across subsequent examinations of an eye. More studies are required to test this new reference plane placement technique rigorously, especially to identify its strengths in glaucoma patients.

7.3.3 Algorithms for Detecting Glaucomatous Damage of the ONH

The stereometric parameters of an ONH can be used to describe the configuration of the ONH. Some of the useful HRT stereometric parameters are cup area, cup volume, rim area, rim volume, cup-to-disc area ratio, rim-to-disc area ratio, vertical cup-to-disc diameter ratio (VCDR), horizontal cup-to-disc diameter ratio, and cup shape measurement [54]. Configuration changes in the ONH due to an onset of glaucoma influence the stereometric parameter estimates. For example, with a rim loss along the vertical poles of the optic disc, the resulting changes in the shape of the cup and rim will be reflected in all of the above-mentioned stereometric parameters. Wide variability in the shape and size of the optic discs limits the usefulness of these stereometric parameters in differentiating normal and glaucomatous ONHs.

7.3.3.1 Moorfields Regression Analysis

Garway-Heath et al. derived a disc size–dependent cutoff for the VCDR parameter using a linear relationship for differentiating glaucoma patients from normal patients [32]. The cutoff was defined using the 95% and 97.5% prediction intervals of the VCDR measurements from a set of normal patients, and the classification scheme was found to be sensitive for detecting changes in discs with smaller diameters (1.38 to 1.63 mm). This VCDR to disc size relationship and the VCDR classification cutoffs agree with a similar derivation in a different study population [33]. In a similar study, the rim area and cup-to-disc area ratio parameters adjusted for disc area were found be sensitive in detecting early glaucoma conditions in 51 glaucoma patients using a 99% prediction interval cutoff estimated from a group of 80 normal study participants [56].

A linear discriminant function defined using a linear regression between the log-neuroretinal rim area and disc area estimates, accounting for the effects of age on the rim area measurements, is available in the HRT software module as Moorfields regression analysis (MRA) [57]. HRT-3 uses ethnic-specific measurements from a normative database composed of 452 Caucasian participants, 111 participants of African origin, and 64 Indian participants for defining 95% and 99% prediction intervals of the normal log-rim area measurements for classification. The log-rim area measurements of topographs above the 95% prediction interval cutoff are considered *normal*; the measurements that lie in the interval between 95% and 99% prediction interval cutoffs are considered to be *borderline* between normal and glaucomatous conditions and the measurements that are less than the 99% prediction interval cutoff are considered *outside normal limits*. The area under the receiver operating characteristic curves (AUROCs) for MRA and stereometric parameters for identifying glaucomatous eyes based on visual function defects range from 0.74 to 0.88 [58–60].

7.3.3.2   Glaucoma Probability Score

Swindale and associates have developed a method of constructing a patient- specific ONH model of an eye from the ONH topographs [61]. The generalized model captures the shape of the optic cup using a Gaussian shape function and models the peripapillary retinal region as a parabolic surface. A given ONH surface topography is modeled as

$$z(x,y) = \frac{z_m}{1+e_0^{((r-r)/s)}} + z_0 + a(x-x_0) + b(y-y_0)^2 + c(x-x_0)^2 + d(y-y_0)^2$$

where z is the estimated topograph height measurements at (x, y), z_m is the topograph height at the optic cup center (x_0, y_0), z_0 is the height of the image plane, r_0 is the radius of the optic cup, s is the slope of the cup wall, a is the nasotemporal slant, b is the vertical slant, c is the horizontal curvature, and d is the vertical curvature. Additionally, a circular region R with center (x_0, y_0) and of radius $r_0 + \log_e(9)s$ is used to derive morphological model indices. The morphological indices are: (1) global radial pixel gradient within the region R, (2) radial pixel gradient within R in the nasal region, (3) radial pixel gradient within R in the temporal region, and (4) an index of maximum cup depth calculated as the average of the 500 deepest points within the region R. The method utilizes an initialization procedure and an iterative procedure to estimate the model parameters described earlier. The initialization procedure computes rough estimates of the parabolic surface parameters a, b, c, and d with initial cup parameters of cup radius $r_0 = 0.5$ mm and cup slant $s = 0.1$ for the cup shape function.

To determine an initial location (x, y), and depth z_m of the cup, the initial parabolic surface is subtracted from the ONH topograph measurements. The cup depth and the location are estimated as the magnitude and the location of the average of the 50 deepest measurements in the parabolic-surface-subtracted image. With this new cup depth and location information, the parabolic surface parameters are once again updated. After the initialization step, the cup and parabolic surface parameters are updated concurrently in an iterative fashion using the Levenburg-Marquardt optimization technique in the iterative procedure. A cost function in the form of an l_2-norm is used to judge the degree of agreement between the model and the original ONH topograph and also to judge the model convergence in the iterative procedure. For identifying glaucomatous ONH topographs, a subset of the model parameters and morphological indices is used as classifying features in a Bayesian framework.

In the HRT-3 software, the ONH modeling approach is available as the Glaucoma Probability Score (GPS) for identifying glaucomatous eyes. The ONH model parameters of cup size, cup depth, rim steepness, horizontal curvature, and vertical curvature are used as input to a relevance vector machine (RVM) for classifying an ONH topograph as glaucomatous. The RVM assigns a glaucoma probability value for the ONH topograph using the same normative database described in Section 7.3.4.1 under MRA. Probability scores between 0 and 0.27 are considered to be within normal limits, the scores between 0.28 and 0.64 are considered to be borderline, and the scores between 0.64 and 1.0 are considered to be outside the normal limits. The AUROCs for the GPS classification for identifying glaucomatous eyes with visual function defects are in the range of 0.7 to 0.93 [58–60].

Other studies have investigated the use of nonlinear machine learning classifiers such as neural networks and *support-vector machines* for efficiently combining multidimensional HRT stereometric parameters and classifying normal and glaucoma patients [62–64]. The performance of the machine learning classifiers was found to be superior to the linear discriminant functions and to have a good agreement with the human qualitative assessment with AUROCs ranging from 0.84 to 0.98. In some participants, the machine learning classifiers can predict a glaucomatous condition before the onset of visual defects.

7.4 Algorithms for Detecting Glaucomatous Progression in the ONH

Due to a wide variability in the morphology of the optic disc structure among healthy and glaucoma patients, there is a large overlap in their stereometric parameter distributions. Also in the early stages of glaucoma, the characteristic changes associated with the structural glaucomatous damages may not be drastic and therefore may not be obvious. Because glaucoma is characterized as a progressive optic neuropathy, the sensitivity of detecting glaucoma patients can be improved by detecting changes in the ONH structure of an eye from a baseline condition. Note that identifying any ONH changes that are characteristic of glaucoma can be useful in identifying an onset of glaucoma, measuring the progression rate in glaucoma patients for planning any medical interventions, and assessing the effect of glaucoma medications and surgery on glaucoma patients. Structural changes that reflect in the ONH topographs can be analyzed using the stereometric parameters that summarize topograph measurements in the regions of interest such as the optic cup, neuroretinal rim, and peripapillary retina, and by analyzing dense pixel-level changes in the ONH region. In progression analysis, the ONH topographs acquired during the longitudinal follow-up exams are compared with a topograph acquired with the eye in a baseline condition to detect any progressive structural changes in the ONH.

7.4.1   Essential Qualities of the Computational and Statistical Methods Applied for Detecting Glaucomatous Structural Changes

Besides any glaucoma-related structural changes observed in the ONH region from a baseline condition, the topograph measurements may also exhibit nonglaucomatous changes due to the topograph height measurement variability associated with: (1) any inherent ONH structure changes due to systemic conditions that could influence the appearance of the structure, for example, due to fluctuations in the IOP, and (2) the imaging instrument. The instrument measurement variability could be due to the Poisson nature of the image formation in the CCD camera and also due to any illumination changes between follow-up exams. Pixel-level HRT height measurements were found to be reproducible for both normal and glaucoma patients with height variability $<50\,\mu m$ [65, 66]. The pixel-level height variabilities were found to be higher in the region within the optic disc margin and also in the locations with a higher height gradient, such as blood vessels and topograph borders, compared to the height measurements in the peripapillary retina. Because the follow-up topographs of the ONH cannot be acquired exactly at the same x, y, and z coordinates of the imaging space with reference to the baseline topograph imaging coordinate (x, y, z), the follow-up topographs should be corrected for imaging artifacts such as shifts along the vertical and horizontal directions, rotational movements, and tilts along the optical axis of the imaging instrument. The recent HRT-3 version of the software corrects the follow-up topographs for the shift, rotational, and tilt imaging artifacts using unique landmark features extracted from baseline and follow-up topographs with respect to a baseline topograph [67]. The general consensus is that the new alignment algorithm is relatively robust compared to the earlier HRT software versions. In summary, the computational and statistical methods developed and applied for detecting glaucomatous progression should account for and characterize the inherent structure variability and the instrument measurement variability to detect the structural changes associated with any glaucomatous progression.

In the following sections, we discuss the use of ONH stereometric parameters in detecting structural progression and two specific pixel-level topograph change analysis methods of topographic change analysis [68] and statistic image mapping of the human retina [69]. For demonstration, we will use the follow-up exams of two participants in the UCSD DIGS database. Within the study period chosen for demonstration, one of the participants, P_{stable}, was assessed to be stable with no glaucomatous progression by photograph and visual function. The other participant, $P_{progressive}$, was assessed with progressive glaucomatous changes both by photographs and visual function defects. The optic disc sizes of both the participants were comparable (optic disc area of $P_{stable} = 1.6\ mm^2$ and of $P_{progressive} = 1.5\ mm^2$). Figures 7.8 and 7.9 show the ONH topographs and reflectance images of both the participants at the baseline condition and on follow-up visits.

7.4.2   Stereometric Parameter-Based Progression Analysis

Figures 7.10(a, c) show the trend of the HRT rim volume estimates in the global and suprotemporal regions, respectively, for the participants P_{stable} and $P_{progressive}$. For the participant $P_{progressive}$, the global and suprotemporal rim volume measurements

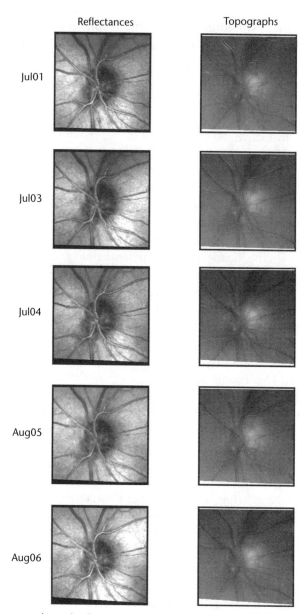

Figure 7.8 HRT topographs and reflectance images of a participant (P_{stable}) in the UCSD DIGS database with no progressive structural changes assessed using optic disc photographs, and no visual function changes assessed using standard automated perimetry Glaucoma Progression Analysis.

decrease progressively and consistently relative to the participant P_{stable}. Also the decrease in the rim volume measurements of $P_{progressive}$ from the baseline exam is more obvious in the suprotemporal sector [Figure 7.10(c)] compared to the global rim volume changes [Figure 7.10(a)]. The HRT software uses the mean stereometric parameter values from a group of normal and glaucoma patients to normalize the respective parameters to a range of −1 to 1 [53]. Normalized parameter values closer

Reflectances   Topographs

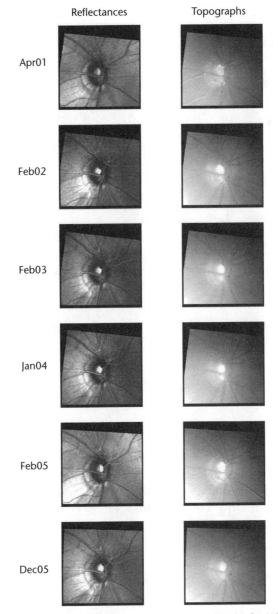

Figure 7.9   HRT topograph and reflectance images of a participant in the UCSD DIGS database ($P_{progressive}$) with progressive structural changes assessed using optic disc photographs, and progressive visual function defects assessed using standard automated perimetry Glaucoma Progression Analysis.

to −1 and +1 indicate extreme conditions of deterioration and improvement, respectively, in the individual eyes. The individual parameters are normalized as

$$Param_{normalized} = \frac{Param_{follow-up} - Param_{baseline}}{Param_{normal} - Param_{glaucoma}}$$

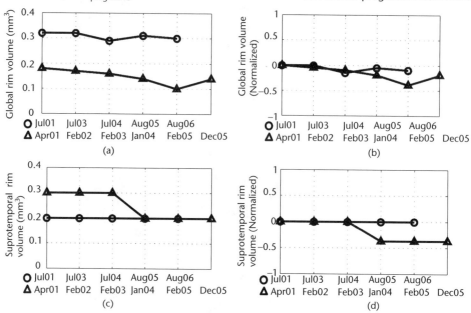

Figure 7.10   (a–d) HRT stereometric parameter trends of the participants P_{stable} and $P_{progressive}$.

Figures 7.10(b, d) show the normalized rim volume trend plots for the global and suprotemporal regions, respectively, for participants P_{stable} and $P_{progressive}$. The normalized rim volume estimates of participant $P_{progressive}$ decrease progressively from 0 from the first follow-up onward, suggesting a progressive deterioration in the neuroretinal rim area with respect to the baseline exam in general [Figure 7.10(b)] and more pronounced changes in the suprotemporal sector [Figure 7.10(d)].

For detecting glaucomatous progression, it is essential: (1) to estimate the direction and magnitude of changes in each follow-up exam from the baseline; for example, a decrease in the rim area and an associated increase in the cup area may indicate optic neuropathy; (2) to estimate the significance of the change from the baseline at each follow-up; and (3) to confirm the detected changes, typically by the observed changes repeating in successive follow-ups.

Burgoyne and associates have applied a variety of statistical approaches to measure glaucomatous structural changes [70], to characterize the variability for identifying the change significance [71, 72], and to confirm the detected changes [71, 72]. Their analysis utilizes at least six topographs per exam to characterize the intrasession variability in the stereometric parameters. For detecting changes in the ONH surface of primates in their experimental glaucoma studies, they used 95% confidence interval (CI) estimates of the stereometric parameters at a baseline condition and experimental glaucoma conditions [70]. The magnitude of the change in a follow-up exam is computed as a difference in the mean parameter values from the baseline to the follow-up visits. The change in a follow-up exam is found to be significant when the 95% CI estimates of the baseline and the follow-up parameters do not overlap. In another study, they have applied a minimum detectable change (MDC) criterion for each of the stereometric parameters for identifying significant changes in a follow-up exam from a baseline using Tukey's W procedure [72], a multiple comparison procedure used to compare a set of 6 follow-up topographs acquired per session with a set of 18 baseline topographs acquired in three different baseline imaging sessions. In brief, the MDC captures the variability in the parameter estimates at a baseline; it also determines the minimum required changes in the respective follow-up parameters from baseline to be statistically significant. The MDC values of each of the parameters of an individual eye were estimated at a baseline condition using a set of three baseline exams with 6 topographs per exam (for a total of 18 topographs). To reduce the false positives during classification and to detect clinically significant changes, they used multiple follow-ups with constraints of significant changes occurring in either two or three consecutive exams [71, 72]. Note that a direct extension of the previously discussed MDC procedure for the HRT parameters may be limited by the large number of topographs required at each imaging session. In addition, several other promising methods have been developed to analyze stereometric parameters to detect change over time, including pointwise linear regression [73] and repeatable rim area change greater than specified limits of variability [74, 75].

7.4.3   Topographic Change Analysis (TCA)

The simplest form of identifying pixel-level changes between a baseline and a follow-up topograph is to estimate the difference between the mean height measurements at each pixel location in the baseline topograph (I_{bl}) and the follow-up topograph (I_{fup}) as ($I_{bl} - I_{fup}$). The significance of the mean height differences at each pixel can be estimated using an analysis of variance (ANOVA) similar to the stereometric parameter change detection method used by Burgoyne and associates [71, 72]. Although the HRT topograph measurements have good reproducibility at the pixel level, estimating the pixel-level MDC thresholds for each pixel location that is specific to an individual eye may not be feasible. Also applying a global threshold cutoff for the mean height difference at each pixel location may not be sensitive enough to detect true changes. For example, consider the mean height difference image of the participant P_{stable} between a baseline visit and the first follow-up shown in Figure 7.11. It can be seen that the locations around vessels have relatively larger mean height differences and this may influence the choice of a global threshold for identifying the locations with significant changes.

7.4.3.1   Mixed-Effect Three-Way ANOVA Model

Chauhan and associates have developed a mixed-effect three-way ANOVA model to detect superpixel-level changes in the ONH topographs from a baseline to a follow-up exam [68], and their ANOVA model is the basis for the TCA method available in the HRT software module [53, 68]. The details of the TCA ANOVA model presented here are based on the original description of the method in [68], so there may be implementation differences in the TCA model available in the HRT software. The model details discussed here were confirmed with one of the authors

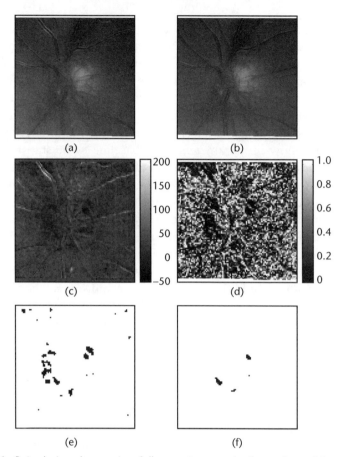

Figure 7.11 (a–f) Analyzing changes in a follow-up topograph of a study participant P_{stable}. The pixel-level change significance map in part (d) was estimated using TCA using HRT software version 3.0. The locations of significant retinal height reduction are shown in parts (e) and (f). (a) Baseline topograph. (b) Follow-up topograph. (c) Difference image. (d) Change probability map. (e) ONH region with significant height reduction. (f) Significant height reduction with the optic disc margin.

of the method [personal communication with David C. Hamilton, February 23, 2007] [76].

In TCA, each superpixel covers a neighborhood of 4×4 individual pixel locations in a topograph for sensitive deduction of localized changes. The topograph height measurements from all the pixel locations within a superpixel are pooled for analysis.

For detecting a superpixel change from a baseline to a follow-up visit, TCA utilizes topograph height measurements from 16 different locations within a superpixel and N different topograph scans acquired in each visit (in HRT, $N = 3$ scans are acquired during each exam). To determine whether a superpixel has changed significantly from a baseline to a follow-up visit, in addition to the effects of the scan time (based on whether the scans were acquired at the baseline or follow-up), the location effects of each of the individual pixel height measurements within a superpixel and the effects of each of the topograph scans (i.e., each of the acquired three scans) within a visit on the height measurements should be taken into account. Hypothetically, the average value of the mean topograph heights within a superpixel could be influenced by: (1) the location of each of the pixels within the superpixel (for example, due to anatomical differences), (2) the effect of each scan/image within the visit, and (3) more importantly could be influenced by whether the scans were acquired at a baseline or a follow-up condition. Thus, the factors due to the scan time T (whether baseline or follow-up), scan/image I, and the locations of the individual height measurements within a superpixel L can be analyzed for significant changes in the mean height measurements of a superpixel. In progression analysis, we are specifically interested in determining whether there is a change in the mean height from a baseline to a follow-up visit and therefore the time factor **T** is a fixed factor. Also the locations or indices of the height measurements within a superpixel are fixed and, therefore, the location factor **L** is also a fixed factor. We are not interested in any of the specific scans acquired during a visit, and these scans are random samples from a large possibility of scans. Therefore, the scan/image factor **I** is a random factor. Because the scans acquired per visit are obviously not the same between the baseline and follow-up visits, the scan/image factor **I** is a factor nested within the time factor **T**. [The effects of factor **I** will be represented as $I(T)$—that is, I nested within T—in the ANOVA model.] The ANOVA model for detecting changes in a superpixel, with each superpixel containing 16 topograph height measurements at locations $l = 1, 2, ..., 16$, from a baseline (at time $t = 1$) to a follow-up visit (at time $t = 2$) using a set of N topographs each acquired at the baseline and follow-up visits is given by

$$h_{tli} = \mu... + T_t + L_l + I(T)_{i(t)} + TL_{tl} + \varepsilon_{tli} \qquad (7.1)$$

where

h_{tli} is the topograph height at location l (within a superpixel) in a topograph i acquired at time t;

μ is the mean topograph height in a superpixel (computed using topograph height measurements from all the t, l, and i locations);

T is the main effect of the time factor **T**;

L is the main effect of the location factor **L**;

$I(T)$ is the main effect of the random image factor **I** with independent $N(0, \sigma_{i(t)}^2)$;

TL is the two-way interaction effect between time factor **T** and location factor **L**;

ε_{tli} is the model error with independent $N(0, \sigma_\varepsilon^2)$.

Table 7.1 shows the height measurements for the factor level combinations of the time, location, and image factors. The sum of squares and degrees of freedom for the model terms [T_t, L_l, $I(T)_{i(t)}$, and TL_{tl}] are shown in Table 7.2. For details refer to [77].

The expected mean squares of the ANOVA terms are

Table 7.1   Three-Way Partially Nested Model for Topograph Height Measurements

		Time Factor **T**					
		Baseline Time t = 1			*Follow-Up Time t = 2*		
		Nested Image Factor I			*Nested Image Factor I*		
		i = 1	*i = 2*	*i = 3*	*i = 1*	*i = 2*	*i = 3*
Location factor **L**	*l = 1*						
	l = 2						
	...	...	...	...	...	...	...
	l = 16						

Note: Factor levels are shown in italics.

Table 7.2   Analysis of Variance

Model term	Sum of Squares (SS)	Degrees of Freedom (DF)
T_t	$16N\sum_{t=1}^{2}\left(\bar{h}_{t..}-\bar{h}_{...}\right)^2$	$(t-1)=1$
L_l	$2N\sum_{l=1}^{16}\left(\bar{h}_{.l.}-\bar{h}_{...}\right)^2$	$(l-1)=15$
$I(T)_{i(t)}$	$16\sum_{t=1}^{2}\sum_{i=1}^{N}\left(\bar{h}_{t.i}-\bar{h}_{t..}\right)^2$	$t(N-1)=2(N-1)$
TL_{tl}	$N\sum_{t=1}^{2}\sum_{l=1}^{16}\left(\bar{h}_{tl.}-\bar{h}_{t..}-\bar{h}_{.l.}+\bar{h}_{...}\right)^2$	$(t-1)(l-1)=15$
Total	$\sum_{t=1}^{2}\sum_{l=1}^{16}\sum_{i=1}^{N}\left(h_{tli}-\bar{h}_{...}\right)^2$	$(tli-1)=(32N-1)$
Error ε_{tli}	$SS_{Total}-SS_T-SS_L-SS_{I(T)}-SS_{TL}=$ $\sum_{t=1}^{2}\sum_{l=1}^{16}\sum_{i=1}^{N}\left(h_{tli}-\bar{h}_{t.i}-\bar{h}_{tl.}+\bar{h}_{t..}\right)^2$	$t(l-1)(i-1)=30(N-1)$

$$E(MS_T)=\frac{SS_T}{DF_T}+16\sigma^2_{i(t)}+\sigma^2_{\varepsilon}$$

$$E(MS_L)=\frac{SS_L}{DF_L}+\sigma^2_{\varepsilon}$$

$$E\left(MS_{I(T)}\right)=16\sigma^2_{i(t)}+\sigma^2_{\varepsilon}$$

$$E(MS_{TL})=\frac{SS_{TL}}{DF_{TL}}+\sigma^2_{\varepsilon}$$

$$E(MS_{\varepsilon})=\sigma^2_{\varepsilon}$$

7.4.3.2   Estimating the Significance of a Superpixel Change

For estimating the significance of a change in a superpixel location from a baseline visit to a follow-up visit, we need to estimate the significance of the main effects T of time factor **T** and time/location interaction effects TL of time factor **T** and location factor **L**. This can be tested using a null hypothesis H_0 of no significance for the main effect T and the interaction effect TL. Because the significance test involves multiple terms from the model, that is, T and TL in (7.1), the *Satterthwaite approximate*

F-test needs to be applied [77]. The numerator of the *F*-statistic can be computed as a linear combination of model terms T and TL as follows.

$$
\begin{aligned}
MS_{NUM} &= \frac{SS_T + SS_{TL}}{DF_T + DF_{TL}} \\
&= \frac{DF_T\, MS_T + DF_{TL}\, MS_{TL}}{(t-1) + (t-1)(l-1)} \\
&= \frac{1}{16} MS_T + \frac{15}{16} MS_{TL}
\end{aligned}
\tag{7.2}
$$

The expected mean squares of the numerator in (7.2) is given as

$$
\begin{aligned}
E(MS_{NUM}) &= E\!\left(\frac{1}{16} MS_T\right) + E\!\left(\frac{15}{16} MS_{TL}\right) \\
&= \frac{1 SS_T}{16 DF_T} + \sigma^2_{i(t)} + \frac{1}{16}\sigma^2_\varepsilon + \frac{15 SS_{TL}}{16 DF_{TL}} + \frac{15}{16}\sigma^2_\varepsilon
\end{aligned}
\tag{7.3}
$$

The denominator of the *F*-statistic is typically chosen as a linear combination of the mean squares of the ANOVA terms (i.e., $MS_{DENOM} = \sum_j a_j MS_j$) such that the *F*-statistic = 1 under H_0. Therefore, under H_0 when there are no time and time/location interaction effects, the expected mean square in (7.3) becomes

$$
E(MS_{NUM}) = \sigma^2_{i(t)} + \sigma^2_\varepsilon
$$

We can see that the denominator of the *F*-statistic can be chosen as the linear combination of $MS_{I(T)}$ and MS_ε as follows:

$$
MS_{DENOM} = \frac{1}{16} MS_{I(T)} + \frac{15}{16} MS_\varepsilon
\tag{7.4}
$$

The degrees of freedom associated with the numerator in (7.2) and the denominator term in (7.4) can be estimated using Satterthwaite's procedure [77] as follows:

$$
DF_{NUM} = \frac{16^2\, MS^2_{NUM}}{MS^2_T + 15 MS^2_{TL}}
$$

$$
DF_{DENOM} = \frac{2(N-1)16^2\, MS^2_{DENOM}}{MS^2_{I(T)} + 15 MS^2_{TL}}
$$

Now the probability of observing the calculated mean height change in the superpixel location due to chance alone can be estimated using the approximate *F*-statistic ($= MS_{NUM}/MS_{DENOM}$) and the associated numerator and denominator degrees of freedom (DF_{NUM} and DF_{DENOM}).

7.4.3.3   Change Detection Strategies Using TCA Pixel Probability Maps

Figure 7.11(d) (earlier in this chapter) shows a map of change probabilities in each superpixel location for testing the null hypothesis of no significant difference in the mean height measurements (within a superpixel) between the baseline and follow-up visits for participant P_{stable}. In the change probability map, the darker pixel locations correspond to lower probabilities (closer to 0) and more significant changes, and the brighter pixel locations correspond to higher probabilities (closer to 1.0) and less significant changes. For each follow-up visit, a height difference image and its associated pixel change probability map can be exported from the HRT software for further analysis [53].

To detect any characteristic glaucomatous changes, we are interested in the locations of significant retinal height decreases, which can be identified as the locations with mean height differences of less than 0 in the difference image map shown in Figure 7.11(c). Now the locations in the ONH region with significant retinal height decreases can be identified by combining the locations of height decrease [Figure 7.11(c)] and their associated change probabilities [Figure 7.11(d)] and by applying a change probability cutoff, for instance $p < \alpha$, where α is the level of significance. Figure 7.11(e) shows the locations of significant retinal height decreases in a follow-up visit from the baseline for participant P_{stable} with change probability < 0.05. Any random, isolated significant change locations are removed from further analysis by applying a spatial filter. (Only the significant change locations connected to at least four significant change locations are retained.)

Glaucomatous structural changes usually affect a contiguous region in the ONH and appear as a group of significant superpixel change locations. For example, mean height measurement changes in isolated locations are most likely due to random measurement variations and usually do not correspond to morphological changes in the ONH regions. Therefore, superpixel change locations with significant reduction in mean retinal height measurements are grouped together as clusters. Although clinically applicable pixel and cluster change cutoff criteria are still under development, glaucomatous changes can be inferred using both the area/size of the clusters and the depth of change [78, 79]. Also, clusters within the optic disc that are 2% to 5% of the size of the optic disc and depths of changes in the range of 20 to 50 μm were observed to be useful TCA superpixel change summary parameters for identifying glaucomatous patients from healthy populations [79].

For detecting only the clinically significant glaucomatous changes, the superpixel locations with significant decreases in mean height measurements in either two or three successive follow-ups are used for estimating the previously discussed TCA superpixel change summary parameters of the cluster size, cluster size measured as a percentage of the optic disc, and the depth of changes [79]. In version 3.0 of the HRT software, the significant ONH topograph change locations based on the TCA are identified using the superpixel locations with significant changes repeated in the same locations: (1) in the first two follow-up exams for the second follow-up, (2) in the first three follow-up exams for the third follow-up, and (3) in at least three out of the four past follow-up exams, including the current follow-up, for the fourth follow-up and onwards.

For participants P_{stable} and $P_{progressive}$, the locations with significant mean height measurements from the baseline to follow-up visits at a 5% level of significance are shown in Figures 7.12 and 7.13, respectively. The size of the largest cluster of superpixels with significant height decreases repeated in three successive follow-ups within their respective optic disc margins was smaller for participant P_{stable} (largest cluster size = 9 superpixels) compared to participant $P_{progressive}$ (largest cluster size = 93 superpixels) as shown in Figures 7.12(d) and 7.13(d), respectively.

7.4.4 Nonparametric Permutation Tests Progression Analysis

The TCA method discussed in the previous section constructs an F-statistic map for estimating the change probabilities p^K for testing the null hypothesis H_0^K of no mean height changes at superpixel location K between baseline and follow-up topographs at a level of significance α. The ANOVA model used for analysis makes parametric assumptions about the distributions of the HRT height measurements. The final step of clustering contiguous and significant superpixel change locations for identifying progression from a baseline condition, however, does not impose a control on the type 1 error of falsely diagnosing a progression at the topograph level. For example, consider a cluster of 20 contiguous superpixel locations with significant changes in their respective mean height measurements at $\alpha^K = 0.05$, which can be considered as progressing based on the cluster-size cutoff criteria of 20 superpixel locations. Because the changes in the superpixel locations within the cluster are significant, the probability of change at each of the superpixel locations K is $p^K < \alpha$ (= 0.05); for discussion, assume a probability of $p^K = 0.03$ ($< \alpha = 0.05$) at each of these locations. Although the superpixels within the cluster are significant, the probability associated with the 20-superpixel cluster will be $\sum_{K=1}^{20} p^K = 0.6$, which is not significant to reject the topograph-level null hypothesis of significant change from the baseline to follow-up. This is because of the fact that a topograph cannot be classified as progressing based on observing a significant change in a single pixel or a superpixel location. It is referred to as the *multiple comparison problem* [80], in which the results of the multiple hypothesis tests H_0^K at each of the pixel or superpixel locations need to be combined to decide on the state of a topograph or the effect of time in the topograph measurements. Therefore, the level of significance (α^K) chosen for testing the superpixel level null-hypothesis tests H_0^K should be adjusted based on the level of significance associated with the topograph-level null hypothesis. Methods such as Bonferroni correction can be used to adjust the α^K for the pixel-level tests H_0^K such as by setting $\alpha^K = \alpha/N$, where α is the level of significance for testing the topograph-level null hypothesis of no change and N is the total number of superpixels in a topograph. However, such a correction might be too conservative and does not account for the spatial correlation that exists between the height measurements.

The nonparametric permutation tests, as opposed to the generalized linear regression models, do not make parametric assumptions about the distribution of the height measurements. In addition, several procedures are available in the nonparametric permutation framework to account for the multiple comparison problem. The main principle of the permutation tests for estimating the effects of various factors stems from the principle of resampling theory or randomization tests.

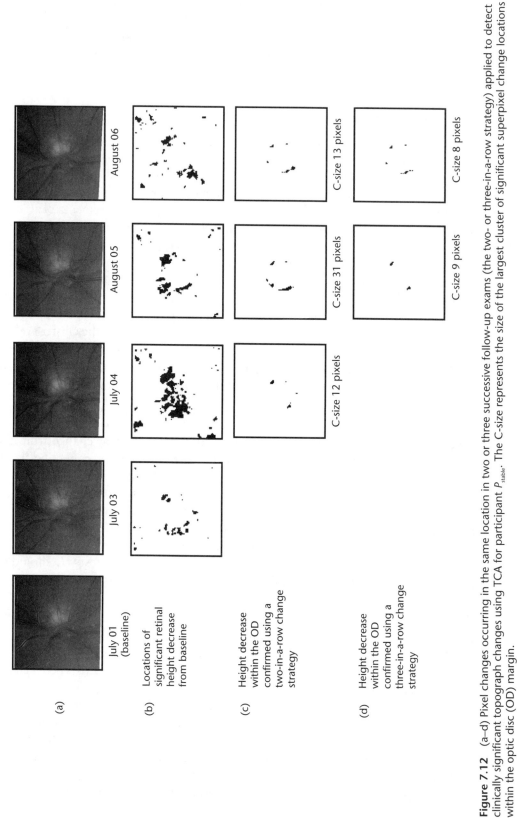

(a) July 01 (baseline)   July 03   July 04   August 05   August 06

(b) Locations of significant retinal height decrease from baseline

(c) Height decrease within the OD confirmed using a two-in-a-row change strategy — C-size 12 pixels, C-size 31 pixels, C-size 13 pixels

(d) Height decrease within the OD confirmed using a three-in-a-row change strategy — C-size 9 pixels, C-size 8 pixels

Figure 7.12 (a–d) Pixel changes occurring in the same location in two or three successive follow-up exams (the two- or three-in-a-row strategy) applied to detect clinically significant topograph changes using TCA for participant P_{stable}. The C-size represents the size of the largest cluster of significant superpixel change locations within the optic disc (OD) margin.

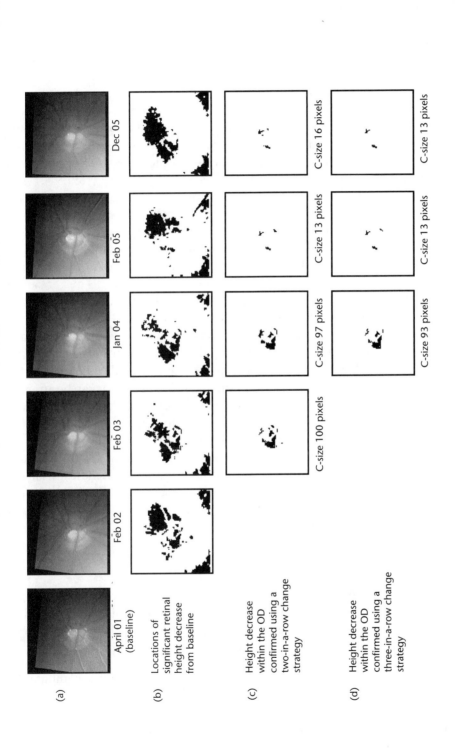

Figure 7.13 (a–d) Application of the TCA two- and three-in-a-row pixel change strategies for participant $P_{progressive}$. The C-size is the size of the largest cluster of significant superpixel change locations within the optic disc (OD).

Consider a set of topograph scans $S_B = \{S_{b1}, S_{b2}, S_{b3}\}$ acquired at a baseline (at time $t = 1$) and a set of scans $S_F = \{S_{f1}, S_{f2}, S_{f3}\}$ acquired at a follow-up visit (time $t = 2$). If the scan time (baseline or follow-up) does not have any effect on the topograph height measurements, then any other random combinations of these scans will not be significantly different from the originally observed scan order. For example, when there is no time effect in the topograph height measurements S_B and S_F, the test statistic estimated from the scans acquired in the order of $\{S_{b1}, S_{f2}, S_{f1}\}$ at baseline and $\{S_{b2}, S_{b3}, S_{f3}\}$ is likely to be same as the test statistic estimated for the original scan order of S_B and S_F. The test statistic in the nonparametric tests is not limited to the conventional parameters such as the t-statistic. A recent application of a nonparametric permutation test for testing the ONH retinal height changes uses the regression parameters of slope and standard error to estimate the test statistic as ($|slope| / SE$) [69, 81].

In the permutation test framework, all possible scan permutations or unique scan orders are utilized to build a *permutation distribution* of a test statistic empirically. The permutation distribution represents a sample distribution of the chosen test statistic. The significance of the time effect in the observed scans at the baseline and follow-up visits is estimated as the proportion of test statistics in the permutation distribution greater than or equal to the observed test statistic. Thus, it can be observed that the significance of an observed test statistic is estimated using the variability information seen in an individual study participant.

The robustness of the permutation test depends on the number of unique resamples or permutations available for building the nonparametric test statistic distribution. To test a null hypothesis H_0 at a level of significance α, a minimum of $1/\alpha$ unique resamples are required, that is, the minimum probability that can be attained from N permutations is $1/N$. Using s topograph scans acquired from n different visits, $(s \times n)! / (s!)^n$ unique scan permutations are possible [69].

Two common permutation tests are available in the neuroimaging literature that can account for the multiple comparison problem: (1) the single threshold test and (2) the suprathreshold test [80]. We briefly discuss the details of these tests in the context of detecting glaucomatous progression. These tests specifically adjust the probability p^K at each of the pixel locations based on the height measurements in all other pixel locations in a topograph.

7.4.4.1   Single Threshold Test

In this method, a maximal test statistic from each of the resamples or permutations is used to build the nonparametric test statistic distribution, which is then used to estimate the significance of the observed test statistic at each of the pixel locations. The following steps summarize the details of the single threshold test:

1. For each unique topograph scan rearrangement or relabeling, compute the test statistic at each of the pixel locations and estimate the maximum of the test statistics from all the pixel locations.
2. Build the permutation distribution for the test statistic using the maximum test statistic values from each of the scan arrangements.
3. With N unique scan arrangements or permutations, the critical test statistic threshold t-$stat_a$ to test the significance of an observed test statistic at a level

of significance α can be estimated to be the $(\lfloor \alpha N \rfloor + 1)$th largest member of the maximum test statistic permutation distribution.

4. Now, each of the topograph pixel locations K in the observed topograph scans can be tested for the null hypothesis H_0^K of no difference between the baseline and follow-up height measurements using the test statistic threshold $t\text{-}stat_\alpha$. A pixel location K is considered significantly changed from the baseline to the follow-up visits if the test statistic observed at pixel location K, $t\text{-}stat_K$, computed using the topographs in the observed scan order, is greater than the critical threshold $t\text{-}stat_\alpha$.

5. Corrected p-values at each of the topograph pixel locations can be estimated using the permutation distribution as $p_{corr} = N_{obs}/N$, where, N_{obs} is the number of maximum test statistic measures in the permutation distribution greater than the observed test statistic $t\text{-}stat_K$ and N is the total number of maximum test statistic measures in the maximum test statistic permutation distribution.

Note that the single threshold test accounts for the multiple comparison problem by using a test statistic that combines information from all the pixel locations using a "maximum" operator in a topograph.

7.4.4.2   Suprathreshold Cluster Tests

In suprathreshold cluster tests, a global primary threshold is initially applied to select the pixel locations with significant changes. To test the significance of topograph-level changes from baseline to follow-up, a permutation distribution of the largest cluster size is built from all of the unique scan order rearrangements. The steps of the suprathreshold cluster tests are summarized here:

1. For all unique scan order arrangements, apply a predetermined primary threshold for the pixel-level test statistic estimates and select all the locations of significant changes. This step is similar to the method of building the pixel change significance map in the TCA method.

2. Determine the size of the largest cluster of significant pixel locations from every scan order arrangement and build a largest cluster-size (test statistic) permutation distribution.

3. For testing the significance of the topograph-level changes from the baseline to follow-up visits at a level of significance α, the largest cluster size (test statistic) threshold $(t\text{-}stat_\alpha)$ will be $(\lfloor \alpha N \rfloor + 1)$th largest element in the permutation distribution.

4. Corrected p-values at each of the topograph pixel locations can be estimated as $p_{corr} = N_{obs}/N$, where N_{obs} is the number of largest cluster size measures in the permutation distribution greater than the largest cluster size in the originally observed topograph sequences and N is the total number of unique permutations used to build the permutation distribution.

Recently, suprathreshold permutation tests were applied for detecting structural glaucomatous progression in ONH topographs from a baseline visit to multiple follow-up visits [69]. As mentioned earlier, the test statistics were computed at each pixel location as a ratio of the absolute value of the slope to the standard error values from the regression analysis using the height measurements from all topograph scans at baseline and follow-up visits in the observed order. Additionally, the standard error values used for computing the test statistic were spatially smoothed using a Gaussian filter to reduce the influence of noise in the variance estimates. To generate a pixel-level change significance map, a 5% global primary threshold is applied to the slope-based test statistic at each pixel location to generate a pixel-level change significance map. Using the pixel-level change significance maps generated at each of the unique scan order rearrangements, the largest cluster-size test statistics were computed. A total of 1,000 unique scan order rearrangements were used: (1) to generate the slope-based test statistic permutation distribution at each pixel location, which is used to determine the 5% global primary threshold; and (2) to build the largest cluster size permutation distribution for inference at the topograph level. The direction of the regression line, estimated as the sign of the slope estimate, is used to determine the pixel locations with decreases in retinal height from baseline. Pixel locations associated with a negative slope are considered to be the locations with a decrease in their retinal heights. The significance of the topograph changes between the baseline and the follow-up visits was estimated using the significance of the largest pixel cluster size estimated for the topograph scans in the originally observed order.

Figure 7.14 shows the permutation distribution of the largest-cluster-size test statistic (t-$stat_{LCS}$) within the optic disc for participant P_{stable}, and Figure 7.17 shows a similar permutation distribution for participant $P_{progressive}$. The results for the ONH topograph sequences from a scan order rearrangement (permutation #300) are chosen for demonstration. We can see that the test statistic t-$stat_{LCS}$ for the observed topograph sequences is smaller (71 pixels) and insignificant ($p = 0.28$) for participant P_{stable} compared to the permuted ONH topograph sequences from permutation #300 (291 pixels; p-value = 0.043) and the same for participant $P_{progressive}$ is larger (1,251 pixels) and significant ($p = 0.001$) compared to the permuted ONH topograph sequences from the permutation #300 (10 pixels; p-value = 0.84). Figures 7.15 and 7.18 show the statistic image maps and cluster maps for the observed ONH topograph sequences for participants P_{stable} and $P_{progressive}$, respectively, and

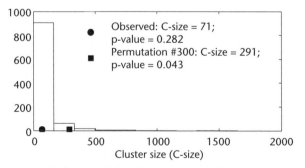

Figure 7.14 Permutation distribution of the cluster size (C-size) statistic for P_{stable}. The statistic image maps and the cluster maps for the observed sequence and for permutation sequence #300 are shown in Figures 7.15 and 7.16.

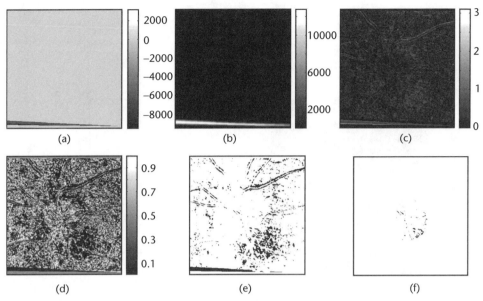

Figure 7.15   Static image maps and cluster maps of the observed sequences of topographs for participant P_{stable}. (a) Slope matrix. (b) Smoothed standard error. (c) Observed test statistic. (d) Change probability map. (e) Locations of retinal height reduction. (f) Locations of significant height reduction within the optic disc (OD); largest cluster inside the OD = 71 pixels with $p = 0.28$.

Figures 7.16 and 7.19 show similar maps for the permuted ONH topograph sequences from the permutation #300 for participants P_{stable} and $P_{progressive}$, respectively.

We can observe that the accuracy and the power of the suprathreshold cluster tests depend on the primary threshold chosen for generating the initial pixel significance map, which is later used for identifying the largest cluster test statistic $t\text{-}stat_{LCS}$. A lower primary threshold may be effective for finding diffuse losses in the ONH, and a higher primary threshold may be effective for finding focal and localized losses in the ONH topographs. A method of combining the single threshold tests and suprathreshold cluster tests for sensitive detection of both the focal and diffuse changes is available [82].

In general, permutation tests are powerful, but at the same time require large memory and computational resources. Our implementation of the suprathreshold cluster tests used in the SIM of the retina [69] in MATLAB, version 7, runs for about 10 minutes for analyzing the topographs from a single participant using 1,000 unique permuted scan rearrangements, with opportunities for further improvement in the MATLAB environment. We used a dual-core Xeon Processor at a 2.33-GHz workstation with 4 Gbyte of RAM for running the analysis. Computationally intensive segments of the analysis were implemented as native-C programs for use in the MATLAB environment. For real-time analysis, an approximate permutation test procedure known as the Monte Carlo procedure can be used, wherein the requirements of the number of permutations are reduced significantly while retaining the strengths of the permutation test.

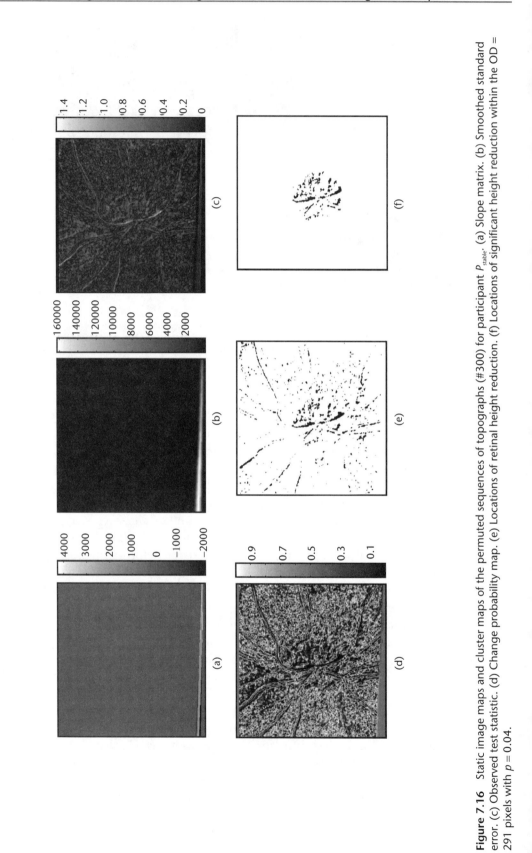

Figure 7.16 Static image maps and cluster maps of the permuted sequences of topographs (#300) for participant P_{stable}. (a) Slope matrix. (b) Smoothed standard error. (c) Observed test statistic. (d) Change probability map. (e) Locations of retinal height reduction. (f) Locations of significant height reduction within the OD = 291 pixels with $p = 0.04$.

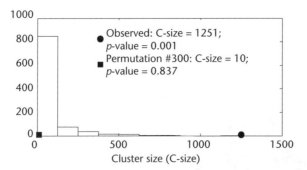

Figure 7.17 Permutation distribution of the cluster size (C-size) statistic for $P_{progressive}$. The statistic image maps and the cluster maps for the observed sequence and for the permutation sequence #300 are shown in Figures 7.18 and 7.19.

7.5 Summary

The visual function changes in a glaucomatous eye are generally asymptomatic in the initial stages of the disease because the brain may compensate for small localized loss of visual functions. Although the IOP of an eye may provide information about the physiological conditions, it cannot be used as a sole indicator to determine the presence or absence of glaucoma, because eyes throughout the range of IOP values can have glaucoma. The effect of an IOP level in an eye is influenced by the eye's CCT. Eyes with thinner CCTs are more susceptible to developing glaucoma. In glaucoma, the death of the RGCs and their nerve axons is reflected in the appearance of the ONH region. Therefore, assessing the appearance of the ONH region provides valuable insights for diagnosing glaucoma.

CSLOs are very useful in capturing the three-dimensional appearance and status of an ONH with relative ease in a clinical setting. Several ONH parameters called stereometric parameters can be automatically estimated from the CSLO topograph height measurements. The CSLO stereometric parameters are generally useful in providing a general indication of the status of an ONH. However, their use in identifying glaucoma patients from healthy individuals is limited due to a large overlap in their values between these two groups, and the fact that parameters can be changing yet still remain in the normal range. Due to this difficulty and also due to the progressive nature of the structural losses in glaucoma, detecting a progression of structural defects from a baseline condition is more useful in identifying a glaucomatous condition in an eye.

The regional height measurements in an ONH topograph are used to estimate the global and regional stereometric parameters; therefore, their sensitivity in detecting a glaucomatous progression depends on the choice of the region of analysis. For example, the VCDR or superior or inferior rim area parameters may be more specific in identifying the neuroretinal rim loss along the vertical poles of an optic disc compared to an overall cup-to-disc diameter or rim-to-disc diameter ratio. Therefore, the stereometric parameters can be more sensitive when physiological a priori information about the pathological condition is available. In HRT, a commercially available CSLO, a standard reference plane placed at a distance of 50 μm in reference to the optic disc margin of an eye, is used to identify the lower extent

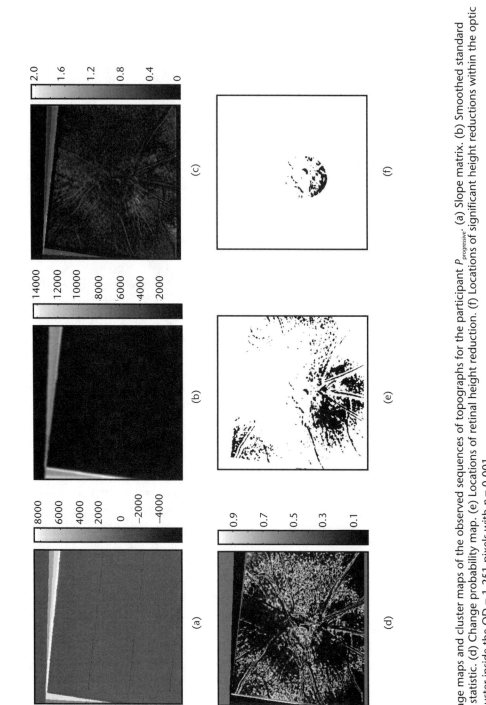

Figure 7.18 Static image maps and cluster maps of the observed sequences of topographs for the participant $P_{progressive}$. (a) Slope matrix. (b) Smoothed standard error. (c) Observed test statistic. (d) Change probability map. (e) Locations of retinal height reduction. (f) Locations of significant height reductions within the optic disc (OD); the largest cluster inside the OD = 1,251 pixels with $p = 0.001$.

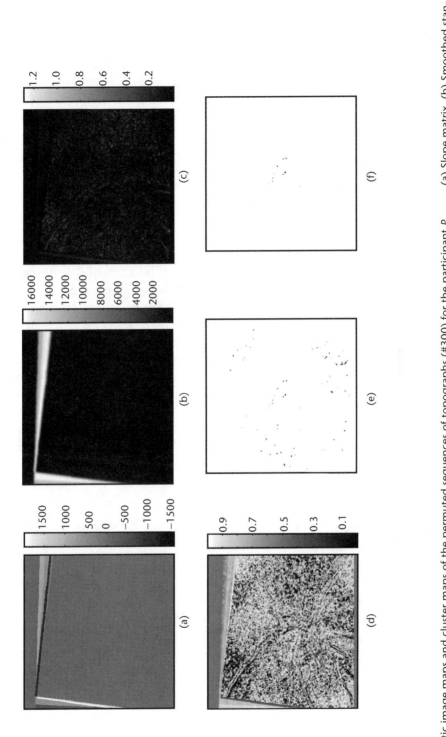

Figure 7.19 Static image maps and cluster maps of the permuted sequences of topographs (#300) for the participant $P_{progressive}$. (a) Slope matrix. (b) Smoothed standard error. (c) Observed test statistic. (d) Change probability map. (e) Locations of retinal height reduction. (f) Locations of significant height reductions within the optic disc (OD); the largest cluster inside the OD = 10 pixels with $p = 0.84$.

of the nerve fiber layer and delineate the neuroretinal rim from the optic cup within the optic disc margin. The stereometric parameter estimates, though proven to be reproducible and repeatable, can be influenced by operator-dependent, manual placement of optic disc margins and by tilts in the ONH topograph scans. The new ONH specific reference plane is a promising approach for locating the lower extent of the nerve fiber layer and requires more investigation using various populations in various stages of glaucoma. This method will be especially useful for analyzing progression in an individual as the location of the reference plane is chosen specific to an individual ONH and is kept at the same location during follow-ups. The GPS available in the HRT-3 software does not require a reference plane for ONH analysis and classification. The performance of the GPS in classifying glaucomatous eyes is similar to that of MRA. In addition, the GPS does not require a manual contour line and is fully automatic.

In contrast to the ONH summary parameters, the pixel-level analysis methods of TCA and nonparametric permutation tests do not require a priori regional information for detecting the progression of structural changes and do not require a manually drawn contour line to identify clusters of changed pixels. However, the criterion to determine clinically significant change has yet to be agreed on. In addition, the current usage of the parametric TCA method for detecting progression using the superpixel cluster-size statistics does not account for the multiple comparison problem. One can see that the suprathreshold permutation tests can be easily applied to the parametric TCA framework to address the multiple comparison problem. Because the permutation tests require significantly more computational resources, this may not be suitable for a real-time analysis. Our implementation of the suprathreshold permutation tests completes in 10 minutes in the MATLAB environment. A more efficient and faster implementation can be accomplished by implementing the analysis using processor-efficient programming languages such as ANSI C and using efficient processor-specific libraries.

Acknowledgments

This publication was supported in part by research support from Heidelberg Engineering, Inc., and NEI EY11008 (LMZ). The contents are solely the responsibility of the authors and do not necessarily represent the official views of the funding agencies. The authors thank Dr. David C. Hamilton with Dalhousie University, Halifax, Canada, for clarifying details of the TCA method and Drs. Andrew J. Patterson and David P. Crabb with City University, London, for clarifying the details of the SIM of the retina method.

References

[1]   Bergman, J., and J. Calkins, "Is the Backwards Human Retina Evidence of Poor Design?" *J. American Scientific Affiliation*, Vol. 52, No. 1, 2000, pp. 18–30.

[2]   Kolb, H., "How the Retina Works—Much of the Construction of an Image Takes Place in the Retina Itself Through the Use of Specialized Neural Circuits," *American Scientist*, Vol. 91, No. 1, 2003, pp. 28–35.

[3] Kolb, H., E. Fernandez, and R. Nelson, "The Organization of the Retina and Visual System," http://retina.umh.es/Webvision/index.html, 2005.

[4] Gangwar, D. N., I. S. Jain, and S. P. Dhir, "Course of the Nerve Fibres in the Temporal Half of the Retina," *Indian J. Ophthalmol.*, Vol. 25, No. 3, 1977, pp. 17–19.

[5] Jonas, J. B., et al., "Human Optic Nerve Fiber Count and Optic Disc Size," *Invest. Ophthalmol. Vis. Sci.*, Vol. 33, No. 6, 1992, pp. 2012–2018.

[6] Quigley, H. A., and A. T. Broman, "The Number of People with Glaucoma Worldwide in 2010 and 2020," *Br. J. Ophthalmol.*, Vol. 90, No. 3, 2006, pp. 262–267.

[7] Lee, P. P., et al., "Cost of Patients with Primary Open-Angle Glaucoma: A Retrospective Study of Commercial Insurance Claims Data," *Ophthalmology*, 2007.

[8] Lee, P. P., et al., "A Multicenter, Retrospective Pilot Study of Resource Use and Costs Associated with Severity of Disease in Glaucoma," *Arch. Ophthalmol.*, Vol. 124, No. 1, 2006, pp. 12–19.

[9] Salm, M., D. Belsky, and F. A. Sloan, "Trends in Cost of Major Eye Diseases to Medicare, 1991 to 2000," *Am J. Ophthalmol.*, Vol. 142, No. 6, 2006, pp. 976–982.

[10] Klein, B. E., et al., "Prevalence of Glaucoma. The Beaver Dam Eye Study," *Ophthalmology*, Vol. 99, No. 10, 1992, pp. 1499–1504.

[11] Rudnicka, A. R., et al., "Variations in Primary Open-Angle Glaucoma Prevalence by Age, Gender, and Race: A Bayesian Meta-Analysis," *Invest. Ophthalmol. Vis. Sci.*, Vol. 47, No. 10, 2006, pp. 4254–4261.

[12] Conforti, L., R. Adalbert, and M. P. Coleman, "Neuronal Death: Where Does the End Begin?" *Trends Neurosci.*, Vol. 30, No. 4, 2007, pp. 159–166.

[13] Bellezza, A. J., R. T. Hart, and C. F. Burgoyne, "The Optic Nerve Head as a Biomechanical Structure: Initial Finite Element Modeling," *Invest. Ophthalmol. Vis. Sci.*, Vol. 41, No. 10, 2000, pp. 2991–3000.

[14] Sigal, I. A., J. G. Flanagan, and C. R. Ethier, "Factors Influencing Optic Nerve Head Biomechanics," *Invest. Ophthalmol. Vis. Sci.*, Vol. 46, No. 11, 2005, pp. 4189–4199.

[15] Weinreb, R. N., and P. T. Khaw, "Primary Open-Angle Glaucoma," *Lancet*, Vol. 363, No. 9422, 2004, pp. 1711–1720.

[16] Foster, P. J., et al., "The Definition and Classification of Glaucoma in Prevalence Surveys," *Br. J. Ophthalmol.*, Vol. 86, No. 2, 2002, pp. 238–242.

[17] Gordon, M. O., et al., "The Ocular Hypertension Treatment Study: Baseline Factors That Predict the Onset of Primary Open-Angle Glaucoma," *Arch. Ophthalmol.*, Vol. 120, No. 6, 2002, pp. 714–720; discussion, pp. 729–730.

[18] Heijl, A., et al., "Reduction of Intraocular Pressure and Glaucoma Progression: Results from the Early Manifest Glaucoma Trial," *Arch. Ophthalmol.*, Vol. 120, No. 10, 2002, pp. 1268–1279.

[19] Collaborative Normal-Tension Glaucoma Study Group, "The Effectiveness of Intraocular Pressure Reduction in the Treatment of Normal-Tension Glaucoma," *Am. J. Ophthalmol.*, Vol. 126, No. 4, 1998, pp. 498–505.

[20] Nickells, R. W., et al., "Surgical Lowering of Elevated Intraocular Pressure in Monkeys Prevents Progression of Glaucomatous Disease," *Exp. Eye Res.*, Vol. 84, No. 4, 2007, pp. 729–736.

[21] Kass, M. A., et al., "The Ocular Hypertension Treatment Study: A Randomized Trial Determines That Topical Ocular Hypotensive Medication Delays or Prevents the Onset of Primary Open-Angle Glaucoma," *Arch. Ophthalmol.*, Vol. 120, No. 6, 2002, pp. 701–713; discussion, pp. 729–730.

[22] Ventura, L. M., and V. Porciatti, "Restoration of Retinal Ganglion Cell Function in Early Glaucoma After Intraocular Pressure Reduction: A Pilot Study," *Ophthalmology*, Vol. 112, No. 1, 2005, pp. 20–27.

[23] Palmberg, P., "What Is It About Pressure That Really Matters in Glaucoma?" *Ophthalmology*, Vol. 114, No. 2, 2007, pp. 203–204.

[24] Repka, M. X., and H. A. Quigley, "The Effect of Age on Normal Human Optic Nerve Fiber Number and Diameter," *Ophthalmology*, Vol. 96, No. 1, 1989, pp. 26–32.

[25] Mikelberg, F. S., et al., "The Normal Human Optic Nerve: Axon Count and Axon Diameter Distribution," *Ophthalmology*, Vol. 96, No. 9, 1989, pp. 1325–1328.

[26] Jonas, J. B., et al., "Histomorphometry of the Human Optic Nerve," *Invest. Ophthalmol. Vis. Sci.*, Vol. 31, No. 4, 1990, pp. 736–744.

[27] Jonas, J. B., and W. M. Budde, "Diagnosis and Pathogenesis of Glaucomatous Optic Neuropathy: Morphological Aspects," *Prog. Retin. Eye Res.*, Vol. 19, No. 1, 2000, pp. 1–40.

[28] Fingeret, M., et al., "Five Rules to Evaluate the Optic Disc and Retinal Nerve Fiber Layer for Glaucoma," *Optometry*, Vol. 76, No. 11, 2005, pp. 661–668.

[29] Harizman, N., et al., "The ISNT Rule and Differentiation of Normal from Glaucomatous Eyes," *Arch. Ophthalmol.*, Vol. 124, No. 11, 2006, pp. 1579–1583.

[30] Zangwill, L., et al., "Agreement Between Clinicians and a Confocal Scanning Laser Ophthalmoscope in Estimating Cup/Disk Ratios," *Am. J. Ophthalmol.*, Vol. 119, No. 4, 1995, pp. 415–421.

[31] Feuer, W. J., et al., "The Ocular Hypertension Treatment Study: Reproducibility of Cup/Disc Ratio Measurements over Time at an Optic Disc Reading Center," *Am. J. Ophthalmol.*, Vol. 133, No. 1, 2002, pp. 19–28.

[32] Garway-Heath, D. F., et al., "Vertical Cup/Disc Ratio in Relation to Optic Disc Size: Its Value in the Assessment of the Glaucoma Suspect," *Br. J. Ophthalmol.*, Vol. 82, No. 10, 1998, pp. 1118–1124.

[33] Crowston, J. G., et al., "The Effect of Optic Disc Diameter on Vertical Cup to Disc Ratio Percentiles in a Population Based Cohort: The Blue Mountains Eye Study," *Br. J. Ophthalmol.*, Vol. 88, No. 6, 2004, pp. 766–770.

[34] Susanna, Jr., R., and F. A. Medeiros, *The Optic Nerve in Glaucoma*, Rio de Janeiro, Brazil: Cultura Medica, 2006.

[35] Sowka, J., "New Thoughts on Normal Tension Glaucoma," *Optometry*, Vol. 76, No. 10, 2005, pp. 600–608.

[36] Artes, P. H., and B. C. Chauhan, "Longitudinal Changes in the Visual Field and Optic Disc in Glaucoma," *Prog. Retin. Eye Res.*, Vol. 24, No. 3, 2005, pp. 333–354.

[37] Zangwill, L. M., et al., "Baseline Topographic Optic Disc Measurements Are Associated with the Development of Primary Open-Angle Glaucoma: The Confocal Scanning Laser Ophthalmoscopy Ancillary Study to the Ocular Hypertension Treatment Study," *Arch. Ophthalmol.*, Vol. 123, No. 9, 2005, pp. 1188–1197.

[38] Zangwill, L. M., et al., "Optic Nerve Head Topography in Ocular Hypertensive Eyes Using Confocal Scanning Laser Ophthalmoscopy," *Am. J. Ophthalmol.*, Vol. 122, No. 4, 1996, pp. 520–525.

[39] Weinreb, R. N., et al., "Effect of Repetitive Imaging on Topographic Measurements of the Optic Nerve Head," *Arch. Ophthalmol.*, Vol. 111, No. 5, 1993, pp. 636–638.

[40] Mikelberg, F. S., K. Wijsman, and M. Schulzer, "Reproducibility of Topographic Parameters Obtained with the Heidelberg Retina Tomograph," *J. of Glaucoma*, Vol. 2, 1993, pp. 101–103.

[41] Rohrschneider, K., et al., "Reproducibility of the Optic Nerve Head Topography with a New Laser Tomographic Scanning Device," *Ophthalmology*, Vol. 101, No. 6, 1994, pp. 1044–1049.

[42] Miglior, S., et al., "Intraobserver and Interobserver Reproducibility in the Evaluation of Optic Disc Stereometric Parameters by Heidelberg Retina Tomograph," *Ophthalmology*, Vol. 109, No. 6, 2002, pp. 1072–1077.

[43] Hatch, W. V., et al., "Interobserver Agreement of Heidelberg Retina Tomograph Parameters," *J. Glaucoma*, Vol. 8, No. 4, 1999, pp. 232–237.

[44] Garway-Heath, D. F., et al., "Inter- and Intraobserver Variation in the Analysis of Optic Disc Images: Comparison of the Heidelberg Retina Tomograph and Computer Assisted Planimetry," *Br. J. Ophthalmol.,* Vol. 83, No. 6, 1999, pp. 664–669.

[45] Zangwill, L. M., et al., "Optic Nerve Imaging: Recent Advances," in *Essentials in Ophthalmology: Glaucoma,* F. Grehn and R. Stamper, (eds.), New York: Springer, 2004, pp. 63–91.

[46] Zangwill, L. M., and C. Bowd, "Retinal Nerve Fiber Layer Analysis in the Diagnosis of Glaucoma," *Curr. Opin. Ophthalmol.,* Vol. 17, No. 2, 2006, pp. 120–131.

[47] Minsky, M., "Microscopy Apparatus," U.S. Patent No. 3013467, issued December 19, 1961.

[48] Minsky, M., "Memoir on Inventing the Confocal Scanning Microscope," *Scanning,* Vol. 10, No. 4, 1988, pp. 128–138.

[49] Kaufman, S. C., "The Development and Application of a White-Light Confocal Microscope in the Morphologic and Anatomic Characterization of Cells Within the Cornea," Ph.D. Thesis, Department of Anatomy, LSU School of Medicine, December 1996.

[50] Chrástek, R., et al., "Automated Segmentation of the Optic Nerve Head for Diagnosis of Glaucoma," *Medical Image Analysis,* Vol. 9, 2005, pp. 297–314.

[51] Strouthidis, N. G., et al., "Improving the Repeatability of Heidelberg Retina Tomograph and Heidelberg Retina Tomograph II Rim Area Measurements," *Br. J. Ophthalmol.,* Vol. 89, No. 11, 2005, pp. 1433–1437.

[52] Burk, R. O., et al., "Development of the Standard Reference Plane for the Heidelberg Retina Tomograph," *Graefes Arch. Clin. Exp. Ophthalmol.,* Vol. 238, No. 5, 2000, pp. 375–384.

[53] "Heidelberg Retina Tomograph: Operating Manual Software Version 3.0," Heidelberg Engineering, GmbH, 2005.

[54] Girkin, C. A., "Principles of Confocal Scanning Laser Ophthalmoscopy for the Clinician," in *The Essential HRT Primer,* M. Fingeret, J. G. Flanagan, and J. M. Liebmann, (eds.), San Ramon, CA: Jocotto Advertising, 2005, pp. 1–9.

[55] Tan, J. C., and R. A. Hitchings, "Reference Plane Definition and Reproducibility in Optic Nerve Head Images," *Invest. Ophthalmol. Vis. Sci.,* Vol. 44, No. 3, 2003, pp. 1132–1137.

[56] Wollstein, G., D. F. Garway-Heath, and R. A. Hitchings, "Identification of Early Glaucoma Cases with the Scanning Laser Ophthalmoscope," *Ophthalmology,* Vol. 105, No. 8, 1998, pp. 1557–1563.

[57] Garway-Heath, D. F., "Moorfields Regression Analysis," in *The Essential HRT Primer,* M. Fingeret, J. G. Flanagan, and J. M. Liebmann, (eds.), San Ramon, CA: Jocotto Advertising, 2005, pp. 31–39.

[58] Zangwill, L. M., et al., "The Effect of Disc Size and Severity of Disease on the Diagnostic Accuracy of the Heidelberg Retina Tomograph Glaucoma Probability Score," *Invest. Ophthalmol. Vis. Sci.,* Vol. 48, No. 6, 2007, pp. 2653–2660.

[59] Burgansky-Eliash, Z., et al., "Glaucoma Detection with the Heidelberg Retina Tomograph 3," *Ophthalmology,* Vol. 114, No. 3, 2007, pp. 466–471.

[60] Coops, A., et al., "Automated Analysis of Heidelberg Retina Tomograph Optic Disc Images by Glaucoma Probability Score," *Invest. Ophthalmol. Vis. Sci.,* Vol. 47, No. 12, 2006, pp. 5348–5355.

[61] Swindale, N. V., et al., "Automated Analysis of Normal and Glaucomatous Optic Nerve Head Topography Images," *Invest. Ophthalmol. Vis. Sci.,* Vol. 41, No. 7, 2000, pp. 1730–1742.

[62] Uchida, H., L. Brigatti, and J. Caprioli, "Detection of Structural Damage from Glaucoma with Confocal Laser Image Analysis," *Invest. Ophthalmol. Vis. Sci.,* Vol. 37, No. 12, 1996, pp. 2393–2401.

[63] Bowd, C., et al., "Comparing Neural Networks and Linear Discriminant Functions for Glaucoma Detection Using Confocal Scanning Laser Ophthalmoscopy of the Optic Disc," *Invest. Ophthalmol. Vis. Sci.,* Vol. 43, No. 11, 2002, pp. 3444–3454.

[64] Bowd, C., et al., "Confocal Scanning Laser Ophthalmoscopy Classifiers and Stereophotograph Evaluation for Prediction of Visual Field Abnormalities in Glaucoma-Suspect Eyes," *Invest. Ophthalmol. Vis. Sci.*, Vol. 45, No. 7, 2004, pp. 2255–2262.

[65] Dreher, A. W., P. C. Tso, and R. N. Weinreb, "Reproducibility of Topographic Measurements of the Normal and Glaucomatous Optic Nerve Head with the Laser Tomographic Scanner," *Am. J. Ophthalmol.*, Vol. 111, No. 2, 1991, pp. 221–229.

[66] Chauhan, B. C., et al., "Test–Retest Variability of Topographic Measurements with Confocal Scanning Laser Tomography in Patients with Glaucoma and Control Subjects," *Am. J. Ophthalmol.*, Vol. 118, No. 1, 1994, pp. 9–15.

[67] Capel, D., *Image Mosaicing and Super-Resolution*, New York: Springer-Verlag, 2004.

[68] Chauhan, B. C., et al., "Technique for Detecting Serial Topographic Changes in the Optic Disc and Peripapillary Retina Using Scanning Laser Tomography," *Invest. Ophthalmol. Vis. Sci.*, Vol. 41, No. 3, 2000, pp. 775–782.

[69] Patterson, A. J., et al., "A New Statistical Approach for Quantifying Change in Series of Retinal and Optic Nerve Head Topography Images," *Invest. Ophthalmol. Vis. Sci.*, Vol. 46, No. 5, 2005, pp. 1659–1667.

[70] Burgoyne, C. F., et al., "Global and Regional Detection of Induced Optic Disc Change by Digitized Image Analysis," *Arch. Ophthalmol.*, Vol. 112, No. 2, 1994, pp. 261–268.

[71] Burgoyne, C. F., et al., "Early Changes in Optic Disc Compliance and Surface Position in Experimental Glaucoma," *Ophthalmology*, Vol. 102, No. 12, 1995, pp. 1800–1809.

[72] Burgoyne, C. F., et al., "Basic Issues in the Sensitive and Specific Detection of Optic Nerve Head Surface Change Within Longitudinal LDT TopSS Images: Introduction to the LSU Experimental Glaucoma (LEG) Study," in *The Shape of Glaucoma*, H. G. Lemij and J. S. Schuman, (eds.), The Hague, the Netherlands: Kugler Publications, 2000.

[73] Strouthidis, N. G., et al., "Optic Disc and Visual Field Progression in Ocular Hypertensive Subjects: Detection Rates, Specificity, and Agreement," *Invest. Ophthalmol. Vis. Sci.*, Vol. 47, No. 7, 2006, pp. 2904–2910.

[74] Tan, J. C., and R. A. Hitchings, "Approach for Identifying Glaucomatous Optic Nerve Progression by Scanning Laser Tomography," *Invest. Ophthalmol. Vis. Sci.*, Vol. 44, No. 6, 2003, pp. 2621–2626.

[75] Tan, J. C., and R. A. Hitchings, "Optimizing and Validating an Approach for Identifying Glaucomatous Change in Optic Nerve Topography," *Invest. Ophthalmol. Vis. Sci.*, Vol. 45, No. 5, 2004, pp. 1396–1403.

[76] Balasubramanian, M., and L. M. Zangwill, *Analysis of Variance Model in the Topographic Change Analysis*, Technical Report No. TR_MBALA_28DEC2006, San Diego, CA: Hamilton Glaucoma Center, Ophthalmology, University of California at San Diego, December 2006.

[77] Neter, J., et al., *Applied Linear Statistical Models*, Burr Ridge, IL: Irwin, 1996.

[78] Chauhan, B. C., et al., "Optic Disc and Visual Field Changes in a Prospective Longitudinal Study of Patients with Glaucoma: Comparison of Scanning Laser Tomography with Conventional Perimetry and Optic Disc Photography," *Arch. Ophthalmol.*, Vol. 119, No. 10, 2001, pp. 1492–1499.

[79] Artes, P. H., and B. C. Chauhan, "Criteria for Optic Disc Progression with the Topographic Change Analysis of the Heidelberg Retina Tomograph," *Invest. Ophthalmol. Vis. Sci.*, Vol. 47, 2006.

[80] Nichols, T. E., and A. P. Holmes, "Nonparametric Permutation Tests for Functional Neuroimaging: A Primer with Examples," *Hum. Brain Mapping*, Vol. 15, No. 1, 2002, pp. 1–25.

[81] Patterson, A. J., "Analysis of Retinal Images in Glaucoma," Ph.D. Thesis, The Nottingham Trent University, March 2006.

[82] Poline, J. B., et al., "Combining Spatial Extent and Peak Intensity to Test for Activations in Functional Imaging," *Neuroimage*, Vol. 5, No. 2, 1997, pp. 83–96.

CHAPTER 8

Fractal Measures for Fungal Keratitis Diagnosis Using a White-Light Confocal Microscope

Madhusudhanan Balasubramanian, A. Louise Perkins, Roger W. Beuerman, and S. Sitharama Iyengar

In this chapter, we present a fractal measure based pattern classification algorithm for automatic feature extraction and identification of fungi associated with infections of the cornea of the eye. A white-light confocal microscope image of a suspected fungus exhibited locally linear and branching structures. The pixel intensity variation across the width of the fungal element was Gaussian. Linear features were extracted using a set of two-dimensional directional matched Gaussian filters. Portions of fungus profiles that were not in the same focal plane appeared relatively blurred. We used Gaussian filters having a standard deviation slightly larger than the width of a fungus to reduce discontinuities. The cell nuclei of cornea and nerves also exhibited locally linear structures. Cell nuclei were excluded by their relatively shorter lengths. Nerves in the cornea exhibited less branching compared to the fungus. Fractal dimensions of the locally linear features were computed using a box-counting method. A set of corneal images with fungal infection was used to generate class-conditional fractal measure distributions of fungus and nerves. The a priori class-conditional densities were built using an adaptive mixtures method to reflect the true nature of the feature distributions and improve the classification accuracy. A maximum-likelihood classifier was used to classify the linear features extracted from test corneal images as "normal" or "with fungal infiltrates," using the a priori fractal measure distributions. We demonstrate the algorithm on the corneal images with culture-positive fungal infiltrates. The algorithm is fully automatic and will help diagnose fungal keratitis by generating a diagnostic mask of locations of the fungal infiltrates.

8.1 Introduction

The cornea is the transparent portion through which light enters the eye [1]. The cornea and the sclera form the outer tunic of the eye and are mechanically strong. They act as a protective shield and prevent foreign objects from entering the eye.

189

The cornea has five main layers: epithelium, Bowman's layer, stroma, Descemet's membrane, and endothelium. The epithelium contributes to approximately 10% of the corneal thickness and has four to six layers of cells. Bowman's layer is approximately 8 to 10 μm thick and is made of fibrous protein called collagen that forms the outer tunic that protects the eye. The stroma forms the other 90% of the cornea and is made of collagen fibrils. The endothelium is the innermost layer of the cornea. It is a single layer of cells approximately 4 to 6 μm thick. The main function of the endothelium is to pump any excess water out of the stroma to preserve its mechanical structure and optical clarity. Descemet's membrane lies between the stroma and endothelium and protects against infections and injuries.

Keratitis is an inflammation in the cornea due to viruses, bacteria, or fungi. Around 50,000 cases of keratitis are reported in the United States every year. Bacterial keratitis, often due to contact lens wear, is more common than fungal keratitis and may be difficult to distinguish at the initial stages [2]. Fungi usually enter the cornea during an agriculture-related corneal injury [3]. In all the cases, the cornea must be injured before the organisms can enter the corneal layers [3]. When left untreated, the organism can infiltrate the sclera and may enter the anterior chamber in the eye, resulting in corneal ulcers and possibly loss of vision. Recently, a worldwide outbreak of fungus infections associated with contact lens wear has alerted the medical community to the pervasive nature of fungi [4].

Fungi can be broadly classified as filamentous (multicellular) and yeast-like (unicellular). Hyphae are tubular structures that form the basic unit of filamentous fungi. Hyphae grow from the tip and branch at an angle. Most cases of fungal keratitis are caused by agriculture/farm-related corneal injury. *Aspergillus fumigatus*, a filamentous fungus, is the most common cause of fungal keratitis in the United States (Minnesota), India, London, Bangladesh, Saudi Arabia, and worldwide [3]. In the initial stages, fungi enter through the epithelium whose function is compromised due to a defect or a corneal injury. The fungi then proliferate in the stroma. However, the recent outbreak of fungal infections is due to another species, called *Fusarium,* that is associated with contact lens wear [4]. If appropriate treatment is not initiated, fungal infections can lead to a loss of vision.

Initial diagnosis is confirmed with corneal scraping and specific laboratory procedures [3]. Culture techniques take from 2 days up to several weeks to positively identify the pathogen. We have previously shown that a white-light confocal microscope (WLCM) can be used to provide a sensitive and rapid diagnosis of bacterial and fungal keratitis [5–7]; for WLCM details refer to [8, 9]. In brief, a WLCM allows imaging of microscopic structures, such as the cornea, at various depths without physically sectioning the structure. This is accomplished by selectively allowing the light to be reflected from the structure at the depth of interest using a finitely small detector pinhole. This optical sectioning ability of a confocal microscope allows imaging a thick tissue such as that of the cornea at various depths in real time [10].

This research builds on our previous work [6] on the use of confocal microscopes for the diagnosis of keratitis due to filamentous fungi. The main motivation of this work was to develop a computer-assisted fungal infection diagnosis toolkit to facilitate easier and faster screening of patients and to assist with the diagnosis. Use of confocal microscope images allows for a noninvasive, rapid, and sensitive diagnosis. The computer-assisted toolkit has three major components: (1) characterizing and extracting the filamentous fungi profiles from the corneal confocal microscope images, (2) developing an appropriate feature metric to describe the morphology of filamentous fungi, and (3) classifying a given confocal microscope image as "normal" or positive for "filamentous fungi."

Objects of interest in an image can be located and extracted using a template matching approach [11, 12]. Two-dimensional matched filters are useful for identifying the locally linear features in the images. They have been successfully used for the extraction of retinal blood vessels [13]. We use a set of two-dimensional directional Gaussian filters to generate the initial seed points of the fungal infiltrates in the images. A fungal infiltrate mask is generated through binary morphological dilation operations of the seed points and the corresponding matched filters. Because the seed points for the fungal infiltrates are identified using a matched filter approach, it will also select: (1) nucleus in the epithelial and endothelial cell layers and (2) corneal nerves. The nucleus appears as a relatively smaller object in the corneal images and can easily be omitted using binary morphological operations. To a certain degree, the morphology of the hyphae and corneal nerves are similar. But the hyphae branch often as opposed to the corneal nerves.

The choice of the feature metric to describe the fungal elements should be invariant to any scale, rotation, and translation of the fungal infiltrates. Fourier-Mellin based descriptors are popular for generating rotation, scale, and translation-invariant feature descriptors [14, 15]. Previously we have attempted to characterize the morphology of the fungal infiltrates in the corneal confocal images by considering the differences in their textures using a fractal dimension–based power-law signature [16]. The fractal dimension of the texture was aimed at differentiating the surface roughness in the portion of the corneal images with fungal infiltrates from the rest of the image. However, upon further investigation, we found that the surface roughness measurement cannot uniquely characterize the fungal infiltrates, especially in the epithelial and endothelial layers where there are prominent cell structures that cause poor sensitivity and specificity in detecting the corneal infiltrates.

We have investigated the use of a box-counting based fractal dimension feature metric to characterize filamentous fungus and to differentiate them from the corneal nerves. The box-counting method has been successfully used for studying microbial growth patterns [17], for characterizing the vascular changes in diabetic retinopathy [18, 19], and for identifying the language from a written script [20] among others. A brief background of fractals and the box-counting method is presented in Section 8.2.2. We use a set of training fungal infiltrates and a set of corneal nerves for estimating the distributions of their fractal dimensions. The density distributions are accurately characterized using an adaptive mixtures method [21]. The two-class a priori densities were used for identifying any fungal infiltrates in the corneal confocal microscope images using a maximum-likelihood classifier [22]. The classifier will be used to classify a given image as being fungal infected or normal using the fractal dimension metrics of the extracted objects in the images. In subsequent sections, we describe the feature extraction procedure using matched filters and binary morphological operations, the box-counting method for estimating the fractal dimension of the objects, and the maximum-likelihood classifier for

image classification. We will show that the fractal dimension metric that characterizes the morphology of the hyphae is a more promising approach in detecting the
fungal infiltrates in the corneal confocal images. We use WLCM corneal images
from our previous study [6]. Images with fungal infiltrates were tested culture
positive.

8.2   A Computational Framework for Identifying Filamentous Fungi in Corneal Confocal Images

We use binary morphological operations of erosion, dilation, closing top hat, and
thinning for preprocessing the corneal images and at various stages in the generation
of the fungal infiltrate mask [23]. Image morphological operators use a special kernel called a *structuring element* (SE) to probe and identify the morphological structures in the images. Given an image I and a structuring element SE, binary erosion
operation is defined as follows:

$$I \ominus SE = \{x : SE_x \subset I\} \tag{8.1}$$

Here, SE_x is the translation of the SE by x pixels. Binary dilation can be defined
using the erosion operator as

$$I \oplus SE = \left(I^c \ominus \breve{SE} \right)^c \tag{8.2}$$

where c is the complement operator and $\breve{SE}$ is the rotation of SE by 180° from its
center. The morphological closing operation is defined as

$$I \bullet SE = (I \oplus SE) \ominus SE \tag{8.3a}$$

and the closing top hat is defined as

$$I \,\hat{\bullet}\, SE = (I \bullet SE) - I \tag{8.3b}$$

8.2.1   Fungal Feature Extraction

Closing top-hat morphological operators are applied to the corneal confocal images
prior to feature extraction to enhance the locally linear features in the images. Figure
8.1 shows a corneal image with fungal infiltrates and after applying the closing
top-hat operator as in (8.3). The pixel brightness distribution across the width of the
fungal infiltrates follows a Gaussian distribution with the mean at the midpoint of
the infiltrate. We used a set of 18 directional Gaussian matched kernels, at an angular interval of 10°, to identify the location of the infiltrates in the confocal images.
The matched filters approach typically uses a spatial convolution operation of the
image with all the directional kernels [13]. The kernels are expected to cover various
possible orientations of the fungal infiltrates. A correlation cube was built using
the individual correlation surfaces resulting from the convolution of the matching
kernels with the corneal image. From the correlation cube, a correlation surface con-

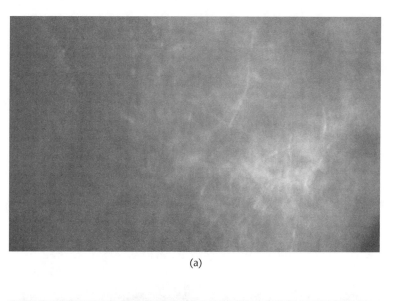

(a)

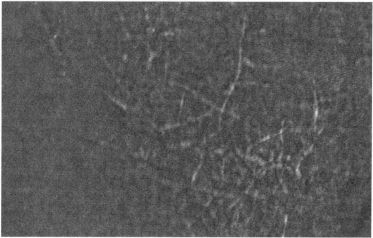

(b)

Figure 8.1 Enhancing the locally linear features in a corneal image using the morphological closing top-hat operation. Images correspond to a corneal region of size 400 μm (height) × 600 μm (width). (a) Acquired original image and (b) after applying morphological closing top-hat operator.

taining the maximal correlation value at each pixel position is constructed. The initial seed points for the infiltrates are determined by applying an adaptive threshold to the correlation surface. Figure 8.2 shows a surface plot of the matching kernels at several orientations. The infiltrate seed points are dilated as in (8.2) to identify the layout of the infiltrates in the image.

The overall mask is further thinned using a binary thinning operation. Figure 8.3 shows a corneal confocal image, its maximal correlation surface, and the fungal infiltrate mask constructed using the two-dimensional directional matched filters. Note that the objects selected using the two-dimensional matched filters method will also include corneal nerves, when they are present in the images, because they have morphological structures similar to filamentous fungi. Therefore, a robust fea-

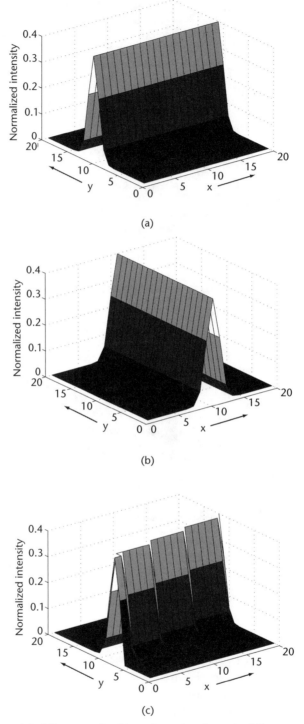

Figure 8.2   Surface plot of the normalized two-dimensional matched filters used for the extraction of locally linear features in the corneal images. (a) 0°; (b) 90°; and (c) 170°.

ture metric is necessary to uniquely characterize the fungal infiltrates using the infiltrate masks created using two-dimensional matched filters.

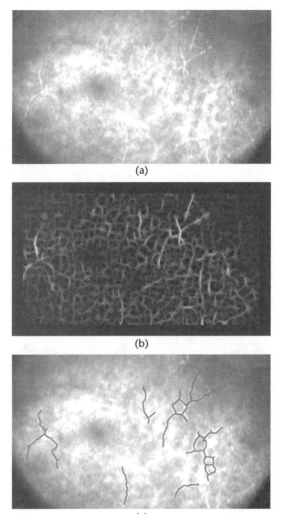

(a)

(b)

(c)

Figure 8.3   A corneal image with fungal infiltrates, its correlation surface, and the fungal infiltrate mask constructed using the matched filter approach. Images correspond to a corneal region of size of 400 μm (height) × 600 μm (width). (a) Corneal image with fungal infiltrates. (b) Correlation surface constructed by convolving various two-dimensional directional matched filters with the corneal image in part (a). (c) Initial seed points constructed by thresholding the maximal correlation surface in part (b) are used to generate the fungal infiltrate mask.

8.2.2   Fractal Dimension Based Feature Metrics

The next step is to extract the individual components present in the fungal infiltrate mask and use a suitable feature metric to describe them. The objects in the masks were formed using groups of pixels that have adjacency with at least one pixel in one of the eight directions (i.e., eight-connectivity). They can be identified and extracted using a connected components algorithm [12].

Pure fractal objects exhibit self-similarity under a change of scale or magnification. Natural objects such as coastlines and clouds do not have this self-similarity property and they are referred to as random fractals. One of the main characteristics used for determining the fractal dimension is the rate of change in the measured values, such as the length of a coastline and surface area of a cloud, when measured using various measuring scales. The relationship between the measured values $M(\varepsilon)$ and the measuring scale ε of a random fractal can be described using the power-law relation:

$$M(\varepsilon) \propto \frac{1}{\varepsilon^D}$$

$$\log(M(\varepsilon)) = -D\log(\varepsilon)$$

(8.4)

where D is the fractal dimension of the object. Box counting is one of the most popular methods used for measuring the fractal dimension [18]. The fractal dimension D is the negative of the slope of the log-log plot between the measure $M(\varepsilon)$ and scale ε as in (8.4) in logarithmic coordinates. Pattern classification algorithms using the fractal dimension as a feature metric could use additional features from the log-log plot, such as the goodness-of-fit (F-ratio) and the y-intercept of the regression line, to support the confidence on the fractal dimension estimate D [16, 24]. Together these features are referred to as a *power-law signature.*

The hyphae grow from the tip, and branch often, at an angle (approximately 45°). Therefore we expect that the fractal dimension of the filamentous fungi in the corneal images computed using the box-counting method will be different from the fractal dimension of the corneal nerves.

8.3 Image Classification

Algorithm 8.1: Algorithm for classifying a given corneal image I

1. **procedure** CornealImageClassifier(Input image I)
2.   Let, fd_I be the fractal dimension of the input image I.
3.   Let N be the class of normal corneal images.
4.   Let F be the class of corneal images with fungal infiltrates.
5.   Let $p(fd_I | N)$ be the a priori probability of observing a fractal dimension fd_I among class N.
6.   Let $p(fd_I | F)$ be the a priori probability of observing a fractal dimension fd_I among class F.
7. **If** $p(fd_I | F) \geq p(fd_I | N)$, **then**
8.   Classify I as a corneal image with fungal infiltrates
9. **else**
10.   Classify I as normal corneal image
11. **end if**
12. **end procedure.**

A set of fungal infiltrates and corneal nerves was used for building a priori fractal dimension distributions of the hyphae and the corneal nerves. The two-class probability distributions were used for identifying the objects in the infiltrate mask generated using the matched filters algorithm from the corneal images. The probability distributions are modeled as a mixture of normal distributions with various parameters and built using the adaptive mixtures method [21]. The mixture densities are expected to represent the true distributions of the fractal dimension metrics of the fungal infiltrates and the corneal nerves. To facilitate corneal image classification, the fractal dimension of the object in I with maximum fractal dimension will be used as the fractal dimension of the corneal image I. Assuming equal priors $p(N)$ and $p(F)$, the image I can be categorized as "normal" or "with fungal infiltrates" using a maximum-likelihood classifier as described earlier in Algorithm 8.1.

Let TP be the number of images with fungal infiltrates classified as "infected," FP the number of normal corneal images classified as "infected," TN the number of normal corneal images classified as "normal," and FN the number of corneal images with fungal infiltrates classified as "normal." The reliability of the fractal dimension metric and the performance of the classifier in correctly identifying the images with infiltrates with fewer false positives is determined using the sensitivity and specificity of the classifier as follows:

$$\text{Sensitivity} = \frac{TP}{TP + FN} \tag{8.5}$$

$$\text{Specificity} = \frac{TN}{TN + FP} \tag{8.6}$$

Sensitivity and specificity closer to 1.0 indicates that the feature metric used for classification is a suitable candidate for characterizing the corneal images.

8.4 Results

Corneal confocal images with filamentous fungi generated from our previous study [6] were used for constructing the two-class a priori fractal dimension distributions of hyphae and corneal nerves. The a priori distributions were estimated using the adaptive mixtures method from a set of eight training culture-positive fungal infiltrates and six training corneal nerves. A total of seven corneal images were used for testing Algorithm 8.1 among which four corneal images contained regions that were culture positive for fungal infiltrates. Figure 8.4 shows two of the corneal images tested using Algorithm 8.1 along with their infiltrate masks generated using the two-dimensional matched filter algorithm discussed in Section 8.2.1. Figure 8.5 shows a confocal image of corneal nerves along with the infiltrate mask and the fractal dimensions of the individual nerves computed using the box-counting method discussed in Section 8.2.2. The a priori fractal dimension distributions of the hyphae and corneal nerves are presented in Figure 8.6.

Based on the test corneal images, we estimated the sensitivity and specificity of the classifier to be 1.0 and 0.67, respectively. All of the images with fungal infiltrates were positively identified by the classifier. Some of the corneal nerves with branching characteristics similar to those of hyphae were misclassified as fungal infiltrates (such as in Figure 8.7), leading to a reduction in the specific detection of fungal infiltrates with the test data. In a clinical setting, this would translate into

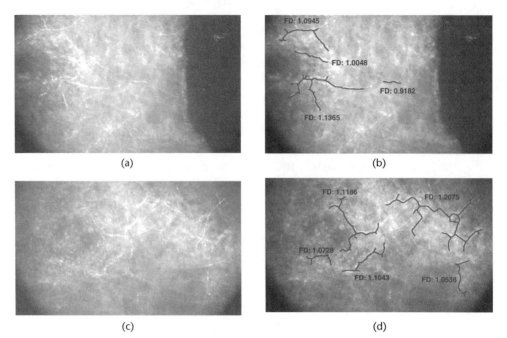

(a)   (b)

(c)   (d)

Figure 8.4 Corneal fungal infiltrate masks generated using two-dimensional matched filters algorithm along with the fractal dimensions of the objects computed using the box-computing method. Images correspond to a corneal region of size 400 μm (height) × 600 μm (width). (a, c) Corneal images with fungal infiltrates. (b, d) Infiltrate masks generated using two-dimensional matched filters.

alerting a physician about a positive fungal infection when it is normal, only in fewer cases (specific detection of 0.67). However, all of the images with culture-positive fungal infiltrates were correctly identified by the classification algorithm (sensitive detection of 1.0). The existing clinical diagnostic tests can then be applied to separate out the false positives; thus, only a smaller population will be required to take the more evasive tests.

8.5 Conclusions and Future Work

We have developed a pattern analysis and image synthesis framework to assist with the diagnosis of fungal keratitis due to filamentous fungi. Initial experimental results suggest that fractal dimension measures using a box-counting method is a suitable characteristic for describing filamentous fungi. The infiltrate mask computed using the two-dimensional matched filter approach can be used as an aid for diagnosing fungal keratitis. Implementing the two-dimensional matched filters algorithm in the frequency domain significantly reduces the computational load and makes the algorithm suitable for real-time analysis of corneal confocal images. Clinically, even a single fungal profile identified is serious.

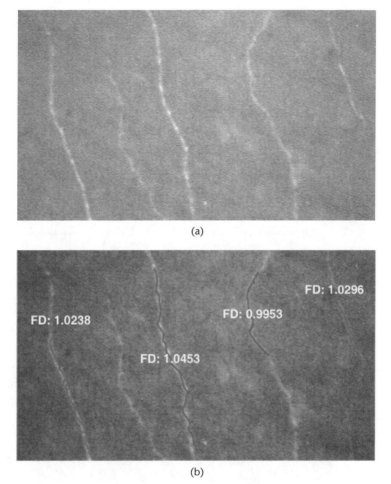

Figure 8.5   A confocal image of corneal nerves superimposed with the infiltrate mask generated using two-dimensional matched filters. Images correspond to a corneal region of size 400 μm (height) × 600 μm (width). (a) Corneal nerve and (b) infiltrate mask.

We have identified the following avenues for improving the specificity of the classifier:

1. WLCM images exhibit scan lines (due to the Nipkow scanning disk), which sometimes give raise to false seed points for the infiltrate mask as in Figure 8.3 and 8.7. Restoration of the confocal microscope images from scan lines and blurring due to the aberrant light contributions from out-of-focus layers can improve the accuracy of quantitative analysis using WLCM images [8, 9].

2. The morphological thinning operation used for creating an infiltrate mask in Section 8.2.1 introduces artificial branches in some cases as in Figure 8.7 and need to be pruned. The artificial branches and fluctuations in the mask (as in Figure 8.7) due to the thinning operation will have a major effect on

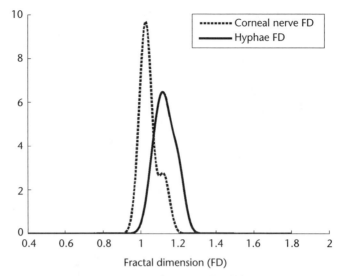

Figure 8.6   A priori probability distributions $p(fd_i | N)$ and $p(fd_i | F)$.

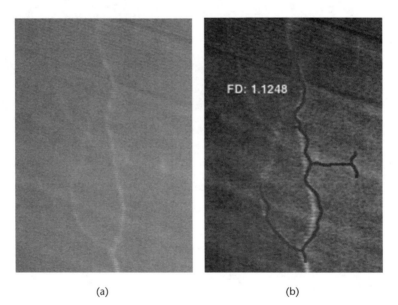

(a)   (b)

Figure 8.7   The infiltrate mask of a corneal nerve showing artifacts due to the confocal scanning lines: (a) corneal nerve and (b) infiltrate mask.

the fractal dimension of the object. Therefore, a better thinning approach needs to be identified or developed.

3. Confocal microscope images can be acquired at various depths in the cornea using the optical sectioning property. Some portions of the elements of interest, such as fungal infiltrates, may not be in the focal plane, leading to discontinuities in the mask. If the detected portions of the infiltrates are small, they may be mistaken for a cell nucleus and may be omitted from further analysis. It may be useful to construct a topographic map of the

WLCM optical serial sections as in the laser confocal microscopy software to generate a continuous map of infiltrates and the objects of interest.

The future direction of this work includes:

1. Using a larger population of human and animal subjects with fungal keratitis to further validate the use of the fractal measures as a reliable metric to characterize and identify filamentous fungi in fungal keratitis.
2. Studying the morphology of various species of fungi common in fungal keratitis and their fractal dimensions. This will have excellent uses in designing the treatment course for the patients. It will be very similar to the study conducted by Obert et al. [17] and will focus more on studying the morphology of various species as seen using a confocal microscope in fungal keratitis.
3. Characterizing the morphological structure of hyphae using fractal dimension measurements from the corneal confocal images at various stages of growth of filamentous fungi. This will be a valuable tool for identifying the fungi at an early stage using confocal images and also for determining the effects of various antifungal drugs during treatment.
4. Fourier-Mellin–based object descriptors have the rotation, scale, and translation-invariance property as with fractal dimension estimates. It may be useful to investigate the choice of Fourier-Mellin descriptors for characterizing the pathogens and compare their performance with the fractal dimension metrics.
5. Investigating the choice of object descriptors, including fractal dimension and Fourier-Mellin descriptors, to identify a broader class of pathogens (bacteria, virus, or fungi) causing the keratitis using the confocal images.

Acknowledgments

This publication was made possible through financial support in part from NEI EY12416 and EY02377 and NIH Grant No. P20 RR16456 from the INBRE Program of the National Center for Research Resources. Its contents are solely the responsibility of the authors and do not necessarily represent the official views of the funding agencies. This work was presented in part as an oral presentation at the Optics & Photonics 2006 SPIE Annual Conference in San Diego, California.

References

[1]   Klyce, S. D., and R. W. Beuerman, "Structure and Function of the Cornea," in *The Cornea*, H. E. Kaufman, (ed.), Boston, MA: Butterworth-Heinemann, 1998.

[2]   Zloty, P., "Diagnosis and Management of Fungal Keratitis," *Focal Points: Clinical Modules for Ophthalmologists*, Vol. 20, No. 6, 2002, pp. 1–13.

[3]   Liesegang, T. J., "Fungal Keratitis," in *The Cornea*, H. E. Kaufman, (ed.), Boston, MA: Butterworth-Heinemann, 1998.

[4]   Khor, W. -B., et al., "An Outbreak of *Fusarium* Keratitis Associated with Contact Lens Wear in Singapore," *J. of Am. Med. Assoc.*, Vol. 295, 2006, pp. 2867–2873.

[5]   Kaufman, S. C., et al., "Diagnosis of Bacterial Contact Lens Related Keratitis with the White-Light Confocal Microscope," *Contact Lens Association of Ophthalmologists (CLAO) J.*, Vol. 22, 1996, pp. 274–277.

[6]   Avunduk, A., et al., "Confocal Microscopy of Aspergillus Fumigatus Keratitis," *Br. J. of Ophthalmol.*, Vol. 87, 2003, pp. 409–410.

[7]   Fan, D., et al., "Comments on Confocal Microscopy of Aspergillus Fumigatus Keratitis," *Br. J. of Ophthalmol.*, Vol. 88, 2003, pp. 849–850.

[8]   Balasubramanian, M., "A Computational Framework for the Structural Change Analysis of 3D Volumes of Microscopic Specimens," Ph.D. Thesis, Louisiana State University, May 2006.

[9]   Balasubramanian, M., et al., "Real-Time Restoration of White-Light Confocal Microscope Optical Sections," *J. of Electronic Imaging,* Vol. 16, No. 3, 2007.

[10] Kaufman, S. C., "The Development and Application of a White-Light Confocal Microscope in the Morphologic and Anatomic Characterization of Cells Within Cornea," Ph.D. Thesis, Louisiana State University School of Medicine, December 1996.

[11] Balasubramanian, M., et al., "Analysis of z-Axis Structural Changes in the Cow Lamina Cribrosa with Changes in Pressure Using Image Matching," *Investigative Opthamology & Visual Science,* Vol. 45, 2004.

[12] Gonzalez, R. C., and R. E. Woods, *Digital Image Processing*, 2nd ed., Upper Saddle River, NJ: Prentice-Hall, 2002.

[13] Chaudhuri, S., et al., "Detection of Blood Vessels in Retinal Images Using Two-Dimensional Matched Filters," *IEEE Trans. on Medical Imaging*, Vol. 8, 1989, pp. 263–269.

[14] Casasent, D., and D. Psaltis, "Position, Rotation and Scale Invariant Optical Correlation," *Applied Optics*, Vol. 65, No. 1, 1976, pp. 1793–1799.

[15] Reddy, B. S., and B. Chatterji, "An FFT-Based Technique for Translation, Rotation and Scale-Invariant Image Registration," *IEEE Trans. on Image Processing*, Vol. 5, 1996, pp. 1266–1271.

[16] Balasubramanian, M., "Computer Assisted Eye Fungal Infection Diagnosis," Master's Thesis, Louisiana State University, December 2003.

[17] Obert, M., P. Pfeifer and M. Sernetz, "Microbial Growth Patterns Described by Fractal Geometry," *J. of Bacteriology*, Vol. 172, 1990, pp. 1180–1185.

[18] Masters, B. R., "Fractal Analysis of the Vascular Tree in the Human Retina," *Ann. Rev. Biomedical Engineering*, Vol. 6, 2004, pp. 427–452.

[19] Avakian, A., et al., "Fractal Analysis of Region-Based Vascular Change in the Normal and Non-Proliferative Diabetic Retina," *Curr. Eye Res.*, Vol. 24, No. 4, 2002, pp. 274–280.

[20] Tao, Y., E. C. Lam, and Y. Y. Tang, "Extraction of Fractal Feature for Pattern Recognition," *Proc. 15th Int. Conf. on Pattern Recognition*, Vol. 2, 2000, pp. 2527–2530.

[21] Silverman, B., *Density Estimation for Statistics and Data Analysis*, New York: Chapman and Hall, 1986.

[22] Martinez, W., *Computational Statistics Handbook with MatLab*, New York: Chapman and Hall/CRC, 2002.

[23] Dougherty, E.R., and R. A. Lotufo, *Hands-On Morphological Image Processing (Tutorial Texts in Optical Engineering)*, Vol. TT59, Bellingham, WA: SPIE Press, 2003.

[24] Solka, J. L., C. E. Priebe, and G. W. Rogers, "An Initial Assessment of Discriminant Surface Complexity for Power Law Features," *Simulation*, Vol. 58, 1992, pp. 311–318.

CHAPTER 9

Vessel Detection Experiments Using a Gaussian Matched Filter

M. Al-Rawi, H. Karajeh, and A. Abu-Dalhoum

Automated inspection of retinal images has become a nontrivial task that helps oph-thalmologists in their work. Pathological changes in the retinal blood vessels can lead to many forms of blindness, in particular, diabetic retinopathy. These patho-logical changes may modify the color and shape of retinal vessels, such as narrowing or widening of some vessels, the occurrence of new vessels, or the occurrence of irregular edges. To estimate retinal changes, a reliable method of vessel detection that takes appropriate measurements of the retinal blood vessels is needed. How-ever, retinal vessels are usually thin with low local contrast and they almost never have ideal step edges.

Due to the complicated structure of retinal vessels, their detection via common edge detection techniques may not give acceptable results. In fact, a special-purpose edge detection technique should be adopted to fulfill the task. Figure 9.1 shows a retinal image, Figure 9.2(a) shows the green band of the retinal image shown in Fig-ure 9.1, and a hand-labeled version of this image is shown in Figure 9.2(b). Figure 9.3 shows another retinal image. The DRIVE data set [1] has been used in most of the work presented in this chapter. The DRIVE data set is a retinal image data set such that each retina image has a corresponding hand-labeled image showing the blood vessels. Hand-labeled images are done by an experienced ophthalmologist. Information about the Hoover data set can be found in [2].

9.1 Spatial Filters for the Detection of Blood Vessels

As shown in Figure 9.4, blood vessels have some form of Gaussian shape profile. Designing a filter to detect the peaks of those vessels may be implemented by con-catenating many Gaussian shape functions. Let the vessel profile be approximated by the function $g_A(x)$; then the concatenation of the many functions can be expressed by the following formula:

$$f_{A,\theta}(x,y) = g_A(u) \quad \forall p_\theta \in N \tag{9.1}$$

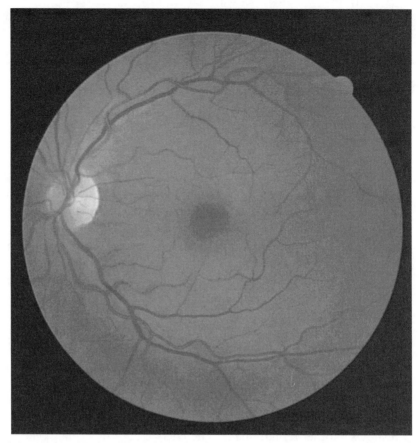

Figure 9.1   A color retinal image of the DRIVE data set, which is a publicly available data set [1].

where A is a set containing some filter parameters, θ indicates the orientation of the filters kernel, u is calculated from (9.2), and p_θ, which is a point in the neighborhood N generated by rotating each (x, y) by an angle θ using the rotation matrix, is given by

$$p_\theta = [u\ v] = [x\ y]\begin{bmatrix} \cos\theta & \sin\theta \\ -\sin\theta & \cos\theta \end{bmatrix} \tag{9.2}$$

where $N = \{(u, v) : |u| \leq T, |v| \leq L\}$, $2L + 1$ is the length of the vessel segment, and $2T + 1$ is width of the profile segment. The filter given in (9.1) is then normalized to have zero mean. The function $g_A(x)$ may have these forms:

1. *Gaussian matched filter.* In this case, the filter function may have a Gaussian profile function given by

$$g_\sigma(u) = \exp(-u^2 / 2\sigma^2) \tag{9.3}$$

2. *Second-order Gaussian matched filter.* In this case, the filter function may have the following form:

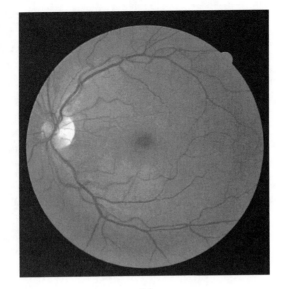

(a)

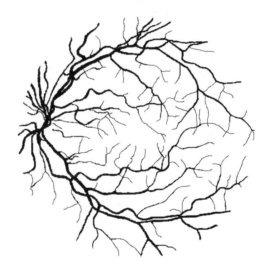

(b)

Figure 9.2   (a) The green band of the retinal image displayed in gray-level mode and (b) a hand-labeled version of the image shown in part (a).

$$g_\sigma(u) = \frac{1}{\sqrt{2\pi}\sigma}\left(u^2 - \sigma^2\right)\ \exp\!\left(-u^2/2\sigma^2\right) \tag{9.4}$$

In this work, all the experiments are performed using the Gaussian matched filter.

An example for some kernels produced from Filter 1 at different rotations is shown in Figure 9.5. These kernels in the convolution process have real values, but for display purposes, the weighting coefficients in the kernels are multiplied by a scale factor of 10 and rounded to their nearest integer. To illustrate how these ker-

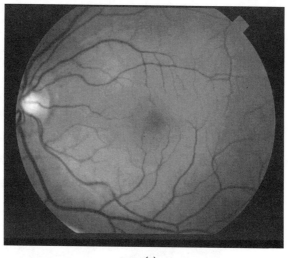

(a)

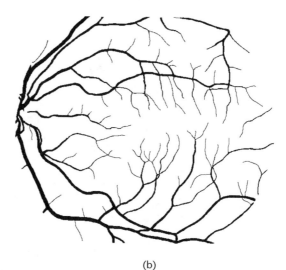

(b)

Figure 9.3   (a) An image from the Hoover data set and (b) a hand-labeled version of the retinal image shown in part (a).

nels have a Gaussian shaped curve, a three-dimensional graph for the kernels of Filter 1 is drawn in Figure 9.6.

9.2   Estimation of Filter Parameters

The estimation of filter parameters for vessel detection is a very sensitive process. The filter to be used in vessel detection contains some parameters that govern its response to the vessels of the image. Each time the filter is used, a measure is needed to tell whether the filter's performance is acceptable or not. This is done as described next.

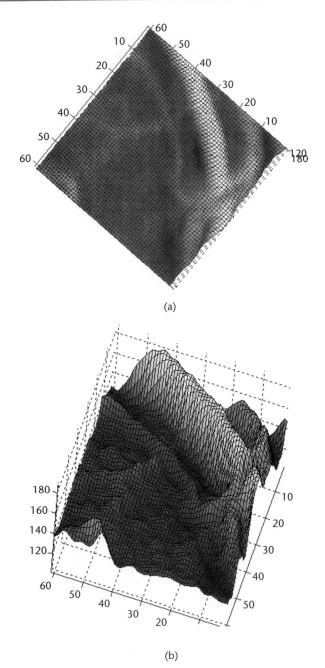

(a)

(b)

Figure 9.4 (a) A magnified vessel segment extracted from the green band of a retina image. (b) A three-dimensional view of the vessel segment shown in part (a). The magnified vessel is displayed as a negative image.

9.2.1 True Versus False Vessels

An algorithm for checking that the detected vessels are correct (true) or not correct (false) is needed. After that, the true vessels and false vessels are used to estimate filter performance by calculating the area under the receiver operating curve (ROC). The area under the ROC is obtained as follows:

0	0	0	0	0	0	0	0	0	0	0	0	0	0	0	0	0
0	0	0	0	0	0	0	0	0	0	0	0	0	0	0	0	0
0	0	0	0	0	0	0	0	0	0	0	0	0	0	0	0	0
0	0	1	1	1	1	1	1	1	1	1	1	1	1	1	0	0
0	0	1	1	1	1	1	1	1	1	1	1	1	1	1	0	0
0	0	1	1	1	1	1	1	1	1	1	1	1	1	1	0	0
0	0	1	1	1	1	1	1	1	1	1	1	1	1	1	0	0
0	0	0	0	0	0	0	0	0	0	0	0	0	0	0	0	0
0	0	-9	-9	-9	-9	-9	-9	-9	-9	-9	-9	-9	-9	-9	0	0
0	0	0	0	0	0	0	0	0	0	0	0	0	0	0	0	0
0	0	1	1	1	1	1	1	1	1	1	1	1	1	1	0	0
0	0	1	1	1	1	1	1	1	1	1	1	1	1	1	0	0
0	0	1	1	1	1	1	1	1	1	1	1	1	1	1	0	0
0	0	1	1	1	1	1	1	1	1	1	1	1	1	1	0	0
0	0	0	0	0	0	0	0	0	0	0	0	0	0	0	0	0
0	0	0	0	0	0	0	0	0	0	0	0	0	0	0	0	0
0	0	0	0	0	0	0	0	0	0	0	0	0	0	0	0	0

(a)

0	0	0	0	0	0	0	0	0	1	0	0	0	0	0	0	0
0	0	0	0	0	0	0	1	1	1	0	0	0	0	0	0	0
0	0	0	0	0	-6	1	1	1	1	1	0	0	0	0	0	0
0	0	0	1	0	-8	-2	1	1	1	1	1	0	0	0	0	0
0	0	1	1	1	-4	-7	1	1	1	1	1	0	0	0	0	0
0	1	1	1	1	0	-8	-3	1	1	1	1	1	0	0	0	0
0	1	1	1	1	1	-2	-8	0	1	1	1	1	0	0	0	0
0	0	1	1	1	1	1	-6	-5	1	1	1	1	0	0	0	0
0	0	1	1	1	1	1	-1	-9	-1	1	1	1	1	0	0	0
0	0	0	1	1	1	1	1	-5	-6	1	1	1	1	0	0	0
0	0	0	0	1	1	1	1	0	-8	-2	1	1	1	1	1	0
0	0	0	0	1	1	1	1	1	-3	-8	0	1	1	1	1	0
0	0	0	0	0	1	1	1	1	1	-7	-4	1	1	1	0	0
0	0	0	0	0	1	1	1	1	1	-2	-8	0	1	0	0	0
0	0	0	0	0	0	1	1	1	1	1	-6	0	0	0	0	0
0	0	0	0	0	0	0	1	1	1	0	0	0	0	0	0	0
0	0	0	0	0	0	0	1	0	0	0	0	0	0	0	0	0

(b)

Figure 9.5   Two kernels are produced from Filter 1 at (a) 0° and (b) 60°.

1. Let the input image be f. Applying a filter to f yields the filtered output image f_{LoT}, which is a continuous image. By thresholding f_{LoT} via different threshold values from 0 to 1 in a step of 0.05 (assuming the use of normalized intensity and that the number of thresholds is 20), we obtain several binary images. Each image corresponds to a certain threshold.

2. Then, we calculate the true_ratio and the false_ratio for each binary image by comparing it with a corresponding hand-labeled retina image denoted h. The hand-labeled image is obtained from the retinal image by an experienced observer to be used for computer comparison purposes. The comparison yields true pixels (pixels detected as vessels yet they appear as vessels in the hand-labeled image) and false pixels (pixels detected as vessels yet they appear as nonvessels in the hand-labeled image). The true_ratio is obtained

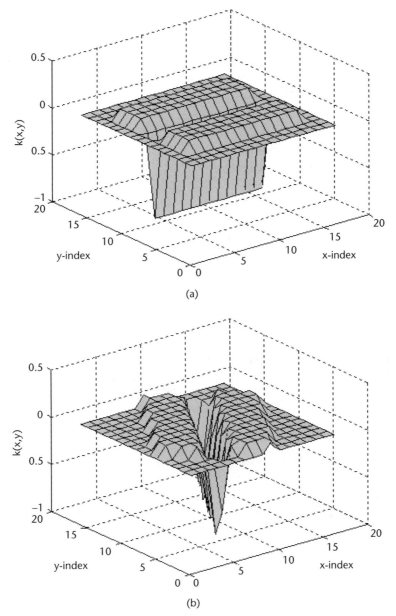

Figure 9.6   (a) A three-dimensional view of the Gaussian matched filter at 0°. (b) Another rotated three-dimensional view of the Gaussian matched filter, at 60° of rotation.

by dividing the true pixels by the number of vessel pixels in h, and the false_ratio is obtained by dividing the false pixels by the number of nonvessel pixels in h.

3. The plot that shows the variation of the false_ratio versus the true_ratio gives the ROC. Then, the area under the ROC is calculated using the trapezoidal method.

Another important measure known in the literature is called the maximum accuracy (MA). The MA is calculated as follows:

1. The image f_{LoT} is thresholded at different levels of threshold between 0 and 1 where the level is a normalized intensity value that lies in the range [0, 1].
2. At each threshold the accuracy value is calculated. The accuracy for one image is calculated by taking the sum for the total number of the pixels correctly classified as vessels and nonvessels at a certain threshold. Then the sum of the pixels is divided by the field of view (FOV), which is the circular area in the retina image.
3. After calculating the set of accuracies at different thresholds, the maximum that gives the MA between all accuracy values is found. The threshold that gives the MA is considered the best threshold. The maximum accuracy average (MAA) is the average over MA values for all images used in the measurements.

9.2.2 Some Filter Parameters

As shown previously, each matched filter is characterized by three parameters: L, σ, and T. The parameters shown in Table 9.1 demonstrate various filters that are rotated in 15° increments. The method used to obtain those parameters is demonstrated in [3]. All experiments for this work were performed on the DRIVE data set [1].

For Image 1, the true vessels and false vessels that Filter 1 yields are shown in Table 9.2.

9.3 Best Rotation for Filter 1

To find the best rotation to use for the filters, a simple experiment was performed using Filter 6 (as a case study), and the ROC area is calculated each time with some rotation. The results are shown in Table 9.3. In this table, the higher the ROC area,

Table 9.1 Parameters That Can Be Used as Filters

Filter Name	T	σ	L
Filter 1	5.2275	0.4942	13.6947
Filter 2	4.2361	1.1575	13.6811
Filter 3	10.5247	1.0485	13.7268
Filter 4	4.5736	1.2105	12.7252
Filter 5	4.6490	1.0880	12.5601
Filter 6	6.2866	0.5745	13.4086
Filter 7	5.7720	0.6119	8.3613
Filter 8	5.8148	0.5702	6.6902
Filter 9	5.9020	0.6374	6.6959
OGMF (Al-Rawi et al. [4])	8	1.9	10.8
GMF (Chaudhuri et al. [5])	6	2	9

Table 9.2 True and False Ratios for Image 1 Using Filter 1

Threshold Levels	True Ratio	False Ratio
0.00	1.0000	1.0000
0.05	0.9940	0.2849
0.10	0.9512	0.0854
0.15	0.8691	0.0396
0.20	0.7775	0.0218
0.25	0.6649	0.0118
0.30	0.5460	0.0066
0.35	0.4418	0.0039
0.40	0.3586	0.0026
0.45	0.2836	0.0018
0.50	0.2161	0.0013
0.55	0.1628	0.0010
0.60	0.1243	0.0009
0.65	0.0965	0.0008
0.70	0.0783	0.0007
0.75	0.0639	0.0006
0.80	0.0509	0.0005
0.85	0.0384	0.0005
0.90	0.0290	0.0004
0.95	0.0220	0.0004

the better the filter. Rotating the filter with 60° gives an average ROC area of 95%, which is acceptable.

9.4 Band Selection of the Retinal Image

Because each retina image is a color red, green, blue (RGB) image, we can choose to work on one of the bands. Let us find the band that gives the highest response to the filter. Performing filtering an each band and calculating the ROC area at each band may provide an answer. The higher the ROC area, the better the filter must be that is used in the vessel and the better the response of the band. The result is shown in Table. 9.4.

9.5 Best Threshold for Segmenting the Vessel Image

It is very important in real-world implementations to know how to automatically find the highest accuracy of the threshold image since no hand-labeled version exists. To find the best threshold that can be used to segment the matched filter output image, a good thresholding procedure is needed. A simple method is performed by calculating the accuracy at each threshold, and the threshold that gives the maxi-

Table 9.3   Area Under the ROC for the 20 Images in the Test Set of the Drive Database Using Filter 6 and Different Degrees of Rotation

Image Number	$\theta = 60°$	$\theta = 30°$	$\theta = 15°$	$\theta = 7.5°$	$\theta = 5°$	$\theta = 2.5°$
1	0.9716	0.9778	0.9780	0.9781	0.9782	0.9783
2	0.9686	0.9749	0.9751	0.9751	0.9753	0.9753
3	0.9540	0.9628	0.9633	0.9632	0.9637	0.9637
4	0.9481	0.9543	0.9546	0.9546	0.9549	0.9549
5	0.9473	0.9525	0.9526	0.9524	0.9528	0.9528
6	0.9410	0.9495	0.9498	0.9497	0.9501	0.9502
7	0.9430	0.9505	0.9509	0.9509	0.9512	0.9512
8	0.9431	0.9506	0.9511	0.9510	0.9518	0.9517
9	0.9441	0.9514	0.9516	0.9515	0.9519	0.9519
10	0.9519	0.9586	0.9589	0.9588	0.9592	0.9592
11	0.9379	0.9456	0.9460	0.9461	0.9467	0.9467
12	0.950c3	0.9570	0.9574	0.9577	0.9579	0.9579
13	0.9463	0.9545	0.9547	0.9549	0.9550	0.9552
14	0.9639	0.9709	0.9709	0.9709	0.9713	0.9713
15	0.9591	0.9646	0.9650	0.9650	0.9654	0.9654
16	0.9510	0.9583	0.9584	0.9583	0.9587	0.9587
17	0.9485	0.9556	0.9555	0.9552	0.9561	0.9560
18	0.9521	0.9575	0.9579	0.9578	0.9582	0.9582
19	0.9628	0.9677	0.9678	0.9677	0.9683	0.9683
20	0.9578	0.9631	0.9634	0.9635	0.9638	0.9639
Mean	0.9521	0.9589	0.9591	0.9591	0.9595	0.9595
Standard deviation	0.0093	0.0088	0.0087	0.0087	0.0087	0.0087

Table 9.4   Average Area Under the ROC for the 20 Retinal Images of the DRIVE Data Set Under Different Bands Using Filter 6

Band	Average ROC Area
Green	0.9591
Blue	0.9227
Red	0.9071

mum accuracy is determined and is considered the best threshold. For all images, the average of the best thresholds that gives the MA is calculated. This *best average threshold value* can then be used to automatically segment any continuous image produced by the matched filter. Using the matched filter that has the parameters (L, σ, T) = (13.4086, 0.5745, 6.2866), the best average threshold values are calculated for the 20 images of the DRIVE test set. Table 9.5 gives the MA and the best average

Table 9.5 Determining the Average Threshold of the Matched Filter with Filter 6 for the DRIVE Test Set

Image Number	Maximum Accuracy	Threshold at Maximum Accuracy	Accuracy at Average Threshold	Error Rate Between the MA and Accuracy Calculated at the Average Threshold
1	0.9548	0.3000	0.9549	0.0001
2	0.9483	0.3500	0.9480	0.0003
3	0.9406	0.2500	0.9373	0.0033
4	0.9394	0.3500	0.9394	0.0000
5	0.9404	0.2500	0.9370	0.0034
6	0.9353	0.2500	0.9323	0.0030
7	0.9344	0.3500	0.9336	0.0008
8	0.9341	0.3000	0.9341	0.0000
9	0.9408	0.2500	0.9386	0.0022
10	0.9447	0.3000	0.9444	0.0003
11	0.9359	0.4000	0.9309	0.0050
12	0.9412	0.3000	0.9412	0.0000
13	0.9360	0.3000	0.9355	0.0005
14	0.9501	0.3500	0.9490	0.0011
15	0.9497	0.4500	0.9435	0.0062
16	0.9420	0.3000	0.9417	0.0003
17	0.9437	0.3000	0.9439	0.0002
18	0.9457	0.3500	0.9458	0.0001
19	0.9573	0.3000	0.9575	0.0002
20	0.9461	0.3000	0.9455	0.0006
Mean	0.9430	0.3150	0.9417	0.0013
Standard deviation	0.0066	0.0516	0.0072	0.0018

thresholds found. Also for the 20 images of the DRIVE test set, Table 9.6 shows the mean and the standard deviation (STD) of the MA, best thresholds, accuracy at the best average threshold, and the error rate. As can be seen in Table 9.6, Filter 1 might be chosen as the best filter because it gives the highest average ROC area and MAA among other matched filters, though some might choose Filter 6.

9.6 Summary

Evaluations performed in this work show that the filter with parameters $(L, \sigma, T) = (13.6947, 0.4942, 5.2275)$ with kernels rotated by $15°$ gives the highest area under the ROC, and the execution time needed to filter one retinal image is less than the to other filters. Experiments under different degrees of rotation showed that rotating the matched filters by $15°$ and $30°$ is adequate to achieve good results. All filters could not detect small vessels to an accurate degree. Performing smoothing with an average filter on each retina image results in lowering the filter's performance; that is, lower ROC and lower MAA compared to nonfiltered

Table 9.6 Difference in MAA Between Thresholding at the Best Threshold and Thresholding at the Average Threshold

Filter Number	Maximum Accuracy	Average Threshold at Maximum Accuracy	Accuracy at Average Threshold	Error Rate Between the MA and Accuracy Calculated at the Average Threshold
Mean (Filter1)	0.9401	0.3675	0.9389	0.0013
STD (Filter1)	0.0067	0.0613	0.0072	0.0005
Mean (Filter2)	0.9427	0.2575	0.9416	0.0012
STD (Filter2)	0.0068	0.0438	0.0074	0.0005
Mean (Filter3)	0.9403	0.3400	0.9391	0.0011
STD (Filter3)	0.0068	0.0576	0.0072	0.0004
Mean (Filter4)	0.9405	0.3650	0.9392	0.0013
STD (Filter4)	0.0068	0.0587	0.0072	0.0004
Mean (Filter5)	0.9411	0.3500	0.9398	0.0012
STD (Filter5)	0.0068	0.0487	0.0072	0.0005
Mean (Filter6)	0.9430	0.3150	0.9417	0.0013
STD (Filter6)	0.0066	0.0516	0.0072	0.0006
Mean (Filter7)	0.9425	0.2175	0.9416	0.0009
STD (Filter7)	0.0075	0.0373	0.0077	0.0002
Mean (Filter8)	0.9421	0.1650	0.9413	0.0008
STD (Filter8)	0.0078	0.0235	0.0080	0.0001
Mean (Filter9)	0.9425	0.1700	0.9417	0.0009
STD (Filter9)	0.0076	0.0299	0.0079	0.0003

retina. One of the astonishing results is that the value of σ, which determines the spread of the Gaussian of the matched filter, is less than 1 (sometimes near 0.5) in most of the performed experiments. This contradicts the $\sigma = 2$ value that originally appeared in [5].

References

[1] Staal, J., et al., "Ridge Based Vessel Segmentation in Color Images of the Retina," *IEEE Trans. on Medical Imaging*, Vol. 23, No. 4, 2004, pp. 501–509.

[2] Hoover, A., V. Kouzntesova, and M. Goldbaum, "Locating Blood Vessels in Retinal Images by Piecewise Threshold Probing of a Matched Filter Response," *IEEE Trans. on Medical Imaging*, Vol.19, No.3, 2000, pp. 203–210.

[3] Karajeh, H., "A Genetic Algorithm Matched Filtered Optimization for Blood Vessels Detection of Digital Retinal Images," Master's Thesis, University of Jordan, 2006.

[4] Al-Rawi, M. S., M. A. Qutaishat, and M. R. Arrar, "An Improved Matched Filter for Blood Vessel Detection of Digital Retinal Images," *Computers Biol. Med.*, Vol. 37, 2007, pp. 262–267.

[5] Chaudhuri, S., et al., "Detection of Blood Vessels in Retinal Images Using Two Dimensional Matched Filters," *IEEE Trans. on Medical Imaging*, Vol. 8, No. 3, 1989, pp. 263–269.

CHAPTER 10

Detection of Retinal Blood Vessels Using Gabor Filters[1]

Rangaraj M. Rangayyan, Fábio J. Ayres, Faraz Oloumi, Foad Oloumi, and Peyman Eshghzadeh-Zanjani

10.1 Introduction

10.1.1 Retinopathy

The structure of the blood vessels in the retina is affected by diabetes, hypertension, arteriosclerosis, and retinopathy of prematurity through modifications in shape, width, and tortuosity [1–6]. Quantitative analysis of the architecture of the vasculature of the retina and changes in the features just mentioned could assist in monitoring disease processes, as well as in evaluating their effects on the visual system. Additionally, images of the retina can reveal pathological features related to retinopathy, such as microaneurysms, hemorrhages, exudates, macular edema, venous beading, and neovascularization [1]. Automated detection and quantitative analysis of retinal features could assist in analyzing the related pathological processes.

Although high sensitivity in the range of 83% to 97% has been achieved in the automated detection of all types of retinopathy using image processing techniques, the corresponding specificity has been low, in the range of 52.8% to 88.9% [1]. In the detection of hemorrhages and microaneurysms, the ranges of sensitivity and specificity have been 73.8% to 82% and 73.8% to 94%, respectively. The sensitivity and specificity reported in the detection of exudates are 88.5% to 96% and 93% to 99.7%, respectively [1]. The performance statistics of the reported methods indicate the need for advanced image processing and pattern analysis methods in order to perform more accurate detection, as well as more precise quantitative analysis of the diagnostic features mentioned earlier.

1. This chapter is an expanded and revised version of the following articles. (1) Rangayyan, R. M., et al., "Detection of Blood Vessels in the Retina with Multiscale Gabor Filters," *J. of Electronic Imaging*, in press, 2008. Reprinted with permission. (2) Rangayyan, R. M., et al., "Detection of Blood Vessels in the Retina Using Gabor Filters," *Proc. 20th Canadian Conf. on Electrical and Computer Engineering (CCECE 2007)*, Vancouver, BC, Canada, April 22–26, 2007, pp. 717–720. Reprinted with permission. (3) Oloumi, F., et al., "Detection of Blood Vessels in Fundus Images of the Retina Using Gabor Filters," *Proc. 29th Annual Intl. Conf. of the IEEE Engineering in Medicine and Biology Society*, Lyon, France, August 22–26, 2007, pp. 6451–6454. Reprinted with permission.

10.1.2 Detection of Retinal Blood Vessels

In many applications of image processing in ophthalmology, the most important step is to detect the blood vessels in the retina [1, 5, 7–14]. We present here a brief review of some of the previously proposed methods and algorithms for the detection of blood vessels in the retina.

- *Matched filters:* Chaudhuri et al. [7] proposed an algorithm based on two-dimensional matched filters, and three assumptions: Vessels can be approximated by piecewise linear segments, the intensity profile of a vessel can be approximated by a Gaussian curve, and the width of vessels is constant. Detection is performed by convolving the given image with the matched filter rotated in several directions, with the maximum response recorded for each pixel.

- *Adaptive local thresholding:* The method of adaptive local thresholding using a verification-based multithreshold probing scheme was used by Jiang and Mojon [15]. In this method, a binary image obtained after applying a threshold is used in a classification procedure to accept or reject any region in the image as a certain object. A series of different thresholds is applied, and the final detection result is a combination of the results provided by the individual thresholds.

- *Ridge-based vessel segmentation:* The assumption that vessels are elongated structures is the basis for the supervised method of ridge-based vessel detection and segmentation, which was introduced by Staal et al. [11]. The image ridges, which roughly coincide with the vessel center lines, are extracted by this algorithm. Then, image primitives are obtained by grouping image ridges into sets that model straight-line elements. Such sets are used to partition the image by assigning each pixel to the closest primitive set. In each partition, a local coordinate is defined by the corresponding line element. Finally, feature vectors are computed for every pixel using the characteristics of the partitions and their line elements, and classified using sequential forward feature selection and a k-nearest-neighbor classifier.

- *Piecewise threshold probing of a matched-filter response:* This method, proposed by Hoover et al. [10], uses local vessel attributes as well as global and region-based attributes of the vascular network structure for the detection and classification of vessels. Different areas and regions in a matched-filter response are probed at several decreasing thresholds. At each level, the region-based attributes are used to determine whether the probing should be continued, and to classify the probed area as a blood vessel.

- *Vessel segmentation using two-dimensional Gabor filters and supervised classification:* An algorithm applying Gabor filters for feature detection and supervised classification of blood vessels was proposed by Soares et al. [16]. A feature vector containing the measurements at several scales obtained from two-dimensional Gabor filters is assigned to each pixel. In the next step, using a Bayesian classifier with class-conditional probability density functions given by a Gaussian mixture model, each pixel is classified as a vessel or nonvessel pixel.

Other methods reported for the detection of blood vessels in the retina include the use of twin snakes or active contour models [17], amplitude-modified second-order Gaussian filters [18], vessel models and the Hough transform [19], the Gabor variance filter with a modified histogram equalization technique [20], mathematical morphology and curvature evaluation [21], and tram-line filtering [22]. Several techniques have been proposed to model and analyze the structure of retinal vasculature, including fractals [23, 24], geometrical models and analysis of topological properties [8, 9, 25], and grading of venous beading [26].

Methods to address the issue of retinal blood vessel detection may take advantage of the fact that blood vessels are elongated, piecewise-linear or curvilinear structures with a preferred orientation. However, most of the directional, fan, and sector filters that have been applied to extract directional elements are not analytic functions; such filters tend to possess poor spectral response and yield images with not only the desired directional elements but also artifacts [27]. One of the fundamental problems with Fourier methods of directional filtering is the difficulty in resolving directional content at the dc point (the origin in the Fourier domain). The design of high-quality fan filters [28] requires conflicting constraints at the dc point. The Gabor function provides a solution to the problem [27, 29–35].

10.2   Methods

10.2.1   Gabor Filters for the Detection of Oriented Patterns

Gabor filters are sinusoidally modulated Gaussian functions that have optimal localization in both the frequency and space domains; a significant amount of research has been conducted on using Gabor functions for texture segmentation, analysis, and discrimination [36]. Gabor functions have been found to provide good models for the receptive fields of simple cells in the striate cortex [37–39]. Gabor functions provide optimal joint resolution in both the Fourier and time (or space) domains, and form a complete basis set through phase shift and scaling or dilation of the original basis function (mother wavelet).

We propose image processing techniques to detect blood vessels in images of the retina based on Gabor filters [34, 35, 40]. The real Gabor filter kernel (or mother wavelet) oriented at the angle $\theta = -\pi/2$ may be formulated as [32, 33]

$$g(x,y) = \frac{1}{2\pi\sigma_x\sigma_y}\exp\left[-\frac{1}{2}\left(\frac{x^2}{\sigma_x^2}+\frac{y^2}{\sigma_y^2}\right)\right]\cos(2\pi f_o x) \tag{10.1}$$

where σ_x and σ_y are the standard deviation values in the x- and y-directions, and f_o is the frequency of the modulating sinusoid. Kernels at other angles are obtained by rotating the mother wavelet. We use a set of 180 kernels, with angles spaced evenly over the range $\theta = [-\pi/2, \pi/2]$.

Gabor filters can be used as line detectors [32, 41]. The parameters in (10.1), namely, σ_x, σ_y, and f_o, need to be derived by taking into account the size of the lines or curvilinear structures to be detected. Let τ be the thickness of the line detector. This parameter is related, in our design, to σ_x and f_o as follows: The amplitude of the exponential (Gaussian) term in (10.1) is reduced to one-half of its maximum at $x =$ $\tau/2$ and $y = 0$; therefore, $\sigma_x = \tau(2\sqrt{2\ln 2})$. The cosine term has a period of τ; hence, f_o $=1/\tau$. The value of σ_y could be defined as $\sigma_y = l\sigma_x$, where l determines the elongation of the Gabor filter in the orientation direction, with respect to its thickness. The value of τ is varied to prepare a bank of filters at different scales for multiresolution filtering and analysis. The Gabor filter designed as above can detect linear features of positive contrast, that is, linear elements that are brighter than their immediate background.

Figure 10.1 shows Gabor filters for various values of the parameters τ, l, and θ, demonstrating the effects of scaling, stretching, and rotation; the corresponding frequency-domain magnitude transfer functions are shown in Figure 10.2. It is evident that a Gabor filter acts as a bandpass filter, with a limited range of response that is dependent on the parameters and orientation of the filter. Increasing the scale factor τ causes the filter to shift to a lower frequency band. Reducing the elongation factor l causes the filter to be less directionally sensitive. Rotating a filter causes a corresponding rotation of the frequency response. Note that there is a $\pi/2$ shift between angles in the space domain (Figure 10.1) and the frequency domain (Figure 10.2).

10.2.2   Procedure for the Detection of Retinal Blood Vessels

Blood vessels in the retina vary in thickness in the range from 50 to 200 μm, with a median of 60 μm [1, 3]. In a comparative analysis of the performance of the Gabor filter and other line detectors [32, 41], the capture range of a given Gabor filter, in terms of detecting lines with an efficiency of more than 90%, in the presence of noise with the normalized standard deviation of 0.2, was determined to be about 0.4–3.2 τ. Although this result implies the adequacy of a single Gabor filter to detect blood vessels in the range previously mentioned, it might be beneficial to use Gabor filters at a few scales for multiresolution analysis, which could lead to improved efficiency of detection of thick and thin blood vessels.

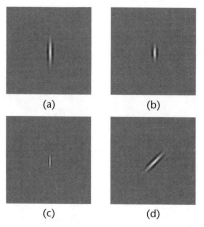

(a)   (b)

(c)   (d)

Figure 10.1   Gabor filters for various values of the parameters τ, l, and θ, demonstrating the effects of scaling, stretching, and rotation. (a) $\tau = 8$, $l = 2.9$, $\theta = -90°$. (b) $\tau = 8$, $l = 1.7$, $\theta = -90°$. (c) $\tau = 4$, $l = 2.9$, $\theta = -90°$. (d) $\tau = 8$, $l = 2.9$, $\theta = 45°$. Each filter was created using a matrix of size 512×512 pixels; however, only the central portion of size 128×128 pixels is shown for each case. See Figure 10.2 for the corresponding frequency-domain magnitude transfer functions.

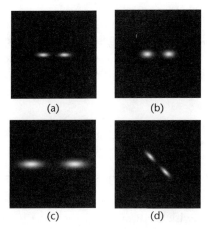

Figure 10.2 Frequency-domain magnitude transfer functions of the Gabor filters shown in Figure 10.1 for various values of the parameters τ, l, and θ, demonstrating the effects of scaling, stretching, and rotation. Each function was created using a matrix of size 512×512 pixels. The dc or $(0,0)$ frequency component is at the center of the frequency domain in each case. (a) $\tau = 8$, $l = 2.9$, $\theta = -90°$. (b) $\tau = 8$, $l = 1.7$, $\theta = -90°$. (c) $\tau = 4$, $l = 2.9$, $\theta = -90°$. (d) $\tau = 8$, $l = 2.9$, $\theta = 45°$.

Figure 10.3(a) shows an image (number 05) from the DRIVE database [11, 42]. The result in Figure 10.3(b) was obtained using a bank of 180 Gabor filters with $\tau = 4$ pixels and $l = 2.9$; that in Figure 10.3(c) was obtained with $\tau = 8$ pixels and $l = 2.9$. The magnitude response image was composed by selecting the maximum response over all of the Gabor filters for each pixel. The result in Figure 10.3(b) indicates that the filters have detected only the edges of the thick vessels, with poor response along their center lines. In contrast, the resulting Figure 10.3(c) shows that, although the thick vessels have been detected well, some of the thinner vessels have not been detected. The results indicate the need for multiscale or multiresolution filtering and analysis, which is easily facilitated by the proposed design of the Gabor filter.

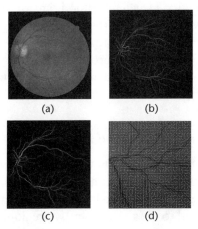

Figure 10.3 (a) Image 05 of the DRIVE database (565×584 pixels). Magnitude response of 180 Gabor filters over $[-\pi/2, \pi/2]$ with: (b) $\tau = 4$ pixels, $l = 2.9$; and (c) $\tau = 8$ pixels, $l = 2.9$. (d) Phase angles related to part (c), shown in the form of needles for every fifth pixel for a part of the image.

The proposed methods also provide the orientation at each pixel, obtained as the angle of the filter with the largest magnitude response. Figure 10.3(d) shows the angle data, in the form of needles for every fifth pixel, for a part of the image in Figure 10.3(a). The angle data exhibit a high level of agreement with the local orientation of the blood vessels.

10.3 Experiments and Results

10.3.1 Data Set of Retinal Images and Preprocessing

The proposed methods were tested with fundus images of the retina from the DRIVE database [11, 42], which contains 40 images (20 for training and 20 for testing). Ground-truth images of blood vessels marked by three observers, and a set of mask images identifying the effective region of each image are also available. The mask images were not used in our work because we have included a procedure to identify the boundary of the effective region for each image.

After converting each pixel in a given image to a vector of color components and normalizing each component (dividing by 255), the result was converted to the luminance component Y, computed as $Y = 0.299R + 0.587G + 0.114B$, where R, G, and B are the red, green, and blue components, respectively, of the color image. Several other works on the detection of retinal vessels have used the green channel only; in the present work, the inverted Y channel was used in order to reduce noise by averaging the three color component images and to obtain an image with positive contrast for the vessels. The effective region of the image was thresholded using the normalized threshold of 0.1. The artifacts present in the DRIVE images at the edges were removed by applying morphological erosion [43] with a disk-shaped structuring element having a diameter of 10 pixels.

To avoid edge artifacts in the results of Gabor filtering, each image was extended beyond the limits of its effective region as follows [16, 34, 35]. First, a four-pixel neighborhood was used to identify the pixels at the outer edge of the effective region. For each of the pixels identified, the mean gray level was computed over all pixels in a 21×21 neighborhood that were also within the effective region. The mean value was assigned to the corresponding pixel location in the gray-scale image. The effective region was merged with the outer edge pixels, forming an extended effective region. The procedure was repeated 50 times, extending the image by a ribbon of pixels having a width of 50 pixels. This procedure for the prevention of edge artifacts has also been used by Soares et al. [16].

10.3.2 Single-Scale Filtering and Analysis

The parameters of the Gabor filters were varied over the range [1, 16] for τ and [1.3, 18.5] for l. For each set of the parameters $\{\tau, l\}$ used, the highest output of the Gabor filters over 180 angles at each pixel was obtained, and a magnitude image was constructed and cropped to obtain the effective region. A sliding threshold was applied to the magnitude images and the result for each threshold was compared with the corresponding ground-truth image to determine the true-positive fraction (TPF) and the false positive-fraction (FPF). Only the effective region of each image was used for

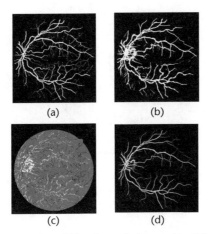

Figure 10.4 (a) Manual segmentation of blood vessels (ground truth) for the image in Figure 10.3(a). (b) Result of thresholding the magnitude response in Figure 10.3(b) with threshold = 0.0023; range of response [0, 0.046]; TPF = 0.88,FPF = 0.11. (c) False-positive (white) and false-negative (black) pixels in the result of part (b) as compared to the image in part (a). (d) Discriminant values produced by the MLP used in multiscale analysis, with τ = {4, 8, 12} and l = 2.9.

the computation of TPF and FPF. Figure 10.4 shows the results of thresholding for a sample image.

The results obtained were combined over the 20 images in the training set, TPF was plotted against the FPF to obtain the receiver operating characteristics (ROC) curve [44], and the area under the ROC curve (A_z) was measured. Table 10.1 presents the A_z values for some of the combinations of the parameters for the 20 images in the training set. The highest blood vessel detection rate of $A_z = 0.94$ was achieved with $\tau = 8$ and $l = 2.9$ (and a few other sets of the parameters) for the training set of 20 images; the corresponding ROC curve is shown in Figure 10.5. Using the same parameters, a detection efficiency of $A_z = 0.95$ was obtained with the 20 images in the test set; the corresponding ROC curve is shown in Figure 10.6.

Table 10.1 Blood Vessel Detection Rate (A_z) for the Training Set (20 Images) of the DRIVE Database [42] for Selected Values of τ and l

Parameters	$l = 1.7$	2.1	2.5	2.9	3.3	3.7
$\tau = 1$	0.62	0.65	0.67	0.70	0.73	0.75
$\tau = 2$	0.69	0.72	0.75	0.77	0.79	0.81
$\tau = 4$	0.85	0.87	0.89	0.90	0.91	0.92
$\tau = 6$	0.91	0.92	0.93	0.94	0.94	0.94
$\tau = 7$	0.92	0.93	0.94	0.94	0.94	0.94
$\tau = 8$	0.93	0.94	0.94	0.94	0.94	0.94
$\tau = 9$	0.94	0.94	0.94	0.94	0.94	0.93
$\tau = 10$	0.94	0.94	0.94	0.93	0.93	0.92
$\tau = 12$	0.93	0.93	0.93	0.92	0.91	0.91

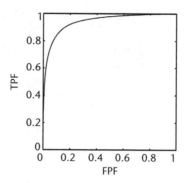

Figure 10.5 ROC curve for the training set of 20 images from the DRIVE database, with $\tau = 8$ and $l = 2.9$. $A_z = 0.94$.

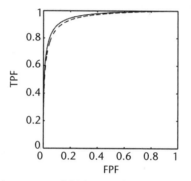

Figure 10.6 ROC curves for the test set of 20 images from the DRIVE database. Solid line: multiscale analysis with $\tau = \{4, 8, 12\}$, $l = 2.9$, and an MLP; $A_z = 0.96$. Dashed line: single-scale analysis with $\tau = 8$, $l = 2.9$; $A_z = 0.95$.

10.3.3 Multiscale Filtering and Analysis

Experiments were conducted with several different methods to combine the magnitude images resulting from Gabor filtering for several combinations of scales for multiscale analysis. In one approach, the maximum value over all scales used was selected for each pixel; this method did not lead to any improvement compared to the use of a single scale. The inclusion of a threshold to minimize false positives did not affect the overall efficiency of detection.

In another approach for multiscale analysis, a classifier was trained using the ground-truth data for the training set of 20 images, and the trained classifier was applied to the test set of 20 images. Several classifiers were studied: Fisher's linear discriminant analysis (FLDA), the Bayesian (quadratic) classifier [45], a generic multilayer perceptron (MLP), and a neural network with radial-basis functions (RBF) [46]. The RBF and MLP classifiers were trained with the data for only 0.1% of the pixels, selected at random from each image in the training set of 20 images. The FLDA and Bayesian classifiers provided no improvement, in terms of the A_z value, with multiscale analysis when compared to single-scale analysis. The RBF network (with 25 hidden nodes) provided marginal improvement, with $A_z = 0.96$ for the multiscale combination $\tau = \{4, 8, 12\}$, with $l = 2.9$. The MLP network (with three hidden nodes) also provided marginal improvement, with $A_z = 0.96$ for the multiscale combination $\tau = \{4, 8, 12\}$, with $l = 2.9$. The corresponding ROC curve is shown in Figure 10.6. Figure 10.4(d) shows the discriminant values generated by the MLP for a sample image.

The efficiency of detection of blood vessels in the retina obtained by several recently reported methods, in terms of A_z values, using the same set of 20 images in the test set of the DRIVE database, are listed in Table 10.2 [16]. The results obtained by the methods proposed in the present work closely match those obtained by Soares et al. [16]. The major differences between the methods proposed in the present work and those of Soares et al. [16] are: (1) the use of real Gabor filters (instead of complex filters), (2) the use of a simple MLP that does not assume a Gaussian mixture model, and (3) the use of the inverted luminance component (Y) instead of the green channel of the color fundus images. In a related work, we applied the same methods to the inverted green channel of the images [40]; the levels of accuracy obtained in the detection of blood vessels, in terms of A_z values, were similar to or marginally lower than those reported in this chapter. The use of the real Gabor filter takes advantage of the fact that the nature of the contrast of the blood vessels in retinal images is known (that is, positive contrast in the inverted luminance channel). Our results also indicate that the advantage to be gained by using multiple scales of Gabor filters is minimal.

10.4   Discussion and Conclusions

Multiscale Gabor filters have provided high efficiency in the detection of retinal blood vessels, with $A_z = 0.96$. However, large numbers of false-positive pixels were observed around the optic nerve head: Methods need to be developed to address this limitation.

Using a semiautomated method, Swanson et al. [3] showed that there is a significant increase in the tortuosity of retinal blood vessels in premature infants with retinopathy of prematurity [47–51]. A few measures of tortuosity of blood vessels have been defined in the literature [49, 52–55]; see Eze et al. [4] for a review and Azegrouz et al. [56] for recent developments. Existing methods are based on the analysis of vessel curvature, or the change in the coordinates or the angle along the vessel segment being analyzed. We propose to develop algorithms to characterize

Table 10.2   Comparison of the Efficiency of Detection of Blood Vessels in the Retina Obtained by Different Methods for the Test Set (20 Images) of the DRIVE Database [42]

Detection Method	A_z
Matched filter; Chaudhuri et al. [7]	0.91
Adaptive local thresholding; Jiang and Mojon [15]	0.93
Ridge-based segmentation; Staal et al. [11]	0.95
Single-scale Gabor filters; Rangayyan et al. [34]	0.95
Multiscale Gabor filters; Soares et al. [16]	0.96
Multiscale Gabor filters; Oloumi et al. [35] and this chapter	0.96

Note: See also [16].

the tortuosity of blood vessels using coherence [57, 58] derived from the angle data provided by the Gabor filters. Gabor filters facilitate the computation of local orientation as well as multiscale analysis of orientation and coherence.

Attention needs to be paid to the diverse and independent information present in the three channels that constitute the color images of the retina [12, 59, 60]. Optimal image processing techniques need to be developed in the most appropriate color representation domain.

Methods need to be designed to facilitate longitudinal quantitative analysis to aid follow-up of patients with retinal pathology and assessment of the effect of treatment [5, 49]. Variations in the width and tortuosity of retinal blood vessels could be analyzed as a function of time to monitor the time course of the pathology, and to evaluate the effects of treatment protocols.

The methods discussed earlier could assist in the detection and diagnosis of retinopathy in premature infants, as well as in patients with diabetes and hypertension. Quantitative techniques for longitudinal analysis could assist ophthalmologists in clinical follow-up of patients, particularly in the analysis of the progression of the disease, evaluation of the effects of treatment, and design of treatment protocols.

Acknowledgments

This work was supported by the Natural Sciences and Engineering Research Council of Canada.

References

[1]   Patton, N., et al., "Retinal Image Analysis: Concepts, Applications and Potential," *Prog. Retinal Eye Res.*, Vol. 25, 2006, pp. 99–127.

[2]   Chapman, N., et al., "Computer Algorithms for the Automated Measurement of Retinal Arteriolar Diameters," *Br. J. Ophthalmol.*, Vol. 85, 2001, pp. 74–79.

[3]   Swanson, C., et al., "Semi-Automated Computer Analysis of Vessel Growth in Preterm Infants Without and with ROP," *Br. J. Ophthalmol.*, Vol. 87, 2003, pp. 1474–1477.

[4]   Eze, C. U., R. Gupta, and D. L. Newman, "A Comparison of Quantitative Measures of Arterial Tortuosity Using Sine Wave Simulations and 3D Wire Models," *Phys. Med. Biol.*, Vol. 45, 2000, pp. 2593–2599.

[5]   Narasimha-Iyer, H., et al., "Robust Detection and Classification of Longitudinal Changes in Color Retinal Fundus Images for Monitoring Diabetic Retinopathy," *IEEE Trans. on Biomedical Engineering*, Vol. 53, No. 6, 2006, pp. 1084–1098.

[6]   Walter, T., et al., "A Contribution of Image Processing to the Diagnosis of Diabetic Retinopathy—Detection of Exudates in Color Fundus Images of the Human Retina," *IEEE Trans. on Medical Imaging*, Vol. 21, No. 10, 2002, pp. 1236–1243.

[7]   Chaudhuri, S., et al., "Detection of Blood Vessels in Retinal Images Using Two-Dimensional Matched Filters," *IEEE Trans. on Medical Imaging*, Vol. 8, No. 3, 1989, pp. 263–269.

[8]   Foracchia, M., E. Grisan, and A. Ruggeri, "Detection of Optic Disc in Retinal Images by Means of a Geometrical Model of Vessel Structure," *IEEE Trans. on Medical Imaging*, Vol. 23, No. 10, 2004, pp. 1189–1195.

[9]   Lowell, J., et al., "Measurement of Retinal Vessel Widths from Fundus Images Based on 2-D Modeling," *IEEE Trans. on Medical Imaging*, Vol. 23, No. 10, 2004, pp. 1196–1204.

[10] Hoover, V., V. Kouznetsova, and M. Goldbaum, "Locating Blood Vessels in Retinal Images by Piecewise Threshold Probing of a Matched Filter Response," *IEEE Trans. on Medical Imaging*, Vol. 19, No. 3, 2000, pp. 203–210.

[11] Staal, J., et al., "Ridge-Based Vessel Segmentation in Color Images of the Retina," *IEEE Trans. on Medical Imaging*, Vol. 23, No. 4, 2004, pp. 501–509.

[12] Li, H., and O. Chutatape, "Automated Feature Extraction in Color Retinal Images by a Model Based Approach," *IEEE Trans. on Biomedical Engineering*, Vol. 51, No. 2, 2004, pp. 246–254.

[13] Niemeijer, M., et al., "Comparative Study of Retinal Vessel Segmentation Methods on a New Publicly Available Database," *Proc. SPIE Intl. Symp. on Medical Imaging*, 2004, pp. 648–656.

[14] Lalonde, M., F. Laliberté, and L. Gagnon, "Retsoft Plus: A Tool for Retinal Image Analysis," *Proc. 17th IEEE Symp. on Computer Based Medical Systems*, 2004, pp. 542–547.

[15] Jiang, X., and D. Mojon, "Adaptive Local Thresholding by Verification-Based Multithreshold Probing with Application to Vessel Detection in Retinal Images," *IEEE Trans. on Pattern Analysis and Machine Intelligence*, Vol. 25, No. 1, 2003, pp. 131–137.

[16] Soares, J. V. B., et al., "Retinal Vessel Segmentation Using the 2-D Gabor Wavelet and Supervised Classification," *IEEE Trans. on Medical Imaging*, Vol. 25, No. 9, 2006, pp. 1214–1222.

[17] Al-Diri, B., and A. Hunter, "A Ribbon of Twins for Extracting Vessel Boundaries," *Proc. 3rd European Medical and Biological Engineering Conf.*, Prague, Czech Republic, November 2005.

[18] Gang, L., O. Chutatape, and S. M. Krishnan, "Detection and Measurement of Retinal Vessels in Fundus Images Using Amplitude Modified Second-Order Gaussian Filter," *IEEE Trans. on Biomedical Engineering*, Vol. 49, No. 2, 2002, pp. 168–172.

[19] Giani, A., E. Grisan, and A. Ruggeri, "Enhanced Classification-Based Vessel Tracking Using Vessel Models and Hough Transform," *Proc. 3rd European Medical and Biological Engineering Conf.*, Prague, Czech Republic, November 2005.

[20] Zhang, M., D. Wu, and J. C. Liu, "On the Small Vessel Detection in High Resolution Retinal Images," *Proc. 27th Annual Intl. Conf. of the IEEE Engineering in Medicine and Biology Society*, Shanghai, China, September 2005, pp. 3177–3179.

[21] Zana, F., and J. C. Klein, "Segmentation of Vessel-Like Patterns Using Mathematical Morphology and Curvature Estimation," *IEEE Trans. on Image Processing*, Vol. 10, No. 7, July 2001, pp. 1010–1019.

[22] Hunter, A., J. Lowell, and D. Steel, "Tram-Line Filtering for Retinal Vessel Segmentation," *Proc. 3rd European Medical and Biological Engineering Conf.*, Prague, Czech Republic, November 2005.

[23] Stosic, T., and B. D. Stosic, "Multifractal Analysis of Human Retinal Vessels," *IEEE Trans. on Medical Imaging*, Vol. 25, No. 8, 2006, pp. 1101–1107.

[24] Kyriacos, S., et al., "Insights into the Formation Process of the Retinal Vasculature," *Fractals*, Vol. 5, No. 4, 1997, pp. 615–624.

[25] Martinez-Perez, E., et al., "Retinal Vascular Tree Morphology: A Semiautomatic Quantification," *IEEE Trans. on Biomedical Engineering*, Vol. 49, No. 8, 2002, pp. 912–917.

[26] Gregson, P. H., et al., "Automated Grading of Venous Beading," *Comput. Biomed. Res.*, Vol. 28, 1995, pp. 291–304.

[27] Rangayyan, R. M., *Biomedical Image Analysis*, Boca Raton, FL: CRC Press, 2005.

[28] Bruton, L. T., and N. R. Bartley, "Using Nonessential Singularities of the Second Kind in Two-Dimensional Filter Design," *IEEE Trans. on Circuits and Systems*, Vol. 36, 1989, pp. 113–116.

[29] Gabor, D., "Theory of Communication," *Journal of the Institute of Electrical Engineers*, Vol. 93, 1946, pp. 429–457.

[30] Ferrari, R. J., et al., "Analysis of Asymmetry in Mammograms Via Directional Filtering with Gabor Wavelets," *IEEE Trans. on Medical Imaging*, Vol. 20, No. 9, 2001, pp. 953–964.

[31] Ayres, F. J., and R. M. Rangayyan, "Characterization of Architectural Distortion in Mammograms," *IEEE Eng. Med. Biol. Mag.*, Vol. 24, No. 1, January/February 2005, pp. 59–67.

[32] Ayres, F. J., and R. M. Rangayyan, "Design and Performance Analysis of Oriented Feature Detectors," *J. of Electronic Imaging*, Vol. 16, No. 2, 2007, pp. 023007:1–12.

[33] Rangayyan, R. M., and F. J. Ayres, "Gabor Filters and Phase Portraits for the Detection of Architectural Distortion in Mammograms," *Med. Biol. Eng. Computing*, Vol. 44, No. 10, October 2006, pp. 883–894.

[34] Rangayyan, R. M., et al., "Detection of Blood Vessels in the Retina Using Gabor Filters," *Proc. 20th Canadian Conf. on Electrical and Computer Engineering (CCECE 2007)*, Vancouver, BC, Canada, April 2007, pp. 717–720.

[35] Oloumi, F., et al., "Detection of Blood Vessels in Fundus Images of the Retina Using Gabor Wavelets," *Proc. 29th Annual Intl. Conf. of the IEEE Engineering in Medicine and Biology Society*, Lyon, France, August 22–26, 2007, pp. 6451–6454.

[36] Manjunath, B. S., and W. Y. Ma, "Texture Features for Browsing and Retrieval of Image Data," *IEEE Trans. on Pattern Analysis and Machine Intelligence*, Vol. 18, No. 8, 1996, pp. 837–842.

[37] Jones, P., and L. A. Palmer, "An Evaluation of the Two-Dimensional Gabor Filter Model of Simple Receptive Fields in Cat Striate Cortex," *J. of Neurophysiology*, Vol. 58, No. 6, 1987, pp. 1233–1258.

[38] Daugman, J. G., "Complete Discrete 2-D Gabor Transforms by Neural Networks for Image Analysis and Compression," *IEEE Trans. on Acoustics, Speech, and Signal Processing*, Vol. 36, No. 7, 1988, pp. 1169–1179.

[39] Daugman, J. G., "Uncertainty Relation for Resolution in Space, Spatial Frequency, and Orientation Optimized by Two-Dimensional Visual Cortical Filters," *J. of Optical Society of America*, Vol. 2, No. 7, 1985, pp. 1160–1169.

[40] Rangayyan, R. M., et al., "Detection of Blood Vessels in the Retina with Multiscale Gabor Wavelets," *J. of Electronic Imaging*, 2008, in press.

[41] Ayres, F. J., and R. M. Rangayyan, "Performance Analysis of Oriented Feature Detectors," *Proc. of SIBGRAPI 2005: XVIII Brazilian Symposium on Computer Graphics and Image Processing*, Natal, Brazil, October 2005, pp. 147–154.

[42] DRIVE: Digital Retinal Images for Vessel Extraction, http://www.isi.uu.nl/Research/Databases/DRIVE/.

[43] Gonzalez, R. C., and R. E. Woods, *Digital Image Processing*, 2nd ed., Upper Saddle River, NJ: Prentice-Hall, 2002.

[44] Metz, C. E., "Basic Principles of ROC Analysis," *Seminars in Nuclear Medicine*, Vol. VIII, No. 4, 1978, pp. 283–298.

[45] Duda, R. O., P. E. Hart, and D. G. Stork, *Pattern Classification*, 2nd ed., New York: Wiley, 2001.

[46] Haykin, S., *Neural Networks: A Comprehensive Foundation*, 2nd ed., Upper Saddle River, NJ: Prentice-Hall, 1999.

[47] Wallace, D. K., et al., "Computer-Automated Quantification of Plus Disease in Retinopathy of Prematurity," *J. of American Association for Pediatric Ophthalmology and Strabismus*, Vol. 7, 2003, pp. 126–130.

[48] Ells, A., et al., "Telemedicine Approach to Screening for Severe Retinopathy of Prematurity: A Pilot Study," *Amer. Acad. Ophthalmology*, Vol. 110, 2003, pp. 2113–2117.

[49] Heneghan, C., et al., "Characterization of Changes in Blood Vessel Width and Tortuosity in Retinopathy of Prematurity Using Image Analysis," *Medical Image Analysis*, Vol. 6, 2002, pp. 407–429.

[50] Gelman, R., et al., "Diagnosis of Plus Disease in Retinopathy of Prematurity Using Retinal Image Multiscale Analysis," *Invest. Ophthalmol. Vis. Sci.*, Vol. 46, No. 12, 2005, pp. 4734–4738.

[51] International Committee for the Classification of Retinopathy of Prematurity, "The International Classification of Retinopathy of Prematurity Revisited," *Archives of Ophthalmology*, Vol. 123, 2005, pp. 991–999.

[52] Hart, W. E., et al., "Automated Measurement of Retinal Vascular Tortuosity," *Proc. of American Medical Informatics Association Annual Fall Conference*, 1997, pp. 459–463.

[53] Hart, W. E., et al., "Measurement and Classification of Retinal Vascular Tortuosity," *Intl. J. Medical Informatics*, Vol. 53, Nos. 2-3, 1999, pp. 239–252.

[54] Grisan, E., M. Foracchia, and A. Ruggeri, "A Novel Method for the Automatic Evaluation of Retinal Vessel Tortuosity," *Proc. 25th Annual Intl. Conf. of the IEEE Engineering in Medicine and Biology Society*, Cancun, Mexico, September 2003, pp. 866–869.

[55] Bullitt, E., et al., "Measuring Tortuosity of the Intracerebral Vasculature from MRA Images," *IEEE Trans. on Medical Imaging*, Vol. 22, No. 9, 2003, pp. 1163–1171.

[56] Azegrouz, H., et al., "Thickness Dependent Tortuosity Estimation for Retinal Blood Vessels," *Proc. of 28th Annual International Conference of the IEEE Engineering in Medicine and Biology Society*, New York, September 2006, pp. 4675–4678.

[57] Mudigonda, N. R., R. M. Rangayyan, and J. E. L. Desautels, "Detection of Breast Masses in Mammograms by Density Slicing and Texture Flow-Field Analysis," *IEEE Trans. on Medical Imaging*, Vol. 20, No. 12, 2001, pp. 1215–1227.

[58] Rao, A. R., and R. C. Jain, "Computerized Flow Field Analysis: Oriented Texture Fields," *IEEE Trans. on Pattern Analysis and Machine Intelligence*, Vol. 14, No. 7, 1992, pp. 693–709.

[59] Salem, N. M., and A. K. Nandi, "Novel and Adaptive Contribution of the Red Channel in Pre-Processing of Colour Fundus Images," *J. of the Franklin Institute*, Vol. 344, 2007, pp. 243–256.

[60] Wang, H., et al., "An Effective Approach to Detect Lesions in Color Retinal Images," *Proc. of IEEE Conf. on Computer Vision and Pattern Recognition*, 2000, pp. 181–186.

Finite Element Simulation of the Eye Structure with Bioheat Analysis: Two- and Three-Dimensional Ocular Surface Temperature Profiles

Eddie Y. K. Ng, E. H. Ooi, and Rajendra Acharya

11.1   Introduction

Infrared (IR) thermography measures only the temperature of a given surface. When the temperature beneath a given surface is desired, IR thermography may not be of much use. This becomes a problem in cases such as the human eye where invasive measurements are not permitted. In such a scenario, mathematical modeling proves to be an alternative for researchers to predict and to conduct analysis on the intraocular temperature.

The use of a mathematical model enables the researcher to virtually dissect any part of the human eye model for further analysis. With the recent advances in computer technology, more sophisticated models with detailed anatomical resemblance of the human eye have been constructed and reported in the literature. In this chapter, we look at the development of two- and three-dimensional models of the human eye using the finite element method. The model is used to simulate the steady-state temperature distribution inside the human eye and results are compared with experimental measurements found in the literature.

In the next section, some of the mathematical models of the human eye that are found in the literature are briefly reviewed. In Section 11.3, we look at the development of the human eye model, which is followed by a mathematical description of the problem in Section 11.4. In Section 11.5, the numerical implementation of the finite element method is briefly discussed. Section 11.6 presents the results from the steady-state analysis, which is followed by results from the sensitivity analysis in Section 11.7. Changes in ocular temperature during electromagnetic (EM) wave radiation are given in Section 11.8, and the chapter is summarized in Section 11.9.

11.2   Previously Developed Models

The use of mathematical models to study the human eye temperature distribution under various conditions has been reported by many researchers in the past. One of the earliest models was developed by Al-Badwaihy and Youssef [1]. They developed a simplified model of the human eye where the eye was assumed to be a homogeneous sphere. The thermal properties of the homogeneous model were obtained by averaging the thermal properties of each individual eye component inside the normal eye. Though crude, such assumptions enabled an analytical solution to the problem to be formulated.

Emery et al. [2] and Guy et al. [3] developed a model of the rabbit eye to investigate the temperature rise during EM wave radiation. The problem was solved numerically using the finite element method. The specific absorption rate of the EM wave inside the rabbit eye during radiation exposure was obtained from experimental measurements. The simulated temperatures were compared with temperature measurements and they generally showed very good agreement. A similar study was carried out by Taflove and Brodwin [4] using a human eye model. In their work, the absorption of heat inside the eye was calculated based on the Maxwell equation using the finite difference time-domain method because experimentation on the human eye can damage it. All three models [2–4] showed an increase in ocular temperature that peaked at the center of the eye.

In 1982, Lagendijk [5] developed an axisymmetric model of both the rabbit and the human eye. The model was solved numerically using the finite difference method. Numerical results obtained using the rabbit eye model were compared with experimental measurements and results were found to agree reasonably well. The data found from experimental measurements on rabbits was fitted into the human eye model, which was later used for predictions of lens and vitreous thermal conductivity. In addition, the convection coefficient between the eye and the surrounding anatomy was also predicted and the value of 65 $Wm^{-2}K^{-1}$ has since been widely used by researchers in the same field.

An axisymmetric finite element model of the human eye was developed by Scott [6] in 1988. The model was used to simulate the steady temperature distribution inside a normal human eye. Sensitivity analysis was carried out to identify the parameters that might have dominant effects on the corneal surface temperature. The model was later used to investigate the changes in ocular temperature during exposure to IR laser radiation for the purpose of studying the mechanism of IR cataractogenesis [7]. Similar studies were also reported by Okuno [8] and Amara [9] in 1991 and 1995, respectively.

In the late 1990s and early 2000s, models of the human eye have included regions surrounding the human eye such as orbital fat, muscles, bones, and brain [10–12]. These studies were focused on the investigations of temperature increase inside the eye during exposure to EM wave radiation. The inclusion of regions such as the orbital fat and bones were necessary since the absorption of heat not only happens inside the eye but in other regions as well, and this may have significant impact on the temperature changes inside the human eye. More recent works on the mathematical modeling of the human eye can be found in [13–17].

11.3   Development of the Human Eye Model

The mathematical model of the human eye in this chapter is constructed based on the dimensions of a six-times-enlarged anatomical human eye model such as that shown in Figure 11.1. The anatomical model consists of six major components: the cornea, the aqueous humor, the lens, the iris, the vitreous, and the sclera. Coordinates of each component are measured using a coordinate measuring machine (CMM); in this case, a Mitutoyo B706 model was used (Figure 11.2).

Two types of probes are available to the CMM: the touch triggering probe and the laser probe. Although the laser probe in general yields a more accurate measurement compared to the touch triggering probe, it is not used in this study due to the curvature of the anatomical eye model (see Figure 11.1). On curved surfaces that are

Figure 11.1   The six-times-enlarged anatomical human eye model. (From: http://www.vicron.com/visnasmt/anaeye.htm. Reproduced with permission.)

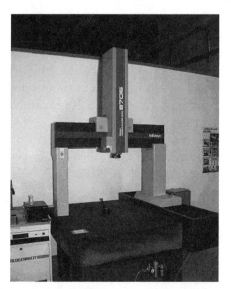

Figure 11.2   The Mitutoyo B706 CMM by Mister, Inc. (From: http://misterinc.com/quality-control-1.shtml. Reproduced with permission.)

too steep, the laser beam from the laser probe that is targeted onto the surface may be reflected in a different direction away from the photo diode detector (Figure 11.3). Consequently, the measured data may not be accurate. The touch triggering probe is used instead.

In the touch triggering probe, the probe is placed onto the model, and the coordinates of the probe with respect to a given origin are measured (Figure 11.4). To enable an accurate measurement on the dimensions of the eye model, large sampling points on the surface of each eye component are taken. During each measuring session, each component is firmly held in its position such that no shifting occurs. Any shifting in the position of the measured eye component will result in an inaccurate measurement (Figure 11.5).

To construct the computational model of the human eye, the measured coordinates are exported into Pro-E [18] where the geometrical structure of each component of the human eye is generated by connecting all of the measured points. Once the structure for each component of the human eye is created, they are assembled to form the complete human eye model. In doing so, we have assumed the human eye to be a deformable structure. We have also ensured that no gaps existed at the interfaces between two contiguous regions. Some fine-tuning work was carried out to "smooth" any sharp edges that existed due to poor selection of sampling points. The model, which was based on a six-times-enlarged anatomical eye model, was then scaled down to its original size.

Heat transfer simulations inside the human eye model are carried out numerically using the finite element method. To do so, we used a commercial software package called COMSOL Multiphysics 3.2 [19], which is hereafter referred to as COMSOL. COMSOL is generally a partial differential equation calculator that utilizes the finite element method to numerically solve any given physical problem.

The model of the human eye that was generated in Pro-E was exported into the workspace of COMSOL. Due to the differences in accuracy supported by Pro-E and COMSOL (1×10^{-6} versus 1×10^{-5}), more additional fine-tuning was required in the workspace of COMSOL. Difficulty arises during the fine-tuning of the three-dimensional model due to the complexity of the geometry. To avoid this, we exported only the two-dimensional geometry; the three-dimensional model was generated in the workspace of COMSOL. This was accomplished by revolving one-half of the two-dimensional model 360° about its pupillary axis.

Figures 11.6 and 11.7 illustrate the graphical image of the two-dimensional and three-dimensional human eye models, respectively, as seen in the workspace of

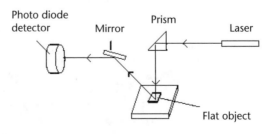

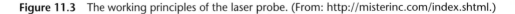

Figure 11.3   The working principles of the laser probe. (From: http://misterinc.com/index.shtml.)

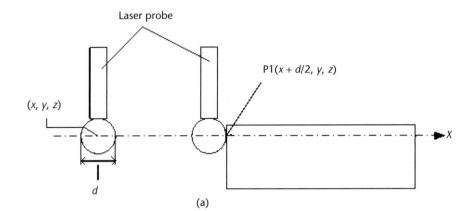

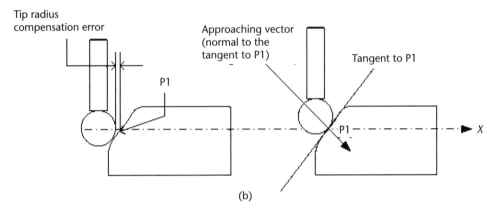

Figure 11.4 (a) Position of a point in space. (Tip radius compensation along a part axis). (From: http://misterinc.com/index.shtml.) (b) Position of a point on the object. (Tip radius compensation along the normal to the tangent of a point.) (From: http://misterinc.com/index.shtml.)

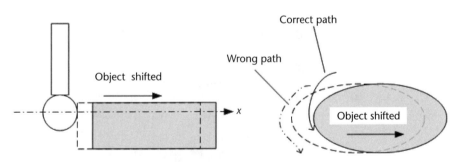

Figure 11.5 Inaccurate measurements as a result of object shifting. (From: http://misterinc.com/index.shtml.)

COMSOL. The pupillary axis is denoted by the horizontal line O–E and is 0.0254m in length. In the actual human eye, one may find two additional layers inside the sclera—the retina and the choroid. The retina and the choroid are relatively thin compared to the sclera and they are thus modeled together with the sclera as one

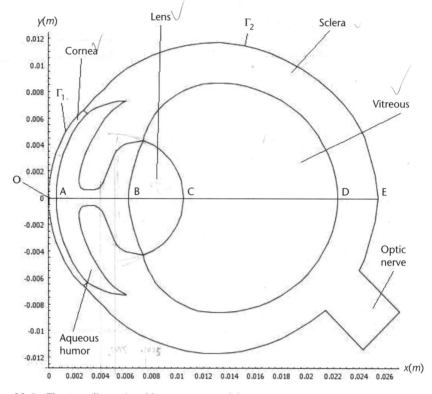

Figure 11.6   The two-dimensional human eye model.

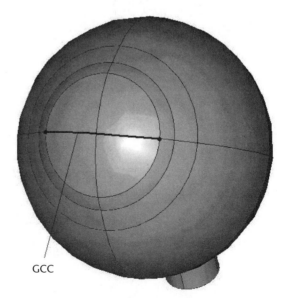

Figure 11.7   The three-dimensional human eye model. (*From:* [14]. © 2007 Elsevier. Reprinted with permission.)

homogeneous region. Similarly, the optic nerve is also assumed to be a part of the sclera. The iris was found to have thermal properties that are similar to those of the sclera [20], so it is, therefore, also modeled as being a part of the sclera.

As a result of these assumptions, the mathematical eye model consists of five major components: the cornea, the sclera, the aqueous humor, the lens, and the vitreous, which we denote as R_1, R_2, R_3, R_4, and R_5, respectively.

11.4 Mathematical Description

11.4.1 Governing Equation

In biological tissues, the equation governing the flow of heat can be described by the Pennes bioheat model [21], which is given by

$$\rho_t c_t \frac{\partial T_t}{\partial t} = \nabla \cdot (k_t \nabla T_t) + \rho_b c_b \omega_b (T_b - T_t) + Q_m \tag{11.1}$$

where ρ and c are the density and specific heat, respectively; T is temperature, t is time, k is thermal conductivity, ω is the blood perfusion rate, and Q_m is the heat generated from metabolic activity of the tissue. Subscripts t and b refer to tissue and blood, respectively.

In the present work, we have modeled the human eye as an organ that is isolated from the human head. The thermal effects of blood flow in the choroidal layer can thus be modeled as part of the boundary conditions, which is discussed in the next section. Consequently, the second term on the right-hand side of (11.1), which models the thermal effects of blood flow, can be omitted. The heat generated from the metabolic activity is neglected in this study [6]. Based on these assumptions and rewriting the governing equation for a steady-state problem, (11.1) reduces to

$$\nabla \cdot (k_i \nabla T_i) = 0 \quad \text{for } i = 1, 2, 3, 4, \text{ and } 5 \tag{11.2}$$

The values of thermal conductivity for each component of the human eye model k_i can be found in the literature and they are summarized in Table 11.1.

11.4.2 Boundary Conditions

Two boundary conditions are required to solve (11.2) and they are specified on the corneal surface, Γ_1, and scleroid surface, Γ_2, respectively (see Figure 11.6).

The surface of the cornea Γ_1 is exposed to the environment. If the ambient temperature is lower than the corneal surface temperature, heat from inside the eye is diffused from the corneal surface into the environment via heat convection and heat radiation. The cooling of the cornea is aided by the evaporation of tears from the corneal surface. These heat loss mechanisms form the boundary condition on the corneal surface, which can be written as

$$-k_1 \frac{\partial T_1}{\partial \mathbf{n}} = h_{amb}(T_1 - T_{amb}) + \varepsilon\sigma(T_1^4 - T_{amb}^4) + E \quad \text{on } \Gamma_1 \tag{11.3}$$

Table 11.1   Properties of the Human Eye for Different Domains

Domains	Index, i	Thermal Conductivity (Wm^{-1}K^{-1})
Cornea	1	0.58 [2]
Sclera	2	1.00 [20]
Aqueous humor	3	0.58 [2]
Lens	4	0.40 [5]
Vitreous body	5	0.60 [6]

where h_{amb} is the convection heat transfer coefficient between the cornea and the environment; ε and σ are the corneal emissivity and the Stefan-Boltzmann constant, respectively; T_{amb} is ambient temperature; E is heat loss due to evaporation rate; and $\partial T_1/\partial \mathbf{n}$ is the rate of change of T_1 in the outward direction normal to surface Γ_1.

As stated in the previous section, the human eye model is assumed to be isolated from the human body. To simulate the thermal effects of blood flow in the choroidal layer, we may employ the approach suggested in [5, 22]. In this approach, scleroid surface, Γ_2 is assumed to be surrounded by an anatomically homogeneous surrounding that has a temperature similar to that of the body core. The transfer of heat between the surrounding area and the human eye is modeled using a constant heat transfer coefficient, written as follows:

$$-k_2 \frac{\partial T_2}{\partial \mathbf{n}} = h_{bl}(T_2 - T_{bl}) \quad \text{on } \Gamma_2 \tag{11.4}$$

where h_{bl} denotes the convection heat transfer coefficient between the surrounding anatomy and the eye, T_{bl} is the body core temperature which we simply refer to as blood temperature, and $\partial T_2/\partial \mathbf{n}$ is the rate of change of T_1 in the outward direction normal to surface Γ_2.

The values of parameters such as ambient temperature, blood temperature, heat transfer coefficients, and heat loss due to tear evaporation can be found in the literature, and they are summarized in Table 11.2. These values are termed the control parameters.

Table 11.2   Control Parameters Used in the Current Study

Control Parameters	Value	Reference
Blood temperature, T_{bl} (°C)	37	[13]
Ambient temperature, T_{amb} (°C)	25	[13]
Emissivity of cornea, ε	0.975	[23]
Blood convection coefficient, h_{bl} (Wm^{-2}K^{-1})	65	[5]
Ambient convection coefficient, h_{amb} (Wm^{-2}K^{-1})	10	[2]
Evaporation rate, E (Wm^{-2})	40	[6]
Stefan-Boltzmann constant, σ (Wm^{-2}K^{-4})	5.67×10^{-8}	[24]

To investigate the importance of each heat loss mechanism, we need to estimate the amount of convective and radiative heat flux on the surface of the cornea. The convective heat flux, q_{conv}, can be estimated using

$$q_{conv} = h_{amb}(T_1 - T_{amb})$$ (11.5)

If we assume the average temperature on the corneal surface to be approximately 34°C (average of 34.8°C [23], 35.4°C [25], 34.3°C [26], 33.0°C [27], 31.9°C [28], and 33.9°C [29]) and using the values of h_{amb} and T_{amb} as given in Table 11.2, the value of convective heat flux is approximated to be 90 Wm^{-2}.

Similarly, for an average corneal surface temperature of 34°C and using the control parameters in Table 11.2, the radiative heat flux, q_{rad}, can be mathematically written as

$$q_{rad} = \varepsilon \times \sigma \times (T_1^4 - T_{amb}^4)$$ (11.6)

which is estimated to have a value of approximately 55.1 Wm^{-2}.

These results show that the heat losses due to convection, radiation, and tear evaporation (40 Wm^{-2}) have the same order of magnitude, and omission of any one of them may have a significant impact on the simulated temperatures.

11.4.3 Additional Assumptions

In addition to the assumptions mentioned in the previous section, a few other assumptions made to simplify the modeling and simulations procedures that are carried out later.

In the actual eye, the aqueous humor containing the aqueous fluid moves in a circular motion as a result of buoyancy that is induced by the presence of thermal gradient across the aqueous humor [30–33]. This circulation has been shown to distort the temperature distribution in the aqueous humor [33]. In this study, however, we have assumed the aqueous fluid to be stagnant. The effects of this circulation on the temperature distribution inside the human eye are discussed in Chapter 14.

The presence of eyelids, which was found to significantly affect the temperature profile on the corneal surface [26, 34], has not been included in this study. This is largely due to the lack of information and the difficulty in modeling the eyelids in a model where the eye is assumed to be isolated from the human head. Blinking of the eyelid, which induces transient temperature responses on the corneal surface, is also neglected in this study.

11.5 Numerical Methodology

This section describes the basic procedures used when implementing the finite element method in the workspace of COMSOL. We have avoided the details on the formulation of the finite element method. Readers are advised to consult textbooks on the finite element method for more information.

To carry out the finite element simulation, the solution domain has to be discretized into small polygonal-shaped elements. In this study, the triangular-shaped elements are used for the two-dimensional model and tetrahedral-shaped elements for the three-dimensional model. When discretizing the model of the human eye, components that are relatively thin such as the cornea are given a finer discretization compared to the other components. This is to ensure that the changes in temperature across the thin region are captured accurately. The meshed structures of both the two- and three-dimensional models are illustrated in Figures 11.8 and 11.9, respectively.

The two-dimensional model consists of 11,255 triangular elements. The geometrical shape order and the element shape function for each of the triangular elements are approximated using the Lagrange quadratic polynomial. In the three-dimensional model, a total of 54,796 tetrahedral elements are generated. Due to the limitations of computer memory, the approximation on the geometrical shape order is reduced to a linear approximation.

All calculations were performed on a Pentium 4, 2.40-GHz, 512-Mbyte RAM personal computer.

11.6 Steady-State Temperature Distribution

The steady-state temperature distribution simulated using the two- and three-dimensional models are presented in this section. In Figures 11.10 and 11.11, the temperature distributions across the entire human eye are plotted for the two- and three-dimensional human eye models, respectively. The isotherms for the two-dimensional model are indicated by the lines across the temperature plot. In the three-dimensional model, temperature at each different reason is presented. The streamlines across the model denote the heat flux, which flows inside the human eye.

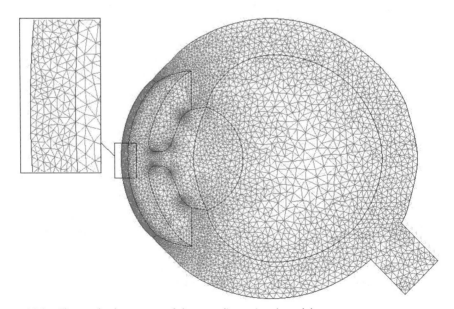

Figure 11.8 The meshed structure of the two-dimensional model.

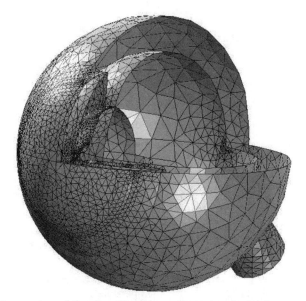

Figure 11.9   A cutaway view of the discretized three-dimensional model.

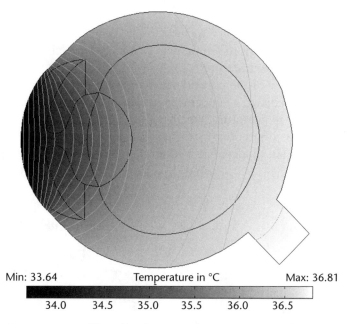

Figure 11.10   Temperature profile and isotherms on the two-dimensional model.

The surface of the cornea appears to be the coolest compared to elsewhere inside the human eye. The temperature is the warmest at the posterior region of the eye.

The isotherms shown in Figure 11.10 are basically symmetrical about its pupillary axis. The presence of optic nerve appears to contribute only very little to the distortion of temperature distribution particularly at the anterior portion of the

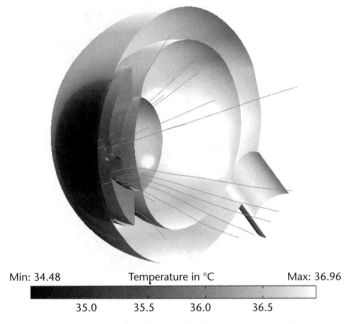

Min: 34.48   Temperature in °C   Max: 36.96

35.0   35.5   36.0   36.5

Figure 11.11   Temperature and heat flux plot for the three-dimensional human eye model.

eye. As such, omitting the optic nerve from the model may not have a significant effect on the temperature predictions inside the eye.

On the center of the corneal surface (point O in Figure 11.6), the two-dimensional model predicted a temperature of 33.64°C. The three-dimensional model simulated a higher temperature, which is found to be 34.48°C at the same location. To investigate the differences in the predicted temperatures between the two- and three-dimensional models, we plotted the variation in temperature at points along the pupillary axis ($y = 0$) for both models. This is shown in Figure 11.12.

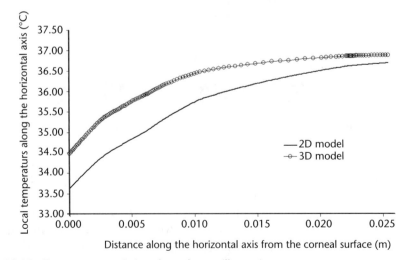

Figure 11.12   Temperature variation along the pupillary axis.

From Figure 11.12, it appears that the use of a two-dimensional model underestimates the temperature distribution inside the human eye. This may be explained by the absence of heat transfer in the direction perpendicular to the plane model shown earlier in Figure 11.6.

In Table 11.3, we compile the temperature measured on the corneal surface by different authors using various measuring techniques. These measurements, which we have sorted chronologically, may be found in the literature. The average of the measured temperature is calculated to be 34.36°C. Comparing the results calculated from both the two- and three-dimensional models, it is clear that the prediction made by the three-dimensional model has a greater accuracy compared to that of the two-dimensional model.

Table 11.3   Summary of Ocular Surface Temperature from Various Literatures Using Various Techniques

	Temperature at the Center of the Cornea (°C)	Technique
Two-dimensional model	33.64	COMSOL
Three-dimensional model	34.48	COMSOL
Author(s)	Ocular Surface Temperature (°C)	Technique
Dohnberg, 1876*	36.60	Mercury bulb
Galezowski, 1877*	36.40	Mercury bulb
Silex, 1893*	35.55	Thermo-element
Gilese, 1894*	35.72	Thermo-element
Hertel, 1900*	35.65	Mercury bulb
Kirisawa, 1942*	34.50	Thermo-element
Holmberg [35]	36.24	Thermo-electric
Hill and Leighton [36]	32.10	Thermister
Mapstone [23]	34.80	Bolometer
Kinn and Tell [37]	35.50	Liquid crystal
Rysä and Sarvaranta [38]	34.80	IR thermography
Hørven [39]	33.67	Contact probe
Fatt and Chaston [40]	34.50	IR thermography
Alio and Padron [41]	32.90	IR thermography
Fielder et al. [42]	33.40	IR thermography
Martin and Fatt [43]	34.50	Heat flow
Efron et al. [26]	34.30	IR thermography
Morgan et al. [34]	33.50	IR thermography
Morgan et al. [28]	31.94	IR thermography
Girardin et al. [44]	33.70	IR thermography
Gugleta et al. [45]	33.65	IR thermography
Craig et al. [46]	33.82	IR thermography
Purslow et al. [47]	35.00	IR thermography

Mean temperature: 34.36°C
* Cited by Holmberg [35].

Such comparisons, however, are crude and unreliable because the temperatures listed in Table 11.3 were obtained using different measuring techniques and may have varying accuracy. Furthermore, the conditions in which each of these measurements was taken may not be the same and may differ from the values of control parameters used in this study.

Figure 11.13 shows the plot of temperature difference between each point along the geometrical center of the cornea (GCC) and the point at the center of the cornea for both the two- and three-dimensional models of the human eye. The GCC for the three-dimensional model is defined by the thick line shown earlier in Figure 11.7. In the two-dimensional model, the corneal surface Γ_1 is assumed as the GCC. In addition to the plots of the two- and three-dimensional models, the results given by Efron et al. [26], which are based on measurements obtained using IR thermography, are also included into Figure 11.13 for better comparison.

The plots obtained for both the two- and three-dimensional models appear to be less widespread compared to the results obtained from experimental measurements. This observation may be attributed to the absence of eyelids in our model. According to Efron et al. [26], the presence of eyelids causes the isotherms on the corneal surface to distribute according to the palpebral aperture, which is elliptical in nature.

11.7   Sensitivity Analysis

The temperature on the surface of the cornea is dependent on many factors, such as age, environmental conditions, and individual variation [23, 34, 48]. In this section, we carry out a sensitivity analysis to identify factors that have dominant effects on

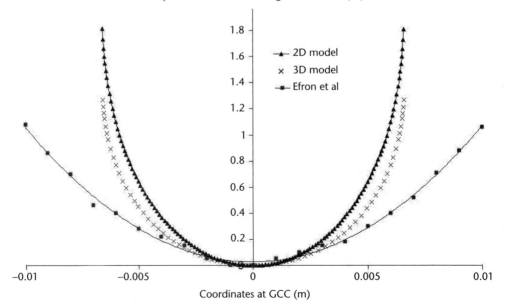

Figure 11.13 Temperature differences between local points and the midcorneal temperature along the GCC line. (*From:* [14]. © 2007 Elsevier. Reprinted with permission.)

the ocular surface temperature. Results from this study may provide researchers, especially ophthalmologists, with information on the parameters of importance during the measurement of corneal temperature.

Six parameters are investigated in this study: lens thermal conductivity, blood convection coefficient, ambient convection coefficient, ambient temperature, blood temperature, and heat loss due to tear evaporation. Temperatures at six points along the pupillary such as those shown in Figure 11.6 are observed. We use only the two-dimensional model for sensitivity analysis because repeated runs using the three-dimensional is computationally expensive.

11.7.1   Effects of Lens Thermal Conductivity

The water content inside the human lens was found to decrease with age [49]. According to Spells [50], the thermal conductivity of a biological tissue increases linearly with its water content. This suggests that the thermal conductivity of the human lens decreases when the age of the individual increases. Among the values of thermal conductivity that have been reported in the literature are 0.21 [2], 0.3 [6], and 0.544 $Wm^{-1}K^{-1}$ [51]. These values are used for investigating the effects of different lens thermal conductivity on the ocular temperature. Results are tabulated in Table 11.4. The highlighted row indicates the values obtained using control parameters.

From the results shown in Table 11.4, we found that the changes in lens thermal conductivity induce only very small changes in the temperature at the center of the cornea (point O). An increase of 0.1 $Wm^{-1}K^{-1}$ increases the temperature by only 0.04°C. As the value of lens thermal conductivity increases, the temperature at point A increases. This suggests that as the age of an individual increases, the temperature on the corneal surface decreases, which agrees with the experimental observations reported in [48].

11.7.2   Effects of Blood Convection Coefficient

The control value of the blood convection coefficient (65 $Wm^{-2}K^{-1}$) was suggested by Lagendijk [5] who assumed the eye to be embedded in an anatomically homogeneous surrounding. When components such as muscles and bones are considered, this value may increase to 110 $Wm^{-2}K^{-1}$. In a recent study, it was suggested that the correct value for the blood convection coefficient lies between 250 and 300 $Wm^{-2}K^{-1}$ [22]. To investigate the effects of different values for the blood convection coef-

Table 11.4   Effects of Lens Thermal Conductivity on Temperature Variation of the Eye

k_4 ($Wm^{-1}K^{-1}$)	Temperature (°C)					
	O	A	B	C	D	E
0.21	33.56	34.06	34.86	35.90	36.62	36.68
0.30	33.60	34.11	34.95	35.84	36.61	36.68
0.40	33.64	34.15	35.03	35.81	36.60	36.67
0.54	33.68	34.19	35.11	35.74	36.59	36.66

ficient, we simulated the problem for blood convection coefficient values of 110, 250, 275, and 300 $Wm^{-2}K^{-1}$. Results are presented in Table 11.5.

Table 11.5 shows that changes in the blood convection coefficient have only very minor effects on the temperature at the corneal surface. Increasing the value of the blood convection coefficient from 65 to 300 $Wm^{-2}K^{-1}$ increases the temperature at the center of the corneal surface by only 0.66°C. Using linear interpolation, this works out to an approximate change of 0.003°C for every 1 $Wm^{-2}K^{-1}$ change in the blood convection coefficient.

11.7.3 Effects of Ambient Convection Coefficient

In the ambient convection coefficient, values of 8 and 15 $Wm^{-2}K^{-1}$, which simulates natural convection heat transfer [24], are considered. In addition, a value of 100 $Wm^{-2}K^{-1}$ is also considered, which simulates the forced convection phenomena. Results from this study are given in Table 11.6.

Temperatures at the posterior region of the eye are not greatly affected by the changes in ambient convection coefficients. However, on the corneal surface, changes in temperature are large. When h_{amb} increases to 100 $Wm^{-2}K^{-1}$, a sharp drop in temperature is observed. The process of forced convection carries away more heat, which leads to an enhanced cooling effect on the cornea. Ultimately, this leads to decreases in the ocular surface temperature such as those shown in Table 11.6.

11.7.4 Effects of Blood Temperature

Changes in blood temperature may be associated with changes in the body core temperature. Different body core temperatures may be caused by the health condition of the individual. For instance, when the subject has a fever, the body's core tempera-

Table 11.5 Effects of Blood Convection Coefficient on Temperature Variation of the Eye

h_{bl} ($Wm^{-2}K^{-1}$)	Temperature (°C)					
	O	A	B	C	D	E
65	33.64	34.15	35.03	35.81	36.60	36.67
110	33.94	34.11	35.35	36.08	36.79	36.84
250	34.25	34.44	35.68	36.36	36.91	36.95
275	34.27	34.46	35.71	36.38	36.92	36.96
300	34.30	34.49	35.73	36.40	36.93	36.96

Table 11.6 Effects of Ambient Convection Coefficient on Temperature Variation of the Eye

h_{amb} ($Wm^{-2}K^{-1}$)	Temperature (°C)					
	O	A	B	C	D	E
8	33.90	34.37	35.18	35.89	36.63	36.70
10	33.64	34.15	35.03	35.81	36.60	36.67
15	33.04	33.64	34.68	35.58	36.53	36.61
100	28.51	29.70	31.88	33.85	35.95	36.14

ture may increase, which may induce a warmer surrounding anatomy around the human eye. To study this parameter, we simulate the model for body core temperature at 38°C and 39°C. Results are given in Table 11.7.

At the control value of 37°C, the ocular surface temperature (OST) is 33.64°C. An increase of 1°C in the body core temperature increases the temperature at the center of the cornea by 0.76°C. The magnitude of temperature elevation is large and can be detected using IR thermography. This suggests the possibility of fever detection based on temperature abnormalities on the corneal surface.

11.7.5 Effects of Ambient Temperature

To investigate the effects of different ambient temperatures on the ocular temperature, we consider two values: one that is lower than the control value and one that is higher. The values of these temperatures are taken to be 20°C and 30°C, respectively. Simulated results are tabulated in Table 11.8.

A lower ambient temperature produces a lower OST. This may be explained by the greater heat loss from the corneal surface due to the larger thermal gradient between the corneal surface and the environment. Based on the results given, an increase of 1°C in the ambient temperature increases the temperature on the corneal surface by approximately 0.2°C, which may be considered to be significant. The changes in temperature become less significant at points inside the human eye (points A to E).

11.7.6 Effects of Tear Evaporation Rate

The control value of 40 Wm^{-2} for heat loss due to tear evaporation was suggested by Scott [6]. Experimental studies have found that this value may vary between individuals, and a range between 20 and 100 Wm^{-2} was proposed for the normal eye [6].

Table 11.7 Effects of Blood Temperature on the Temperature Variation of the Eye

T_{bl}(°C)	Temperature (°C)					
	O	A	B	C	D	E
37	33.64	34.15	35.03	35.81	36.60	36.67
38	34.40	34.95	35.89	36.71	37.57	37.65
39	35.17	35.75	36.75	37.62	38.54	38.62

Table 11.8 Effects of Ambient Temperature on the Temperature Variation of the Eye Model

T_{amb} (°C)	Temperature (°C)					
	O	A	B	C	D	E
20	32.51	33.19	34.37	35.39	36.47	36.56
25	33.64	34.15	35.03	35.81	36.60	36.67
30	34.79	35.12	35.70	36.21	36.74	36.78

In individuals with dry eye syndrome, the absence of the lipid layer on the corneal surface, which helps to prevent excessive evaporation of tears, causes the value of E to be exceptionally high. In this case, a value of 320 Wm^{-2} is considered. Table 11.9 summarizes the results obtained for this study.

It appears that changes in the tear evaporation rate greatly affect the temperature on the corneal surface. A greater heat loss from the evaporation of tears reduces the temperature on the corneal surface. When the value of E is extremely high (320 Wm^{-2}), the temperature on the corneal surface drops to a level lower than 30°C. This suggests the possibility of diagnosing dry eye syndrome based on the abnormally low OST.

11.7.7 Summary from Sensitivity Analysis Study

From the sensitivity analysis conducted, the eye temperature is found to be particularly sensitive towards changes in blood temperature, ambient temperature, and evaporation rate. The OST is also greatly affected when exposed to the environment, which induces forced convection on the corneal surface. The variations in eye temperature are not critically by changes in the blood convection coefficient compared to blood temperature, which agrees with results from previous studies [6, 9]. Scott [6] suggested that the blood convection coefficient will only play a significant role in the eye temperature when it is subjected to an extrenal heat source that increases the temperature inside the eye. When this happens, the blood flow inside the choroidal layer acts as a cooling medium to transfer the absorbed heat away.

11.8   Effects of EM Wave Exposure on the Human Eye

In this section, the three-dimensional model of the human eye is used to simulate the temperature rise during EM wave radiation. Our purpose here is not to study the biological response of the human eye when exposed to EM wave radiation, but rather to examine the robustness of the three-dimensional model under different applications. Simulations conducted in this section are similar to those carried out by Taflove and Brodwin [4].

When the human eye is exposed to EM wave radiation, energy from the wave radiation that is absorbed by the eye is converted into heat. This increases the temperature inside the human eye. The magnitude of the temperature increase depends

Table 11.9 Effects of Evaporation Rate on Temperature Variations in the Eye

E (Wm^{-2})	Temperature (°C)					
	O	A	B	C	D	E
20	33.93	34.09	35.21	35.91	36.64	36.70
40	33.64	34.15	35.03	35.81	36.60	36.67
70	33.21	33.78	34.78	35.64	36.55	36.63
100	32.78	33.41	34.53	35.48	36.50	36.59
320	29.62	30.73	32.67	34.35	36.12	36.28

on the amount of heat that is absorbed, which in turn, depends on the frequency of the EM wave to which the eye is exposed [4, 10–12].

In this study, two EM wave frequencies are considered: 750 MHz and 1.5 GHz. These frequencies are similar to those modeled by Taflove and Brodwin [4]. To obtain the value of the amount of heat that is absorbed inside the eye, we make a simple assumption based on the values calculated by Taflove and Brodwin [4].

According to Taflove and Brodwin [4], for an incident irradiation of 1,000 W m^{-2}, the total dissipated power inside an eye exposed to a 750-MHz EM wave is approximately 0.13W with a peak of 7.3 times its average. When exposed to a 1.5-GHz EM wave, the total power dissipated is 0.19W with a peak of 3.3 times the average. The location of the peak absorption is at the cornea for the 750-MHz radiation and at the center for the 1.5-GHz radiation.

If we assume the distribution of the absorbed power to be uniform over the entire eye [4], the average volumetric heat generation inside the eye can be easily calculated. The volume of the three-dimensional model in the present work is 7.19 cm^3. Using this information, the average volumetric heat absorption inside the eye for the 750-MHz and 1.5-GHz radiation is 18,081 and 26,426 W m^{-3}, respectively. This information together with the peak volumetric absorption is summarized in Table 11.10.

11.8.1   Governing Equations and Boundary Conditions

To account for the heat absorption inside the eye, the governing equation in (11.2) has to be rewritten as

$$\nabla \cdot \left(k_i \nabla T_i \right) + Q_i = 0 \quad \text{for } i = 1, 2, 3, 4, \text{ and } 5 \tag{11.7}$$

where Q_i is the volumetric heat absorption (W m^{-3}) inside region R_i. Boundary conditions and the values of control parameters are similar to those given in (11.3) and (11.4) and in Table 11.2, respectively.

11.8.2   Results and Discussion

The temperature profiles of the human eye subjected to 750-MHz and 1.5-GHz EM wave radiation are shown in Figures 11.14 and 11.15, respectively. The figures show peak temperatures at the center of the eye, and this is true for radiation in both frequencies.

Table 11.10 Summary of the Data Calculated for the 750-MHz and 1.5-GHz Radiation

Parameter	EM Wave Frequency	
	750 MHz	1.5 GHz
Average density (Wm^{-3})	18,081	24,426
Peak density (Wm^{-3})	131,991	80,606
Location	Front of the eye	Center of the eye
Domain involved	Cornea	Vitreous

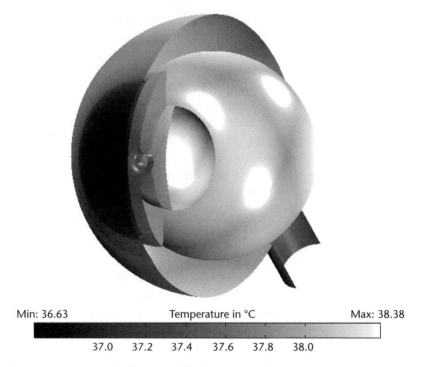

Min: 36.63   Temperature in °C   Max: 38.38

37.0   37.2   37.4   37.6   37.8   38.0

Figure 11.14   Temperature distribution of the human eye when exposed to 750-MHz EM wave radiation.

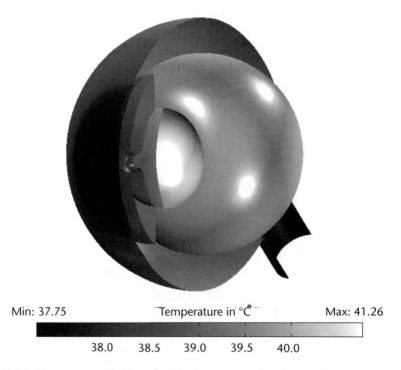

Min: 37.75   Temperature in °C   Max: 41.26

38.0   38.5   39.0   39.5   40.0

Figure 11.15   Temperature distribution of the human eye when exposed to 1.50-GHz EM wave radiation.

When exposed to 750-MHz radiation, the maximum temperature rise is approximately 38.38°C. A maximum temperature of 41.26°C is found for exposure to 1.5-GHz EM wave radiation. These results agree qualitatively with the numerical predictions found in [4]. Similar results were obtained for rabbit eye models found in [3].

In both conditions, the temperature inside the eye increases beyond the body core temperature of 37°C. As a result, the blood flow in the choroidal layer now acts as a cooling medium. The excessive heat generated inside the eye as a result of energy absorption is carried away by the flow of blood. It is expected that when the blood convection coefficient is increased, the cooling effects will be much greater [6]. The cooling of the human eye is aided by the heat loss at the surface of the cornea, as described in Section 11.4.2.

11.9 Summary

Two- and three-dimensional models of the human eye have been developed and presented in this chapter. The model was constructed based on the coordinates of a six-times-enlarged anatomical human eye model. The equation governing the heat flow inside the eye and the respective boundary conditions are solved numerically using the finite element method. This is accomplished using the commercial software package COMSOL, which enables simulations to be carried out effectively.

The models presented in this study complement the studies carried out in the past by other researchers such as Scott [6], Amara [9], and Ridouane and Campo [17]. The use of COMSOL, which enables a color-coded display of eye temperature to be plotted, is helpful particularly when direct comparisons between the numerical predictions and IR thermograms are desired.

The numerical predictions from both the two- and three-dimensional models were compared with experimental data found in the literature and reasonable qualitative agreement is observed. Our results also showed that the two-dimensional model underestimates the heat that is flowing through the human eye, and the use of a three-dimensional model yields results that have a greater accuracy. The numerical accuracy of the three-dimensional model was further justified through the simulation of EM wave exposure conditions, in which results were in good agreement with those reported in the literature.

Results from sensitivity analysis also showed that the temperature on the surface of the cornea is sensitive toward changes in ambient temperature, blood temperature, and tear evaporation rate. These results provide information to researchers, especially ophthalmologists, on the parameters of importance during the measurement of corneal temperature. The results shown here are, however, only preliminary because any quantitative results will require proper statistical analysis coupled with ophthalmologists' opinion on the degree of changes in temperature that should be considered as critical.

References

[1] Al-Badwaihy, K. A., and A. A. Youssef, "Biological Thermal Effect of Microwave Radiation on Human Eyes," in *Biological Effects of Electromagnetic Waves*, C. C. Johnson and M. L. Shore, (eds.), Washington, D.C.: DHEW Publication, 1976, pp. 61–78.

[2] Emery, A. F., et al., "Microwave Induced Temperature Rises in Rabbit Eyes in Cataract Research," *J. of Heat Transfer*, Vol. 97, 1975, pp. 123–128.

[3] Guy, A., et al., "Effect of 2450-MHz Radiation on the Rabbit Eye," *IEEE Trans. on Microwave Theory and Techniques*, Vol. MTT-23, No. 6, 1975, pp. 492–498.

[4] Taflove, A., and M. Brodwin, "Computation of the Electromagnetic Fields and Induced Temperatures Within a Model of the Microwave-Irradiated Human Eye," *IEEE Trans. on Microwave Theory and Techniques*, Vol. MTT-23, No. 11, 1975, pp. 888–896.

[5] Lagendijk, J. J. W., "A Mathematical Model to Calculate Temperature Distributions in Human and Rabbit Eyes During Hyperthermic Treatment," *Phys. Med. Biol.*, Vol.27, No. 11, 1982, pp. 1301–1311.

[6] Scott, J. A., "A Finite Element Model of Heat Transport in the Human Eye," *Phys. Med. Biol.*, Vol. 33, No. 2, 1988, pp. 227–241.

[7] Scott, J. A., "The Computation of Temperature Rises in the Human Eye Induced by Infrared Radiation," *Phys. Med. Biol.*, Vol. 33, No. 2, 1988, pp. 243–257.

[8] Okuno, T., "Thermal Effect of IR Radiation on the Eye: A Study Based on a Model," *Ann. Occupational Hygiene*, Vol. 35, No. 1, 1991, pp. 1–12.

[9] Amara, E. H., "Numerical Investigations on Thermal Effects of Laser Ocular Media Interaction," *Int. J. Heat Mass Transfer*, Vol. 38, No. 13, 1995, pp. 2479–2488.

[10] Bernardi, P., et al., "SAR Distribution and Temperature Increase in an Anatomical Model of the Human Eye Exposed to the Field Radiated by the User Antenna in a Wireless LAN," *IEEE Trans. on Microwave Theory and Techniques*, Vol. 46, No. 12, 1998, pp. 2074–2082.

[11] Hirata, A., S. Matsuyama, and T. Shiozawa, "Temperature Rises in the Human Eye Exposure to EM Waves in the Frequency Range 0.6–6 GHz," *IEEE Trans. on Electromagnetic Compatibility*, Vol. 44, No. 4, 2002, pp. 594–596.

[12] Hirata, A., "Temperature Increase in Human Eyes Due to Near-Field and Far Field Exposures at 900 MHz, 1.5 GHz and 1.9 GHz," *IEEE Trans. on Electromagnetic Compatibility*, Vol. 47, No. 1, 2005, pp. 68–76.

[13] Ng, E. Y. K., and E. H. Ooi, "FEM Simulation of the Eye Structure with Bioheat Analysis," *Comput. Meth. Programs in Biomedicine*, Vol. 82, No. 3, 2006, pp. 268–276.

[14] Ng, E. Y. K., and E. H. Ooi, "Ocular Surface Temperature: A 3D FEM Prediction Using Bioheat Equation," *Comput. Biol. Med.*, Vol. 37, No. 6, 2007, pp. 829–835.

[15] Ooi, E. H., et al., "Variations of Corneal Surface Temperature with Contact Lens Wear," *Proc. Institution of Mechanical Engineers, Part H, J. of Engineering in Medicine*, Vol. 221, No. 4, 2007, pp. 337–350.

[16] Ooi, E. H., W. T. Ang, and E. Y. K. Ng, "Bioheat Transfer in the Human Eye: A Boundary Element Approach," *Eng. Anal. Boundary Elements*, Vol. 31, No. 6, 2007, pp. 494–500.

[17] Ridouane, E. H., and A. Campo, "Numerical Computation of the Temperature Evolution in the Human Eye," *Heat Transfer Res.*, Vol. 37, No. 7, 2006, pp. 607–617.

[18] PTC/Pro-Engineer, http://www.ptc.com.

[19] COMSOL Multiphysics 3.2, http://www.comsol.com.

[20] Cicekli, U., "Computational Model for Heat Transfer in the Human Eye Using the Finite Element Method," M.Sc. Thesis, Department of Civil & Environmental Engineering, Louisiana State University, 2003.

[21] Pennes, H. H., "Analysis of Tissue and Arterial Blood Temperatures in the Resting Forearm," *J. of Applied Physiology*, Vol. 1, No. 2, 1948, pp. 93–122.

[22] Flyckt, V. M. M., B. W. Raaymakers, and J. J. W. Lagendijk, "Modeling the Impact of Blood Flow on the Temperature in the Human Eye and the Orbit: Fixed Heat Transfer Coefficient Versus the Pennes Bioheat Model Versus Discrete Blood Vessels," *Phys. Med. Biol.*, Vol. 51, No. 19, 2006, pp. 5007–5021.

[23] Mapstone, R., "Measurement of Corneal Temperature," *Exper. Eye Res.*, Vol. 7, No. 2, 1968, pp. 237–243.

[24] Incroprera, F., and D. P. DeWitt, *Fundamentals of Heat and Mass Transfer*, 5th ed., New York: John Wiley & Sons, 2002.

[25] Mori, A., et al., "Use of High Speed, High Resolution Thermography to Evaluate the Tear Film Layer," *Am. J. Ophthalmol.*, Vol. 124, No. 6, 1997, pp. 729–735.

[26] Efron, N., G. Young, and N. Brennan, "Ocular Surface Temperature," *Curr. Eye Res.*, Vol. 8, No. 9, 1989, pp. 901–906.

[27] Kocak, I., S. Orgul, and J. Flammer, "Variability in the Measurement of Corneal Temperature Using a Non-Contact Infrared Thermometer," *Ophthalmologica*, Vol. 213, No. 6, 1999, pp. 345–349.

[28] Morgan, P. B., A. B. Tullo, and N. Efron, "Infrared Thermography of the Tear Film in Dry Eye," *Eye*, Vol. 9, No. 5, 1995, pp. 615–618.

[29] Mori, A., et al., "Efficacy and Safety of Infrared Warming of the Eyelids," *Cornea*, Vol. 18, No. 2, 1999, pp. 188–193.

[30] Aihara, M., J. D. Lindsey, and R. N. Weinreb, "Aqueous Humor Dynamics in Mice," *Invest. Ophthalmol. Vis. Sci.*, Vol. 44, No. 12, 2003, pp. 5168–5173.

[31] Maurice, D., "The Von Sallmann Lecture 1996: An Ophthalmological Explanation of REM Sleep," *Exper. Eye Res.*, Vol. 66, No. 2, 1998, pp. 139–145.

[32] Canning, C. R., et al., "Fluid Flow in the Anterior Chamber of a Human Eye," *IMA J. Math. Appl. Med. Biol.*, Vol. 19, No. 1, 2002, pp. 31–60.

[33] Heys, J. J., and V. H. Barocas, "A Boussinesq Model of Natural Convection in the Human Eye and the Formation of Krukenberg's Spindle," *Ann. Biomed. Eng.*, Vol. 30, No. 3, 2002, pp. 392–401.

[34] Morgan, P. B., et al., "Potential Applications of Ocular Thermography," *Optometry Vis. Sci.*, Vol. 70, No. 7, 1993, pp. 568–576.

[35] Holmberg, A., "The Temperature of the Eye During Application of Hot Packs and After Milk Injections," *Acta Ophthalmologica*, Vol. 30, No. 4, 1952, pp. 347–364.

[36] Hill, R. M., and A. J. Leighton, "Temperature Changes of a Human Cornea and Tears Under a Contact Lens 1. The Relaxed Open Eye and the Natural and Forced Closed Eye Conditions," *Am. J. Optom. Arch. Am. Acad. Optometry*, Vol. 42, No. 2, 1965, pp. 9–16.

[37] Kinn, J. B., and R. A. Tell, "A Liquid-Crystal Contact Lens Device for Measurement of Corneal Temperature," *IEEE Trans. on Biomedical Engineering*, Vol. 20, No. 5, 1973, pp. 387–388.

[38] Rysa, P., and J. Sarvaranta, "Corneal Temperature in Man and Rabbit. Observations Made Using an Infra-Red Camera and a Cold Chamber," *Acta Ophthalmologica*, Vol. 52, No. 6, 1974, pp. 810–816.

[39] Hørven, I., "Corneal Temperature in Normal Subjects and Arterial Occlusive Disease," *Acta Ophthalmologica*, Vol. 53, No. 6, 1975, pp. 863–874.

[40] Fatt, I., and J. Chaston, "Temperature of a Contact Lens on the Eye," *Int. Contact Lens Clin.*, Vol. 7, 1980, pp. 195–198.

[41] Alio, J., and Padron, M., "Influence of Age on the Temperature of the Anterior Segment of the Eye," *Ophthalmic Res.*, Vol. 14, No. 3, 1982, pp. 153–159.

[42] Fielder, A. R., et al., "Problems with Corneal Arcus," *Trans. Ophthalmol. Soc. United Kingdom*, Vol. 101, No. 1, 1981, pp. 22–26.

[43] Martin, D., and I. Fatt, "The Presence of a Contact Lens Induces a Very Small Increase in the Anterior Corneal Surface Temperature," *Acta Ophthalmologica*, Vol. 64, No. 5, 1986, pp. 512–518.

[44] Girardin, F., et al., "Relationship Between Corneal Temperature and Finger Temperature," *Arch. Ophthalmol.*, Vol. 117, No. 2, 1999, pp. 166–169.

[45] Gugleta, K., S. Orgul, and J. Flammer, "Is Corneal Temperature Correlated with Blood-Flow Velocity in the Ophthalmic Artery?" *Curr. Eye Res.*, Vol. 19, No. 6, 1999, pp. 496–501.

[46] Craig, J. P., et al., "The Role of Tear Physiology in Ocular Surface Temperature," *Eye*, Vol. 14, No. 4, 2000, pp. 635– 641.

[47] Purslow, C., J. Wolffsohn, and J. Santodomingo-Rubido, "The Effect of Contact Lens Wear on Dynamic Ocular Surface Temperature," *Contact Lens and Anterior Eye*, Vol. 28, No. 1, 2005, pp. 29–36.

[48] Morgan, P. B., M. P. Soh, and N. Efron, "Corneal Surface Temperature Decrease with Age," *Contact Lens & Anterior Eye*, Vol. 22, No. 1, 1999, pp. 11–13.

[49] Siebinga, I., et al., "Age-Related Changes in Local Water and Protein Content of Human Eye Lenses Measured by Raman Microspectroscopy," *Exper. Eye Res.*, Vol. 53, No. 2, 1991, pp. 233–239.

[50] Spells, K. E., "The Thermal Conductivities of Some Biological Fluids," *Phys. Med. Biol.*, Vol. 5, No. 2, 1960, pp. 139–153.

[51] Neelakantaswamy, P. S., and K. P. Ramakrishnan, "Microwave-Induced Hazardous Non-linear Thermoelastic Vibrations of the Ocular Lens in the Human Eye," *J. Biomechanics*, Vol. 12, No. 3, 1979, pp. 205–210.

CHAPTER 12

Variations of the Corneal Surface Temperature with Contact Lens Wear

E. H. Ooi, Eddie Y. K. Ng, C. Purslow, and Rajendra Acharya

12.1 Introduction

In Chapter 11, we discussed the development of both the two-dimensional and three-dimensional human eye models. A heat transfer simulation was carried out using the commercial software package COMSOL Multiphysics 3.2 [1]. This chapter is dedicated to the application of the two-dimensional human eye model simulating the changes in ocular temperature distribution during contact lens wear. In addition, the thermal effects of different contact lens storage temperatures are investigated.

Many studies have been conducted regarding the effects of contact lens wear on the human cornea. These studies focused mainly on the effects of poor oxygen permeability and wettability of the contact, which may lead to corneal hypoxia, a condition commonly associated with irritation experienced by contact lens wearers [2]. Investigations on corneal surface temperature during contact lens wear, however, remain scarce. The importance of understanding how the corneal temperature changes during contact lens wear outlines the objectives of this study which are twofold:

1. The corneal threshold for pain induced by temperature rise was reported to be 2.8°C [3]. Contact lenses may or may not cause an increase in the corneal surface temperature. If an increase in temperature does occur and the increase is beyond this threshold, the irritation that is felt by a contact lens wearer may be caused by the thermal response of the cornea to the elevation in temperature.

2. An in vitro study conducted by Willcox et al. [4] discovered that bacteria have a tendency to bind onto the contact lens at low temperatures (25°C compared to 37°C). This may cause an infection on the corneal surface, which should be prevented.

These statements imply that the corneal surface temperature during contact lens wear should be kept below a certain level to avoid any discomfort that may result from the thermal responses of the cornea. The corneal temperature, however, should not be too low such that it stimulates the binding of bacteria.

12.2 A Brief History of Contact Lens

Contact lenses are used mainly for vision correction. Lately though, application of contact lenses can be traced to areas such as cosmetic, therapeutics [5–7], and blood glucose detection [8]. The first contact lens was introduced in 1887 and was made of glass. It was developed by the Swiss physician A. E. Fick and the French optician Edouard Kalt to correct nearsightedness and farsightedness.

William Feinbloom constructed the first plastic contact lens in 1936. The center of the lens was made of glass while the periphery was made of plastic that covered the sclera. Soon after, opticians such as Kevin Tuohy and George Butterfield introduced fully plastic contact lenses, which were smaller and thinner but still covered the whole cornea. These lenses were called *hard contacts*. Throughout 1950 to 1960, research and development of contact lens was focused on the manufacturing of thinner and smaller lenses.

Soft contact lens was first introduced in the early 1960s by Otto Wichterle using soft, water-absorbing plastic called hydroxyethyl methacrylate (HEMA). The success of developing soft contact lens was credited to the construction of the spin casting machine by Wichterle. In 1971, the first commercialized soft contact lens was made available by Bausch & Lomb in the United States and has since been widely used by almost all contact lens wearers in the world.

The soft contact lens, however, had poor permeability towards oxygen, which resulted in corneal hypoxia among long-term wearers. This had led to the development of the rigid gas permeable (RGP) contact lens, which was made of polymethyl methacrylate (PMMA) and silicone in 1979. In 1981, the RGP lens became available for commercial distribution. Between 1981 and 1998, research on contact lens resulted in various disposable contact lenses. Finally in 1999, the first silicone hydrogel lens—better known as the extended-wear contact lens—was introduced. The silicone hydrogel lenses permitted wearers to wear the contact lens for an extended period without causing corneal hypoxia.

12.3 Tear Film and Tear Evaporation

On the surface of the cornea, one may find a layer of tear film. The thickness of the tear film is between 7 and 10 μm [9]. It consists of three layers: the lipid layer, the aqueous layer, and the mucous layer. The lipid layer is an oily layer that is at the top most layer of the tear film. It helps to reduce evaporation of the tear film from the corneal surface.

When a contact lens is placed onto the cornea, the tear film breaks into two layers: the pre-lens tear film (PLTF), which lies on top of the contact lens, and the post-lens tear film (PoLTF), which sits between the posterior contact lens surface and the corneal surface. This is illustrated in Figure 12.1. The PLTF and PoLTF were measured to be 2.31 and 2.34 μm, respectively [10]. The thin layer of the PLTF was

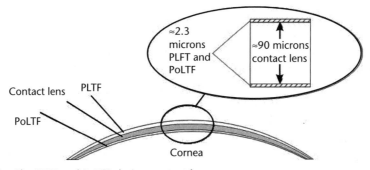

Figure 12.1   The PLTF and PoLTF during contact lens wear.

found to contribute insignificantly to the measurement of the contact lens surface temperature [11].

As the tear film breaks into the PLTF and the PoLTF, the lipid layer that retards evaporation destabilizes. This, in turn, results in an increase in tear evaporation rate [12–14]. The increase in tear evaporation during contact lens wear is independent of the water content inside the contact lens [13], the type of contact lens worn (hydrogel and silicone hydrogel) [15], and the conditions of the contact lens (old and new) [10].

12.4   Contact Lens and Corneal Surface Temperature

The terms *hot* and *burning* are usually used by contact lens wearers to describe the irritation felt when wearing contact lens [16]. Such descriptions have also been used by dry eye patients [17–19]. Although there may not be any immediate link between these descriptions and the temperature changes during contact lens wear, it remains possible that the irritation in eyes caused by contact lens wear may be attributed to the changes in its temperature.

Early attempts to measure the corneal surface temperature during contact lens wear involved the use of a scleral contact lens with a temperature probe embedded in the center of the lens [20, 21]. Using this method, a drop in temperature was found (31.3°C versus 32.1°C), which was suggested to be insignificant. Similar results were reported by Dixon and Blackwood [22] using a microthermistor probe (33.3°C versus 33.7°C).

Fatt and Chaston [23] used a noncontact bolometer to measure the temperature on the corneal surface during contact lens wear for both hard and soft contact lenses. Similarly, they found a decrease in temperature, and this decrease was not significantly affected by the types of contact lens worn (hard or soft contact lens).

Ophthalmologists began using IR thermography when it became available to study the temperature changes during contact lens wear where irregular thermal patterns were reported on the corneal surface [24]. A more extensive study was conducted by Purslow et al. [25]. In their work, three types of soft contact lens were studied: Lotrafilcon A, Balafilcon A, and Etafilcon A. The first two are silicone hydrogels, and the latter, conventional hydrogels. These lenses differ mainly in their percentage of water content as well as the presence of a silicone compound inside the silicone hydrogels, both of which produce different thermal properties. For all types of lenses studied, an increase in corneal temperature was recorded; this was attributed to the insulating properties of the contact lens [25]. Modality of wear (daily wear and continuous wear) was found to contribute insignificantly to the corneal surface temperature. The largest elevation in corneal temperature was reported to be 3.6°C. Based on the temperature threshold for corneal pain, which was reported at 2.8°C [3], the increase of 3.6°C may cause discomfort to wearers.

Martin and Fatt [26] developed a simplified mathematical model to study the changes in corneal temperature during contact lens wear. Only the cornea and contact lens were considered. Simulations were carried out for various contact lens' thicknesses of and percentage of water content inside the contact lens. Using this model, they found that the temperature on the corneal surface increases when the thicknesses and the percentage of water content inside the contact lens increase.

12.5   The Contact Lens Model

12.5.1   Model Development

The two-dimensional human eye model presented in Chapter 11 is used in this study. Three types of contact lens are simulated, namely, Lotrafilcon A, Balafilcon A, and Etafilcon A (similar to those used by Purslow et al. [25]). The dimensions of these lenses were obtained from [25] and are listed in Table 12.1 along with the water composition for each lens. The dimensions given in Table 12.1 are shown graphically in Figure 12.2. Due to the different base curves between the contact lens and the cornea, a gap exists between the cornea and the contact lens. To maintain the dimensions of the eye model and the contact lens, this gap is treated as the PoLTF despite its large thickness, which is approximately 481 μm. This is shown in Figure 12.3.

12.5.2   Governing Equation and Boundary Conditions

Similar to the work presented in Chapter 11, the metabolic activity inside the human eye and the blood flow inside the iris and ciliary body are neglected. This results in a

Table 12.1   Specifications and Properties of Contact Lenses Used in this Study

Contact Lens Material	Lotrafilcon A	Balafilcon A	Etafilcon A
Lens type	Silicone hydrogel	Silicone hydrogel	Hydrogel
Water content (%)*	24	36	58
Base curve (mm)*	8.4	8.6	9.0
Diameter (mm)*	14.2	14.0	13.8
Center thickness (mm)*	0.080	0.090	0.084
Thermal conductivity (Wm^{-1}K^{-1})	0.281	0.331	0.432
Specific heat (J kg^{-1}K^{-1})	2,259.90	2,543.00	3,321.60
Density (kg m^{-3})	1,080	1,064	1,050
Emissivity	0.955	0.958	0.966

* Data obtained from Purslow et al. [25].

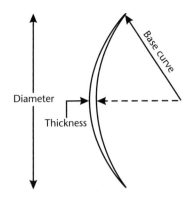

Figure 12.2   Typical parameters in classifying the size of contact lenses. (*From:* [11]. © 2007 Professional Engineering Publishing. Reprinted with permission.)

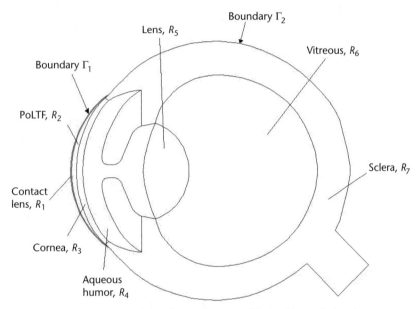

Figure 12.3   The contact lens model with post-lens tear film and boundaries shown.

governing equation that is similar to the classical heat diffusion equation, which is given by

$$\rho_i c_i \frac{\partial T_i}{\partial t} = \nabla \cdot (k_i \nabla T_i) \quad \text{for } i = 1, 2, 3, 4, 5, 6, \text{ and } 7 \tag{12.1}$$

where ρ is density, c is specific heat, k is thermal conductivity, T is temperature, and t is time. Subscripts i refer to each of the regions shown in Figure 12.3.

The values of thermal conductivity, density, and specific heat are required for simulating the propagation of heat inside the human eye. The values of thermal conductivity for each of the contact lens are obtained based on the method used by Martin and Fatt [26], which states that the thermal conductivity of a contact lens can be computed by summing the weight fractions (see Table 12.1) of polymer and water contained in the hydrated lens. Mathematically, this can be written as

$$k_{CL} = \frac{\left((100 - \%\text{water}) \times k_{poly} \times \rho_{poly}\right) + \left(\%\text{water} \times k_{water} \times \rho_{water}\right)}{\rho_{CL}} \qquad (12.2)$$

where k_{CL}, k_{poly}, and k_{water} are the thermal conductivity for the contact lens, dehydrated contact lens polymer, and water, respectively; ρ_{CL}, ρ_{poly}, and ρ_{water} are the density for the contact lens, dehydrated contact lens polymer, and water, respectively; and %water refers to the percentage of water content inside the contact lens.

The thermal conductivity and density of water are chosen as 0.6232 W m^{-1} K^{-1} and 1,000 kg m^{-3} (at 34°C), respectively. For the dry contact lens polymer, the value of thermal conductivity and density are chosen as 0.167 W m^{-1}K^{-1} [26] and 1,210 kg m^{-3} [27], respectively.

The same approach used to approximate the thermal conductivity of contact lens is also applied to calculate the values of specific heat [28]. This would, however, require the knowledge of the specific heat of dry polymer, which is not available to us [11]. To enable predictions of contact lens specific heat, we assume that a linear relationship exists between specific heat and percentage of water content inside the contact lens. Therefore, if the specific heat of water is chosen as 4,178 J kg^{-1} K^{-1} and using the value of specific heat of polyacrylamide (a tissue-equivalent polymer with water content of 71.8% [29]), the specific heat for each contact lens may be evaluated using linear interpolation.

The values of density for each type of contact lens can be found from the data sheets that are available on the Internet. The estimated thermal conductivity and specific heat values for each type of contact lens are summarized in Table 12.1 along with the values of density obtained from the Internet.

Values of thermal conductivity, specific heat, and density for each of the eye components are similar to those used in [30]. These values are tabulated in Table 12.2. In this study, the PoLTF is assumed to have properties similar to those of water. The values of thermal conductivity, specific heat, and density for water are 0.6232 Wm^{-1} K^{-1}, 1,000 kg m^{-3}, and 4,178 J kg^{-1} K^{-1}, respectively.

To solve (12.1), the initial boundary conditions of the problem must be specified. On the surface of the contact lens that is exposed to the environment, we may write

$$-k_1 \frac{\partial T_1}{\partial \mathbf{n}} = h_{amb}\left(T_1 - T_{amb}\right) + \varepsilon\sigma\left(T_1^4 - T_{amb}^4\right) + E \quad \text{on } \Gamma_1 \qquad (12.3)$$

Table 12.2 Thermal Properties for Each Component of the Human Eye

Domains	Index, i	Thermal Conductivity (W m^{-1}K^{-1})	Specific Heat (J kg^{-1} K^{-1})	Density (kg m^{-3})
Cornea	3	0.58	4,178	1,050
Aqueous humor	4	0.58	3,997	996
Lens	5	0.40	3,000	1,050
Vitreous body	6	0.60	4,178	1,000
Sclera	7	1.00	3,180	1,100

where h_{amb} is the ambient convection coefficient; T_{amb} is ambient temperature; ε and σ are emissivity and the Stefan-Boltzmann constant, respectively; E is the heat loss due to tear evaporation; and $\partial T_1/\partial \mathbf{n}$ is the rate of change of T_1 in the outward direction normal to surface Γ_1.

Similarly, on the surface of the sclera, we can write

$$-k_7 \frac{\partial T_7}{\partial \mathbf{n}} = h_{bl}\left(T_7 - T_{bl}\right) \quad \text{on } \Gamma_2 \tag{12.4}$$

where h_{bl} is the blood convection coefficient, T_{bl} is blood temperature, and $\partial T_7/\partial \mathbf{n}$ is the rate of change of T_7 in the outward direction normal to surface Γ_2. Details on the boundary conditions in (12.3) and (12.4) are described in Chapter 11. Values for each of the parameters used in (12.3) and (12.4) are listed in Table 12.3.

The values for contact lens emissivity are not available in the literature. To enable simulations to be carried out, we use the same approach that is used to approximate the specific heat of contact lenses to predict the emissivity on the surface of each contact lens. To do so, we have chosen the emissivity of water to be 0.981 at 298K, and the reference emissivity is chosen to be that of skin, which is 0.971 at 293.5K [31]. The approximated emissivity values are listed in Table 12.1.

The tear evaporation rate values merit some discussion. Observations made using experimental work have found that the evaporation of tears increases during contact lens wear. Furthermore, the type of contact lens was found to produce no effects on the increase in tear evaporation [12–15]. Measurements conducted by different researchers have resulted in different values of evaporation; this may be due to individual variation. To simplify our work, we use the measurement reported by Mathers [14], which has a value of 23.0×10^{-7} g cm^{-2} s^{-1}. If the latent heat of evaporation is chosen to be 2,426.7 kJ kg^{-1} at body temperature [32], the heat loss due to tear evaporation during contact lens wear is calculated to be 55.8 W m^{-2}.

On the interface between two contiguous regions, we assume continuity in temperature and heat flux such that

$$T_i = T_j \text{ and } k_i \frac{\partial T_i}{\partial \mathbf{n}} = k_j \frac{\partial T_j}{\partial \mathbf{n}} \text{ an } I_{ij} \tag{12.5}$$

Table 12.3   Constant Parameters Used in Boundary Conditions

Parameter	Value
Ambient temperature,* T_{amb} (°C)	21
Blood temperature, T_{bl} (°C)	37
Ambient convection coefficient, h_{amb} (Wm^{-2}K^{-1})	10
Blood convection coefficient, h_{bl} (Wm^{-2}K^{-1})	65
Heat loss due to tear evaporation, E (Wm^{-2})	55.8
Stefan-Boltzmann constant (Wm^{-2}K^{-4})	5.67×10^{-8}

*Similar to the experimental study conducted in [25].

where T_i and T_j are the temperatures in regions R_i and R_j, respectively, $\partial T_i/\partial \mathbf{n}$ and $\partial T_j/\partial \mathbf{n}$ are the rate of change of T_i and T_j in the outward direction normal to surface I_{ij}, and I_{ij} is the interface between regions R_i and R_j.

The initial condition used in the transient analysis is obtained from the steady-state solution of (12.1) when no contact lens is worn, which is written as

$$\nabla \cdot (k_i \nabla T_i) = 0 \quad \text{for } i = 3, 4, 5, 6, \text{ and } 7 \tag{12.6}$$

12.5.3 Numerical Methodology

All simulations are conducted in the workspace of COMSOL Multiphysics 3.2 [1], which utilizes the finite element method. Discretizations of the eye model are carried out using triangular elements in which the quadratic Lagrangian polynomial is used to approximate the geometrical order and the shape function of each element.

Domains and boundaries that are relatively thin (contact lens, PoLTF, and cornea) are given extra refinement. As a result, 47,899, 52,367, and 51,661 elements for the Lotrafilcon A, Balafilcon A, and Etafilcon A models, respectively, are generated. The differences in the number of elements for different types of contact lenses are attributed to the different dimensions of each of the contact lens. Numerical experiments indicated convergence using these numbers of elements. Figure 12.4 illustrates the meshed structure of the human eye model fitted with a Lotrafilcon A contact lens. The enlarged view of the contact lens shows the contact lens layer, which has been finely discretized.

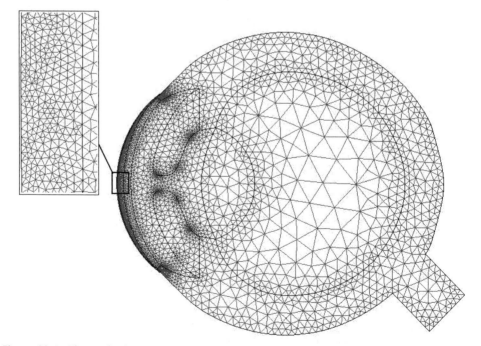

Figure 12.4 The meshed structure of the human eye model with enlarged portion for contact lens.

12.6   Numerical Predictions and Analysis

12.6.1   Steady-State Analysis

Using the values of control parameters given in Table 12.2, the temperature at the center of the corneal surface for a bare cornea (without contact lens) is found to be 32.734°C. When a contact lens is placed onto the cornea, a drop in temperature is observed on the corneal surface (see Table 12.3). As the percentage of water content inside the contact lens increases, the drop in temperature becomes greater. The average decrease in temperature is approximately 0.52°C ± 0.05°C. This observation agrees with the reports given in [19–23] but contradicts with the results found by Purslow et al. [25] and Martin and Fatt [26].

The difference between the simulated results from our model and the one reported by Martin and Fatt [26] may be due to the different modeling approaches employed and the different boundary conditions used. In our model, the presence of PoLTF, which has a thickness of 481 μm, is nearly 200 times greater than the actual thickness of the PoLTF. Therefore contain, the results predicted by our model may inaccuracies.

To investigate the effects of the PoLTF assumption, each contact lens is again remodeled with base curves matching that of the cornea, such as that shown in Figure 12.5. In this case, the presence of PoLTF is eliminated from the model. Using the values of the control parameters given in Table 12.2, the simulations are repeated and results are presented in Table 12.4. A decrease in temperature is still observed. The magnitude of the decrease, however, is smaller than the case when the layer of PoLTF is present.

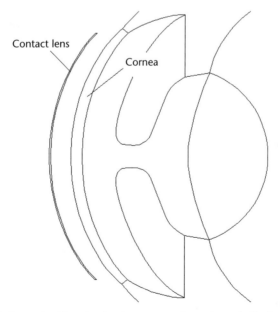

Contact lens

Cornea

Figure 12.5   Model of a contact lens having base curve dimensions similar to those of the cornea. (*From:* [11]. © 2007 Professional Engineering Publishing. Reprinted with permission.)

Table 12.4 Results Simulated for Steady-State Solution

Bare Cornea Temperature: 32.7340°C	*Lotrafilcon A*	*Balafilcon A*	*Etafilcon A*
Water content (%)	24	36	58
Contact lens surface temperature (°C)	32.0025	31.9769	31.8683
Corneal surface temperature (°C)	32.2436	32.2352	32.1554
Temperature difference with the control eye (°C)	0.4904	0.4988	0.5786

12.6.2   Transient Analysis

Simulations with different initial contact lens temperatures may provide information to ophthalmologists that may help them understand the thermal effects of different contact lens storage temperatures on the corneal surface. Five different initial contact lens temperatures were investigated: 16°C, 21°C, 25°C, 30°C, and 35°C. The transient temperature responses at the center of the cornea with time for each of these initial temperatures are plotted in Figures 12.6 through 12.10 using a logarithmic timescale.

When the contact lens is stored at temperatures below that of the bare cornea (32.734°C), the corneal surface temperature experiences a drop immediately after the insertion of contact lens (see Figures 12.6 through 12.9). On the other hand, insertion of contact lens stored above the bare cornea temperature causes a temperature rise (Figure 12.10). Steady state is reached approximately 1,800 seconds after the contact lens is placed on the cornea.

Figure 12.11 shows the variation of corneal temperature with time when the contact lens is stored at a temperature that is the same as the bare cornea tempera-

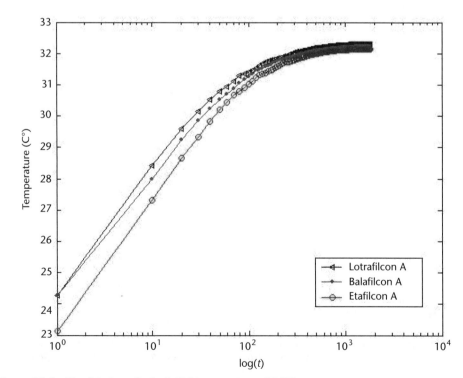

Figure 12.6   Simulated results for initial temperature of 16°C.

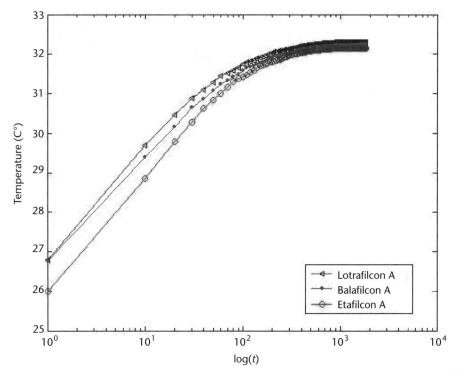

Figure 12.7   Simulated results for initial temperature of 21°C.

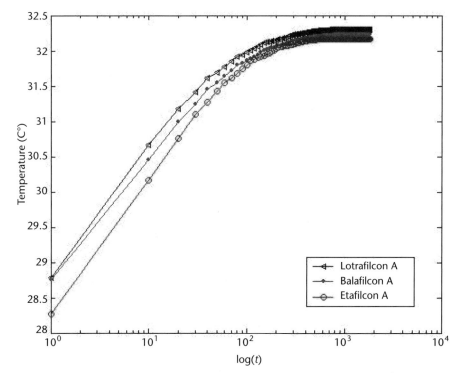

Figure 12.8   Simulated results for initial temperature of 25°C.

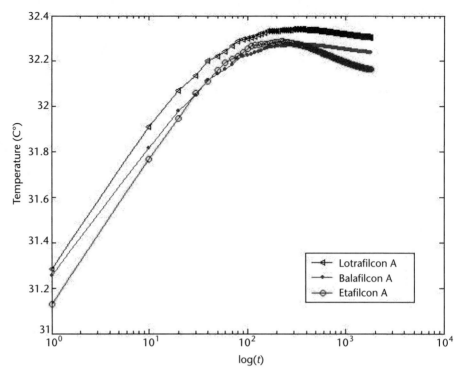

Figure 12.9   Simulated results for initial temperature of 30°C.

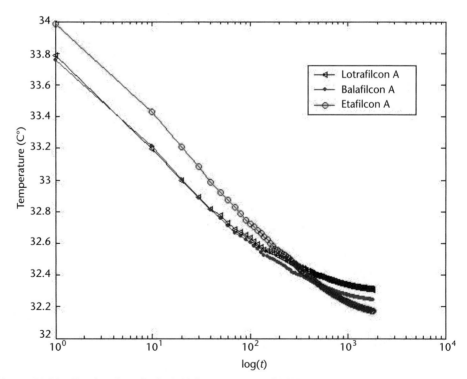

Figure 12.10   Simulated results for initial temperature of 35°C.

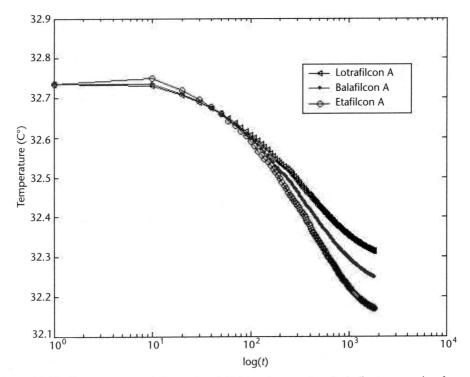

Figure 12.11   Temperature variations when initial lens temperature is similar to corneal surface temperature.

ture (32.734°C). No changes in temperature are found on insertion of the contact lens. As time progresses, the corneal temperature decreases and proceeds towards steady state.

Results from these transient studies show that there is a rapid change in temperature during the first few hundred seconds after the insertion of contact lenses. This behavior, however, is not present when the initial temperature of the contact lens is similar to the temperature on the corneal surface before insertion (see Figure 12.11). Based on these results, it appears that if measurements of temperature on the contact lens are to be taken, sufficient time should be given for the temperature to reach a stable condition. Failure to do so may result in an inaccurate measurement.

12.7   Summary and Discussion

The results found in this study, which show a decrease in temperature during contact lens wear, may be explained by the increase in tear evaporation. Insertion of contact lens into the eye not only disrupts the tear film but also destroys the stability of the lipid layer. Without the lipid layer, excessive loss of tears occurs through evaporation on the surface of the cornea and this leads to excessive amount of heat loss.

The constant value of 55.8 W m^{-2} used to simulate heat loss due to the evaporation from contact lens, which we have assumed to be constant, may not be entirely accurate. In reality, the tear film is a dynamic structure [31]. During each blink of the eyelids, fresh tears are deposited onto the corneal surface. As a result, a new layer of lipid is spread across the corneal surface, thus retarding the evaporation of tears. This implies that the heat loss due to tear evaporation may change with time. This issue must be given particular attention in future studies.

As mentioned in the preceding paragraph, the blinking of eyelids results in the deposition of fresh tears across the corneal surface. The temperature of the newly formed tear layer is greater than the temperature of the tear layer before blinking. The periodical blinking of eyelids results in a cyclic variation of the corneal surface temperature, which has been reported in [33, 34].

The difficulty in incorporating blinking effects into the eye model lies in the scarce anatomical and physiological information pertaining to its heat transfer. During the closure of eyelids, heat loss due to convection, radiation, and tear evaporation is prevented. The eyelids, which contain small blood vessels [9], may, however, warm up the surface of the cornea. This process may be modeled as a boundary condition that is similar to (12.4). Alternatively, we may consider the eyelids as a heat retarding layer such that no transfer of heat takes place during the phase when the eyelids are covering the corneal surface. These suggestions remain speculative and may be investigated in future studies.

Comparisons between the results obtained from the present work disagree with the results presented by Purslow et al. [25]. No experimental work is carried out in this study and, thus, no benchmarking of the numerical results is conducted. Without proper validation of the numerical predictions, it is difficult to evaluate the accuracy of the model, particularly the assumptions that have been made.

Investigations on the effects of contact lens wear on the corneal surface temperature may still require extensive studies, both numerical and experimental. Information obtained from these studies may be useful to ophthalmologists and optometrists in understanding the importance of thermal factors when designing contact lenses as well as prescribing the duration of contact lens wear.

References

[1]   COMSOL Multiphysics 3.2, http://www.comsol.com.

[2]   Hamano, H., and H. E. Kaufman, *Corneal Physiology and Disposable Contact Lenses*, Boston, MA: Butterworth-Heinemann, 1997.

[3]   Beuerman, R. W., and D. L. Tanelian, "Corneal Pain Evoked by Thermal Stimulation," *Pain*, Vol. 7, No. 1, 1979, pp. 1–14.

[4]   Willcox, M. D. P., et al., "Bacterial Interactions with Contact Lenses; Effects of Lens Material, Lens Wear and Microbial Physiology," *Biomaterials*, Vol. 22, No. 24, 2001, pp. 3235–3247.

[5]   Andrew, N. C., and E. G. Woodward, "The Bandage Lens in Bullous Keratopathy," *Ophthalmic Physiolog. Opt.*, Vol. 9, No. 1, 1989, pp. 66–68.

[6]   Montero, J., J. Sparholt, and R. Mely, "Retrospective Case Series of Therapeutic Applications of a Lotrafilcon A Silicone Hydrogel Soft Contact Lens," *Eye Contact Lens*, Vol. 29 (Supplement), 2003, pp. S54–S56.

[7]   Kanpolat, A., and O. Ucakhan, "Therapeutic Use of Focus® Night and Day™ Contact Lenses," *Cornea*, Vol. 22, No. 8, 2003, pp. 726–734.

[8]   Badugua, R., J. R. Lakowicza, and C. D. Geddes, "A Glucose Sensing Contact Lens: A New Approach to Non-Invasive Continuous Physiological Glucose Monitoring," *Proc. SPIE,* I. Gannot, (ed.), Vol. 5317, 2004, pp. 234–245.

[9]   Remington, L. A., *Clinical Anatomy of the Visual System,* 2nd ed., Boston, MA: Elsevier Butterworth Heinemann, 2005.

[10] Nichols, J. J., and P. E. King Smith, "The Effects of Eye Closure on the Post Lens Tear Film Thickness During Silicone Hydrogel Contact Lens Wear," *Cornea,* Vol. 22, No. 6, pp. 539–544, 2001.

[11] Ooi, E. H., et al., "Variations of Corneal Surface Temperature with Contact Lens Wear," *Proc. Institution of Mechanical Engineers, Part H, J. Eng. Med.,* Vol. 221, No. 4, 2007, pp. 337–350.

[12] Hamano, H., "The Change of Precorneal Tear Film by the Application of Contact Lenses (Japanese)," *Contact and Intraocular Lens Med. J.,* Vol. 7, No. 3, 1981, pp. 205–209.

[13] Cedarstaff, T. H., and A. Tomlinson, "A Comparative Study of Tear Evaporation Rates and Water Content of Soft Contact Lenses," *American Journal of Optometry and Physiological Optics,* Vol. 60, No. 3, 1983, pp. 197–174.

[14] Mathers W., "Evaporation from the Ocular Surface," *Exper. Eye Res.,* Vol. 78, No. 3, 2004, pp. 389–394.

[15] Thai, L., M. Doane, and A. Tomlinson, "Effect of Different Soft Contact Lens Materials on the Tear Film," *Invest. Ophthalmol. Vis. Sci.,* Vol. 43, 2002, p. 3083.

[16] Du Toit, R., et al., "The Effects of Six Months of Contact Lens Wear on the Tear Film, Ocular Surfaces and Symptoms of Presbyopes," *Optom. Vis. Sci.,* Vol. 78, No. 6 2001, pp. 455–462.

[17] Begley, C. G., et al., "Use of the Dry Eye Questionnaire to Measure Symptoms of Ocular Irritation in Patients with Aqueous Deficient Dry Eye," *Cornea,* Vol. 78, No. 7, 2002, pp. 664–670.

[18] Bandeen-Roche, L., et al., "Self-Reported Assessment of Dry Eye in a Population Based Setting," *Invest. Ophthalmol. Vis. Sci.,* Vol. 38, No. 12, 1997, pp. 2469–2475.

[19] McMonnies, C. W., "Key Questionnaire in a Dry Eye History," *J. of Am. Optometric Assoc.,* Vol. 57, No. 7, 1986, pp. 512–517.

[20] Hill, R. M., and A. J. Leighton, "Temperature Changes of a Human Cornea and Tears Under a Contact Lens 1: The Relaxed Open Eye and the Natural and Forced Closed Eye," *Am. J. Optometry Arch. Am. Acad. Optometry,* Vol. 42, 1965, pp. 9–16.

[21] Hill, R. M., and A. J. Leighton, "Temperature Changes of Human Cornea and Under a Contact Lens 2: Effects of Intermediate Lid Apertures and Gaze," *Am. J. Optometry Arch. Am. Acad. Optometry,* Vol. 42, 1965, pp. 71–77.

[22] Dixon, J., and L. Blackwood, "Thermal Variations of the Human Eye," *Trans. Am. Ophthalmol. Soc.,* Vol. 89, No. 1, 1991, pp. 183–193.

[23] Fatt, I., and J. Chaston, "Temperature of a Contact Lens on the Eye," *Int. Contact Lens Clin.,* Vol. 7, 1980, pp. 195–198.

[24] Montoro, J., et al., "Use of Digital Infrared Imaging to Objectively Assess Thermal Abnormalities in the Human Eye," *Thermology,* Vol. 3, 1991, pp. 242–248.

[25] Purslow, C., J. Wolffsohn, and J. Santodomingo-Rubido, "The Effect of Contact Lens Wear on Dynamic Ocular Surface Temperature," *Contact Lens Anterior Eye,* Vol. 28, No. 1, 2005, pp. 29–36.

[26] Martin, D., and I. Fatt, "The Presence of a Contact Lens Induces a Very Small Increase in the Anterior Corneal Surface Temperature," *Acta Ophthalmologica,* Vol. 64, No. 5, 1986, pp. 512–518.

[27] Refojo, M. F., "Contact Lenses," in *Encyclopedia of Polymer Science and Technology,* Supplement 1, New York: Wiley, 1976, pp. 195–219.

[28] Ishikiriyama, K., and M. Todoki, "Heat Capacity of Water in Poly(Methyl Methacrylate) Hydrogel Membrane for an Artificial Kidney," *J. of Polymer Sci., Part B: Polymer Phys.*, Vol. 33, No. 5, 2003, pp. 791–800.

[29] Suriowec, A., et al., "Utilization of a Multilayer Polyacrylamide Phantom for Evaluation of Hyperthermia Applicators," *Int. J. Hyperthermia*, Vol. 8, No. 6, 1992, pp. 795–807.

[30] Ng, E. Y. K., and E. H. Ooi, "FEM Simulation of the Eye Structure with Bioheat Analysis," *Comput. Meth. Programs in Biomedicine*, Vol. 82, No. 3, 2006, pp. 268–276.

[31] Togawa, T., "Non-Contact Skin Emissivity: Measurement from Reflectance Using Step Change in Ambient Radiation Temperature," *Clin. Phys. Physiological Measurement*, Vol. 10, No. 1, 1989, pp. 39–48.

[32] Phase Changes, http://hyperphysics.phy-astr.gsu.edu/HBASE/thermo/phase.html.

[33] Mapstone, R., "Determinants of Ocular Temperature," *Br. J. Ophthalmol.*, Vol. 52, No. 10, 1968, pp. 729–741.

[34] Efron, N., G. Young, and N. Brennan, "Ocular Surface Temperature," *Curr. Eye Res.*, Vol. 8, No. 9, 1989, pp. 901–906.

CHAPTER 13

An Axisymmetric Boundary Element Model for Bioheat Transfer in the Human Eye

E. H. Ooi, W.-T. Ang, and Eddie Y. K. Ng

13.1 Introduction

In Chapter 11, finite element models of the human eye were discussed. Comparisons between the predicted ocular surface temperature from finite element models and thermographic measurements reported in the literature showed very good agreement. A disadvantage of the finite element method is that the solution domain has to be divided into many small elements. This may be a complicated task for a multilayered structure like the human eye.

The boundary element method is a relatively new numerical technique for analyzing problems in engineering and physical sciences. One of the main features of the boundary element method, which perhaps is the reason behind its superiority, is that it requires only the boundary of the solution domain to be discretized. Consequently, the system of linear algebraic equations generated by the method is much smaller than that of the finite element method. This helps ease requirements on computer memory.

The boundary element method has several advantages as a numerical tool for modeling heat flow inside the human eye. For one, as mentioned earlier, less computer memory is used because only the boundaries of the different regions of the human eye have to be discretized. The boundary element method solves for the unknown temperature and heat flux on the boundaries and interfaces between different regions. Interior temperature and heat flux are calculated as a postprocessing step. For certain purposes, it may be sufficient to calculate the boundary temperature on the cornea for validation with measurements obtained using infrared thermography. Another advantage of the boundary element method is that it is generally known to compute secondary variables, such as heat flux, more accurately than the finite element method [1].

Recently, bioheat transfer in a two-dimensional human eye model was solved using the boundary element method [2]. The geometries and boundary conditions closely followed those given in [3]. Temperature predicted by the boundary element model showed close agreement with the finite element calculation in [3]. The boundary element method employed only 470 elements on the boundary, while as many as 8,557 domain elements were used in the finite element method to obtain comparable results. A smoothly varying heat flux across the corneal surface was obtained by the boundary element method. This was not the case in the finite element model where wild fluctuations of the heat flux were observed at the vicinity of the corneal edges [2].

The models in [2, 3] are essentially two-dimensional ones based on the assumption that the temperature variation in the direction perpendicular to the cross section of the eye being analyzed is comparatively small. The two-dimensional model may underestimate the warming effects of blood flow in the choroid. The temperature distribution inside the human eye may best be simulated using a geometrically three-dimensional model. However, efforts may be hampered by the requirement of larger computer memory. Even for the boundary element method, a full three-dimensional model would require surfaces to be discretized, a task that may not be easy in general. Nevertheless, if the optic nerve is ignored, the human eye can be modeled as geometrically axisymmetric about the pupillary axis of the eye. Furthermore, the boundary conditions can also be considered axisymmetric about the pupillary axis. With such assumptions, it is possible to carry out analysis that is essentially two dimensional, while simulating three-dimensional physical phenomena.

This chapter focuses on the development of an axisymmetric human eye model. The equation governing the flow of heat inside the eye is solved numerically using the axisymmetric boundary element method. Numerical results obtained from the axisymmetric model are compared with those obtained from a similar three-dimensional finite element model. The finite element model includes the optic nerve and is not axisymmetric. This allows us to examine whether it is justifiable to neglect the optic nerve in the axisymmetric boundary element model. Computation of the finite element model is carried out using a commercial software package, COMSOL Multiphysics 3.2 [4]. Sensitivity analyses are carried out using the boundary element model to identify parameters having dominant effects on the temperature distribution inside the eye.

13.2 The Axisymmetric Human Eye Model

The axisymmetric human eye model is obtained by rotating the regions in Figure 13.1 about the z-axis. Internal structures such as the aqueous humor, the lens, the vitreous, and the sclera are modeled based on anatomical data given in [5–7]. As in [2, 8], the iris, the retina, the choroid, and the sclera are modeled as a single homogeneous region. The eye model consists of five regions: the cornea, the aqueous humor (both anterior and posterior chamber), the lens, the vitreous, and the sclera. The regions are denoted by R_i, as shown in Figure 13.1. For example, R_1 denotes the cornea.

Each region is considered to be homogeneous. The heat flow in it is assumed to be isotropic, following the Fourier law of heat conduction. Consequently, the steady-state temperature field satisfies the Laplace equation throughout the entire

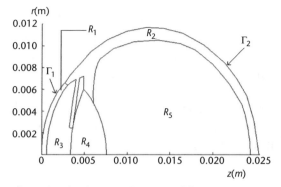

Figure 13.1   The two-dimensional axisymmetric eye model.

eye. The values of the thermal conductivities are obtained from the literature as recorded in Table 13.1.

13.2.1   The Governing Equation

The steady-state heat flow inside the human eye is governed by the Laplace equation:

$$\nabla\left(k_i\nabla T_i(r,z)\right) = 0 \text{ in } R_i\left(\text{for } i = 1, 2, 3, 4, \text{ and } 5\right) \tag{13.1}$$

where k_i is the thermal conductivity and T_i is temperature in R_i. In writing (13.1), the heat generated from the metabolic activity inside the eye and the blood flow inside the iris and ciliary body are neglected [9].

For solving (13.1), appropriate boundary conditions must be specified on Γ_1 and Γ_2. At Γ_1 (the corneal surface), heat from inside the eye is transferred away from the corneal surface via convection and radiation if the ambient temperature is lower than the corneal temperature. The tear film sits on top of the corneal surface and is constantly evaporated and refreshed after each eye blink. The evaporation of tears aids in the cooling of the corneal surface. The boundary condition on the corneal surface may be written as

$$-k_1\frac{\partial T_1}{\partial \mathbf{n}} = h_{amb}\left(T_1 - T_{amb}\right) + \varepsilon\sigma\left(T_1^4 - T_{amb}^4\right) + E_{vap} \text{ on } \Gamma_1 \tag{13.2}$$

Table 13.1 Thermal Conductivity of Each Eye Region

Domains	Index, i	Thermal Conductivity, k (Wm^{-1}K^{-1})
Cornea	1	0.58 [9]
Sclera	2	1.0042 [8]
Aqueous humor	3	0.58 [9]
Lens	4	0.40 [10]
Vitreous	5	0.603 [10]

where the first, second, and third terms on the right-hand side correspond to the heat loss due to convection, radiation, and tear evaporation, respectively; $\partial T_1/\partial \mathbf{n}$ is the rate of change in T_1 in the outward unit normal vector to the external corneal surface Γ_1; h_{amb} and T_{amb} are the ambient convection coefficient and the ambient temperature, respectively; and E_{vap} is the heat loss due to tear evaporation. The nonlinear term appearing in (13.2) is modeled based on the Stefan-Boltzmann law.

The boundary condition on Γ_2 (the external surface of the sclera), where heat from blood flow in the choroid enters the eye via convection [10], can be written as

$$-k_2 \frac{\partial T_2}{\partial \mathbf{n}} = h_{bl}(T_2 - T_{bl}) \text{ on } \Gamma_2 \tag{13.3}$$

where $\partial T_2/\partial \mathbf{n}$ denotes the rate of change in T_2 in the outward normal direction vector to the external scleroid surface, Γ_2 of R_2; h_{bl} denotes the blood convection coefficient; and T_{bl} is the blood temperature. Values of control parameters used in (13.2) and (13.3) are similar to those in [2] as listed in Table 13.2.

The interfaces between different contiguous regions are given by

$$
\begin{aligned}
T_1 &= T_2 \text{ and } k_1 \frac{\partial T_1}{\partial \mathbf{n}} = k_2 \frac{\partial T_2}{\partial \mathbf{n}} \text{ on } I_{12} \\
T_1 &= T_3 \text{ and } k_1 \frac{\partial T_3}{\partial \mathbf{n}} = k_3 \frac{\partial T_3}{\partial \mathbf{n}} \text{ on } I_{13} \\
T_2 &= T_3 \text{ and } k_2 \frac{\partial T_2}{\partial \mathbf{n}} = k_3 \frac{\partial T_3}{\partial \mathbf{n}} \text{ on } I_{23} \\
T_2 &= T_4 \text{ and } k_2 \frac{\partial T_2}{\partial \mathbf{n}} = k_4 \frac{\partial T_4}{\partial \mathbf{n}} \text{ on } I_{24} \\
T_2 &= T_5 \text{ and } k_2 \frac{\partial T_2}{\partial \mathbf{n}} = k_5 \frac{\partial T_5}{\partial \mathbf{n}} \text{ on } I_{25} \\
T_3 &= T_4 \text{ and } k_3 \frac{\partial T_3}{\partial \mathbf{n}} = k_4 \frac{\partial T_4}{\partial \mathbf{n}} \text{ on } I_{24} \\
T_4 &= T_5 \text{ and } k_4 \frac{\partial T_4}{\partial \mathbf{n}} = k_5 \frac{\partial T_5}{\partial \mathbf{n}} \text{ on } I_{45}
\end{aligned}
\tag{13.4}
$$

where k_i is the thermal conductivity of R_i, I_{ij} denotes the interface between R_i and R_j, and $\partial T_i/\partial \mathbf{n}$ (at the interface) denotes the rate of change of temperature T_i in the direction of a normal vector to the interface.

Table 13.2 Control Parameters Used for the Boundary Condition

Control Parameters	Value
Blood temperature, T_{bl} (°C)	37
Ambient temperature T_{amb} (°C)	25
Blood convection coefficient, h_{bl} (Wm^{-2}K^{-1})	65
Ambient convection coefficient, h_{amb} (Wm^{-2}K^{-1})	10
Evaporation rate, E_{vap} (Wm^{-2})	40
Emissivity of the cornea, ε	0.975
Stefan-Boltzmann constant, σ (Wm^{-2}K^{-4})	5.67×10^{-8}

13.3   The Axisymmetric Boundary Element Method

This section outlines the axisymmetric boundary element method for solving (13.1) subject to (13.2) and (13.3). Only the main equations necessary for the implementation of the method are presented. Further details in the derivations of the equation can be found in, for example, [11, 12].

The axisymmetric boundary element method is based on the boundary integral equation of (13.1) given in the form

$$\lambda(\xi,\eta)T_i(\xi,\eta) = \int_{\Lambda_i}\left[T_i(r,z)\frac{\partial}{\partial\mathbf{n}}\Phi(r,z;\xi,\eta)\right]\cdot r\cdot ds(r,z)$$

$$-\int_{\Lambda_i}\left[\Phi(r,z;\xi,\eta)\frac{\partial}{\partial\mathbf{n}}T_i(r,z)\right]\cdot r\cdot ds(r,z) \qquad (13.5)$$

$$\text{for }(\xi,\eta)\in R_i\cup\Lambda_i(i = 1,2,3,4,\text{ and }5)$$

where Λ_i denotes the boundary of the region R_i, $ds(r,z)$ denotes the length of an infinitesimal part of curve Λ_i, $\lambda(\xi,\eta) = 1$ if (ξ,η) lies in the interior of R_i, $\lambda(\xi,\eta) = 0.5$ if (ξ,η) lies on a smooth path of Λ_i, and $\Phi(r,z,\xi,\eta)$ and $\partial\Phi(r,z,\xi,\eta)/\partial n$ are the axisymmetric fundamental solution of the Laplace equation and its normal derivative, respectively, as given by

$$\Phi(r,z;\xi,\eta) = -\frac{K\big(m(r,z;\xi,\eta)\big)}{\pi\sqrt{a\big((r,z;\xi,\eta)\big)+b(r;\xi)}}$$

$$\frac{\partial}{\partial n}\Phi(r,z;\xi,\eta) = -\frac{1}{\pi\sqrt{a\big((r,z;\xi,\eta)\big)+b(r;\xi)}}$$

$$\times\left\{\frac{n_r}{2}\left[\frac{\xi^2-r^2+(\eta-z)^2}{a(r,z;\xi,\eta)-b(r;\xi)}E\big(m(r,z;\xi,\eta)\big)\right.\right.$$

$$\left.-K\big(m(r,z;\xi,\eta)\big)\right] \qquad (13.6)$$

$$\left.+n_z\frac{\eta-z}{a(r,z;\xi,\eta)+b(r;\xi)}E\big(m(r,z;\xi,\eta)\big)\right\}$$

$$m(r,z;\xi,\eta) = \frac{2b(r;\xi)}{a\big((r,z;\xi,\eta)\big)+b(r;\xi)}$$

$$a(r,z;\xi,\eta) = \xi^2+r^2+(\eta-z)^2\text{ and }b(r;\xi) = 2r\xi$$

In (13.6), n_r and n_z are the components of the unit normal vector pointing outward of R_i in the r and z directions, respectively; $K(m)$ and $E(m)$ are the complete elliptic integral of the first and second kind, respectively, as given in [13]. Details on the derivation in (13.6) can be found in [11].

If the boundary Λ_i of the region R_i in (13.5) is discretized into N number of straight-line segments, we can write $\Lambda_i = \Lambda_i^{(1)}\cup...\Lambda_i^{(N_i)}$. For a typical element

$\Lambda_i^{(m)}$ (for $m = 1, 2, .., N_i - 1, N_i$) with starting and ending points denoted by $(r_i^{(m)}, z_i^{(m)})$ and $(r_i^{(m+1)}, z_i^{(m+1)})$, respectively, two points within $\Lambda_i^{(m)}$ can be chosen such that

$$\left(\xi_i^{(m)}, \eta_i^{(m)}\right) = \left(r_i^{(m)}, z_i^{(m)}\right) + \tau\left(r_i^{(m+1)} - r_i^{(m)}, z_i^{(m+1)} - z_i^{(m)}\right),$$

$$\left(\xi_i^{(m+N_i)}, \eta_i^{(m+N_i)}\right) = \left(r_i^{(m)}, z_i^{(m)}\right) + (1-\tau)\left(r_i^{(m+1)} - r_i^{(m)}, z_i^{(m+1)} - z_i^{(m)}\right)$$

(13.7)

where τ is a preselected number in the range of $0 < \tau < 0.5$. The numerical results presented in subsequent sections are obtained using $\tau = 0.25$. Using the approximations in (13.7), temperature along the element $\Lambda_i^{(m)}$ can be approximated by

$$T_i^{(m)}(r,z) = \frac{\left[s_i^{(m)}(r,z) - (1-\tau)\ell_i^{(m)}\right]T_i^{(m)}\left[s_i^{(m)}(r,z) - \tau\ell_i^{(m)}\right]T_i^{(m+N_i)}}{(2\tau - 1)\ell_i^{(m)}}$$

(13.8)

$$\text{for } (r,z) \in \Lambda_i^{(m)}$$

where $T_i^{(m)}$ and $T_i^{(m)}$ are temperature at $(\xi_i^{(m)}, \eta_i^{(m)})$ and $(\xi_i^{(m+N_i)}, \eta_i^{(m+N_i)})$, respectively; $\ell_i^{(m)}$ is the length of element $\Lambda_i^{(m)}$; and $s_i^{(m)}(r,z)$ is given by

$$s_i^{(m)}(r,z) = \sqrt{\left(r - r_i^{(m)}\right)^2 + \left(z - z_i^{(m)}\right)^2}$$

(13.9)

Similarly, the normal derivatives of temperature $(\partial T/\partial n = q)$ can be approximated by

$$q_i^{(m)}(r,z) = \frac{\left[s_i^{(m)}(r,z) - (1-\tau)\ell_i^{(m)}\right]q_i^{(m)} - \left[s_i^{(m)}(r,z) - \tau\ell_i^{(m)}\right]q_i^{(m+N_i)}}{(2\tau - 1)\ell_i^{(m)}}$$

(13.10)

$$\text{for } (r,z) \in \Lambda_i^{(m)}$$

where $q_i^{(m)}$ and $q_i^{(m+N_i)}$ denote the normal derivatives of temperature at $(\xi_i^{(m)}, \eta_i^{(m)})$ and $(\xi_i^{(m+N_i)}, \eta_i^{(m+N_i)})$, respectively.

Using the approximations in (13.8) and (13.10) and by letting $(\xi, \eta) = (\xi_i^{(n)}, \eta_i^{(n)})$, the boundary integral equation in (13.4) can now be written as

$$\lambda\left(\xi_i^{(n)}, \eta_i^{(n)}\right) T\left(\xi_i^{(n)}, \eta_i^{(n)}\right) = \sum_{m=1}^{N_i} \frac{1}{(2\tau - 1)\ell_i^{(m)}}$$

$$\times \left\{ \left[-(1-\tau)\ell_i^{(m)} G_{2i}^{(m)}\left(\xi_i^{(n)}, \eta_i^{(n)}\right) + G_{4i}^{(m)}\left(\xi_i^{(n)}, \eta_i^{(n)}\right) \right] T_i^{(m)} \right.$$

$$+ \left[\tau\ell_i^{(m)} G_{2i}^{(m)}\left(\xi_i^{(n)}, \eta_i^{(n)}\right) + G_{4i}^{(m)}\left(\xi_i^{(n)}, \eta_i^{(n)}\right) \right] T_i^{(m+N_i)}$$

$$- \left[-(1-\tau)\ell_i^{(m)} G_{1i}^{(m)}\left(\xi_i^{(n)}, \eta_i^{(n)}\right) + G_{3i}^{(m)}\left(\xi_i^{(n)}, \eta_i^{(n)}\right) \right] q_i^{(m)} \qquad (13.11)$$

$$\left. - \left[\tau\ell_{3i}^{(m)} G_{1i}^{(m)}\left(\xi_i^{(n)}, \eta_i^{(n)}\right) - G_{3i}^{(m)}\left(\xi_i^{(n)}, \eta_i^{(n)}\right) \right] q_i^{(m+N_i)} \right\}$$

for $n = 1, 2, \ldots, 2N_i - 1, 2N_i$

and $i = 1, 2, 3, 4,$ and 5

where

$$G_{1i}^{(m)}(\xi, \eta) = \int_{\Lambda_i^{(m)}} \Phi(r, z; \xi, \eta) \cdot r \cdot ds(r, z)$$

$$G_{2i}^{(m)}(\xi, \eta) = \int_{\Lambda_i^{(m)}} \frac{\partial}{\partial n} \Phi(r, z; \xi, \eta) \cdot r \cdot ds(r, z)$$

$$\qquad (13.12)$$

$$G_{3i}^{(m)}(\xi, \eta) = \int_{\Lambda_i^{(m)}} s(r, z) \Phi(r, z; \xi, \eta) \cdot r \cdot ds(r, z)$$

$$G_{4i}^{(m)}(\xi, \eta) = \int_{\Lambda_i^{(m)}} s(r, z) \Phi(r, z; \xi, \eta) \cdot r \cdot ds(r, z)$$

The integrals in (13.12) can be computed numerically using standard numerical integration rules such as the Gaussian quadrature.

To solve (13.1), the point $(\xi_i^{(n)}, \eta_i^{(n)})$ is collocated at all $2N_i$ boundary points defined in (13.7) for each eye region. Collocation at these points results in five sets of equations corresponding to the five different eye regions. Note that the interface between two contiguous regions may form part of the boundary, Λ_i. For example, Λ_1 consists of the boundary Γ_1 and the interfaces I_{12} and I_{23}. When $(\xi_i^{(n)}, \eta_i^{(n)})$ is collocated at corneal surface, Γ_1, and the surface of the sclera, Γ_2, the boundary conditions in (13.2) and (13.3) can be rearranged and substituted into (13.11) to give a function that contains only the unknown boundary temperatures. Following the collocation step, a global system of linear algebraic equation containing $2 \times (N_1 + N_2 + N_2 + N_4 + N_5)$ equations and $2 \times (N_1 + N_2 + N_2 + N_4 + N_5)$ unknowns is generated.

If the nonlinear radiation term in (13.2) is ignored as in most studies, the global system of linear algebraic equation presents no difficulty and can be solved for the unknown T_i and $\partial T_i/\partial n$ at all exterior boundaries of $\Gamma_1 \cup \Gamma_2$ and the interfaces. However, if the nonlinear term is to be considered, the complexity involved with solving the system of linear algebraic equations may be made easier by the use of an iterative scheme such as the one suggested in [2]. The iterative scheme basically solves T_i through consecutive approximations, $T_i^{(p-1)}$ where $p = 1, 2, .., P$ where P is the number of iteration steps required for the criterion

$$\left|T_i^{(P)} - T_i^{(P-1)}\right| \le \delta \tag{13.13}$$

to be satisfied where δ is a small preselected positive number. In here, δ is chosen to be 10^{-5}. If we make an initial guess of the temperature at the corneal surface Γ_1 for the first iteration step, $(p = 1)$ to be such that $T_1^{(0)} = T_{amb}$, the boundary condition in (13.2) can be rewritten as follows:

$$-k_1 \frac{\partial T_1}{\partial n} = h_{amb}\left(T_1 - T_{amb}\right) + \varepsilon\sigma\left[\left(T_i^{(0)}\right)^4 - T_{amb}^4\right] + E_{vap} \tag{13.14}$$

Note that the nonlinear term in the boundary condition of (13.2) is approximated as known. With this approximation, the values of $T_i^{(1)}$ and $\partial T_i^{(1)}/\partial n$ on all boundaries $\Gamma_1 \cup \Gamma_2$ and the interfaces can be solved. This procedure may be repeated as many times as necessary until the convergence criterion in (13.13) is satisfied.

To obtain numerical results, the boundaries of the human eye are discretized into 193 elements with 32 of them on the cornea. The discretization at the cornea has to be sufficiently refined to enable the thermal gradient over the relatively thin structure to be properly simulated. The discretized human eye model is shown in

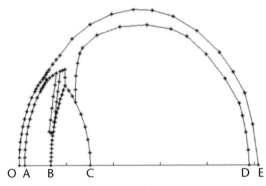

Figure 13.2 The discontinuous linear elements of the human eye model.

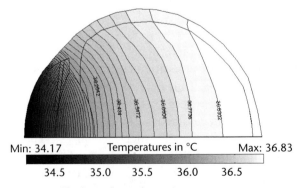

Figure 13.3 Temperature profile throughout the entire eye.

Figure 13.2. The system in (13.11) was set up using MATLAB 6.5 [14] and the simulation was executed on a Pentium 2.4-GHz, 512-Mbyte RAM personal computer.

13.4   Numerical Results and Analysis

The temperature profile throughout the entire eye as obtained from the axisymmetric boundary element model is given in Figure 13.3 together with the isothermal lines. The human eye is coolest at the corneal surface and warmest at the back. This is to be expected from the boundary conditions in (13.2) and (13.3) as

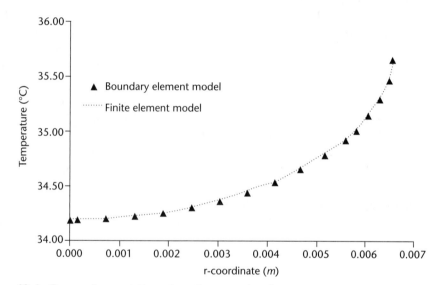

Figure 13.4   Temperature variations along the corneal surface.

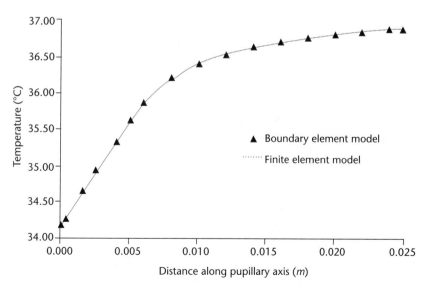

Figure 13.5   Temperatures at various points along the pupillary axis.

heat is transferred by blood through the eye and taken away from the corneal surface through convection and radiation.

The temperature is lowest at the center of the corneal surface, that is, at the point $(r, z) = (0, 0)$. At that point, the temperature is numerically found to be 34.1868°C. The three-dimensional finite element method gives a slightly different temperature of 34.1952°C at the same spot. The difference of 0.0084°C may be regarded as insignificant.

In Figure 13.4, the temperatures obtained using the boundary element and the finite element models are plotted against the r-coordinate of points on the corneal surface. To compare the temperature at other points inside the eye, temperature at $r = 0$ along the pupillary axis is plotted against the z-coordinate in Figure 13.5.

Results from both models agree very closely. The sharp temperature gradient between $z = 0.003$ and $z = 0.008$ is caused by the presence of the human lens, which has a lower thermal conductivity compared to the rest of the ocular regions.

Perhaps to justify the use of the boundary element model, which ignores the optic nerve, temperatures at some of the points in the vicinity where the optic nerve is located are computed and compared to the temperatures found by the finite element model. This is presented in Table 13.3. The results in Table 13.3 show an average difference of approximately 0.025°C. The optic nerve appears not to play a significant role in influencing the temperature at those points.

In Table 13.4, the numerical value of the temperature at the center of the corneal surface, as obtained using the boundary element model here, is compared with some experimental values given in the literature. The predicted numerical value is shown to lie within the range of temperature obtained from experimental studies (i.e.,

Table 13.3 Temperatures at Selected Points at the Vicinity of the Optic Nerve

(r, z)	Temperature (°C)		
	Boundary Element Model	Finite Element Model	Difference
(0.022, 0.007)	36.8739	36.9023	0.0284
(0.021, 0.008)	36.8659	36.8913	0.0254
(0.024, 0.006)	36.8835	36.9060	0.0225

Table 13.4 Temperature Values at the Center of the Cornea

Method	Temperature (°C)
Boundary element model (present work)	34.2
Mapstone [20]	34.8
Fielder et al. [21]	33.4
Efron et al. [22]	34.3
Morgan et al. [17]	33.5
Girardin et al. [23]	33.7
Gugleta et al. [24]	33.7
Purslow et al. [25]	35.0

between 33.7°C and 35.0°C). Differences could be caused by factors such as individual variation (age and eye dimensions) [15–17] and environmental conditions (temperature and humidity) [18, 19].

13.5   Sensitivity Analysis

The purpose of sensitivity analysis is to identify parameters that have dominant effects on the temperature inside the human eye. Of particular interest is the temperature at the corneal surface where validation of numerical values with thermographic measurements can be done. Similar studies have been carried out in the past [3, 9, 26, 27]. Six parameters—the lens thermal conductivity, ambient temperature, blood temperature, ambient convection coefficient, blood convection coefficient, and tear evaporation rate—are investigated [3, 9]. Temperatures at six different points—O, A, B, C, D, and E along the pupillary axis, as shown in Figure 13.2—are compared. The thermal conductivities and other relevant parameters presented in Tables 13.1 and 13.2, respectively, are taken as the control parameters for observing how variation in each parameter influences the temperature.

13.5.1   Effects of Lens Thermal Conductivity

The thermal conductivity of the human lens is investigated for two reasons. First, the water content of the human lens changes with age [28]. Variation in the water content may affect the thermal conductivity [29]. Based on these observations, we could argue that age may be a contributing factor to the individual variation of corneal temperature. Second, the thermal conductivity of the human lens is lower than that of the other eye components. The formation of cataract was hypothesized to originate from increases in lens temperature due to the absorption of heat when the eye is exposed to heating sources such as electromagnetic waves and infrared radiation [30–32]. Increases in the lens temperature may be attributed to its lower thermal conductivity.

The control value for lens thermal conductivity is $0.4 \text{ W m}^{-1}\text{K}^{-1}$ as given in [10]. Other values from the literature include 0.21, 0.3, and $0.54 \text{ W m}^{-1}\text{K}^{-1}$ [9, 30, 33]. Simulations are conducted for these values to analyze the effects of changes in lens thermal conductivity on the human eye. Results obtained are shown in Table 13.5.

Changes in lens thermal conductivity appear not to produce significant effects on the corneal temperature. When the thermal conductivity of the lens is smaller

Table 13.5   Effects of Lens Thermal Conductivity on Eye Temperature

k_4 (Wm^{-1}K^{-1})	Effects of Lens Thermal Conductivity Temperature (°C)					
	O	A	B	C	D	E
0.21		34.02	34.22	34.91	36.28	36.89	36.90
0.30		34.11	34.31	35.05	36.22	36.89	36.89
0.40		34.16	34.38	35.16	36.18	36.88	36.89
0.54		34.26	34.46	35.29	36.14	36.88	36.89

Table 13.6   Effects of Ambient Temperature on Eye Temperature

T_{amb} (°C)	Effects of Ambient Temperature					
	Temperature (°C)					
	O	A	B	C	D	E
20	33.25	33.51	34.55	35.90	36.85	36.85
25	34.19	34.38	35.16	36.18	36.88	36.89
30	35.15	35.27	35.79	36.46	36.92	36.93

than the control value, the temperature gradient across the lens increases. Lower thermal conductivity may tend to retard the flow of heat. Consequently, as explained earlier, human lenses having lower thermal conductivity may be prone to the formation of cataract.

13.5.2   Effects of Ambient Temperature

Changes in ambient temperature have been found to affect the temperature in the cornea [3, 9, 18, 19]. The control value for ambient temperature is taken to be 25°C (the average temperature in an air-conditioned laboratory in Singapore [3]). Two other ambient temperatures, 20°C and 30°C, were also investigated. Numerical results obtained are given in Table 13.6.

In Table 13.6, it is apparent that the temperature at the corneal surface is greatly affected by the changes in ambient temperature. When the ambient temperature is low, the temperature gradient created at the corneal surface becomes greater, thus inducing a larger amount of heat loss via convection and radiation. Considerable changes in the ocular temperature are observed from the corneal surface to the anterior lens surface. The changes in ambient temperature are not effective at depths beyond the lens. This is perhaps due to the relatively low thermal conductivity of the lens. The results in Table 13.6 agree well with experimental data obtained from rabbit [19] and human eyes [18].

13.5.3   Effects of Blood Temperature

Blood temperature is associated with body temperature. Thus, changes in body temperature lead to changes in the blood temperature. The body temperature for a healthy human body is 37°C. This is taken as the control body temperature. Two

Table 13.7   Effects of Blood Temperature on Eye Temperature

T_{bl} (°C)	Effects of Blood Temperature					
	Temperature (°C)					
	O	A	B	C	D	E
37	34.19	34.38	35.16	36.18	36.88	36.89
38	34.99	35.20	36.04	37.12	37.88	37.88
39	35.79	36.01	36.91	38.06	38.87	38.88

additional body temperatures, 38°C and 39°C, are considered. Both represent temperatures in an unhealthy human body, such as one experiencing a fever. Numerical temperatures obtained are shown in Table 13.7.

The corneal surface temperature is evidently affected by changes in blood temperature. According to Table 13.7, an increase of 1°C in the body temperature induces an increase of 0.8°C at the corneal surface. This suggests the possibility of using corneal surface temperature as a diagnostic tool for fevers. However, clinical trials and experimental validations have to be conducted before any definite conclusion can be drawn.

13.5.4   Effects of Ambient Convection Coefficient

The control value for the ambient convection coefficient ($10 \ Wm^{-2} K^{-1}$) lies in the range of natural convection (2 to 25 $Wm^{-2} K^{-1}$). A forced convection coefficient has values between 25 and 250 $Wm^{-2} K^{-1}$ [34]. To study the effects of the ambient convection coefficient on eye temperature, values of 2, 25, and 250 $Wm^{-2} K^{-1}$ were simulated. Results are given in Table 13.8.

A higher convection coefficient implies that more heat is transferred, hence reducing the temperature at the corneal surface. At the extreme value of the forced convection coefficient (i.e., 250 $Wm^{-2} K^{-1}$), the corneal temperature at point O drops to 26.85°C. The temperature at points inside the eye is kept below 30°C up to the anterior lens surface (B). The extreme cooling of the cornea during forced convection may explain the cooling sensation one feels when gusts of air are blown directly at the cornea of an opened eye.

13.5.5   Effects of Blood Convection Coefficient

The human eye model in this study is assumed to be isolated from the human head. Therefore, we can omit the blood perfusion term in the Pennes bioheat equation [35]. The effects of blood flow are modeled by assuming the human eye to be embedded in a homogeneous surrounding anatomy where heat transfer from blood flow in the choroid is described by a single heat transfer coefficient, h_{bl} [10, 36]. The first value suggested for h_{bl} was 65 $Wm^{-2} K^{-1}$, which has been widely used by many researchers [3, 9, 26, 27, 32].

Table 13.8   Effects of Ambient Convection Coefficient on Eye Temperature

$h_{amb} (Wm^{-2}K^{-2})$	Temperature (°C)					
	O	A	B	C	D	E
2	35.18	35.31	35.82	36.47	36.92	36.93
10	34.19	34.38	35.16	36.18	36.88	36.89
25	32.71	33.00	34.18	35.73	36.82	36.83
250	26.85	27.39	29.88	33.65	36.53	36.55

Table 13.9 Effects of Blood Convection Coefficient on Eye Temperature

$h_{bl}(Wm^{-2}K^{-2})$	Effects of Blood Convection Coefficient					
	Temperature (°C)					
	O	A	B	C	D	E
65	34.19	34.38	35.16	36.18	36.88	36.89
110	34.36	34.56	35.35	36.34	36.95	36.96
250	34.56	34.77	35.56	36.51	36.99	36.99
275	34.58	34.79	35.58	36.53	36.99	36.99
300	34.60	34.80	35.60	36.54	36.99	37.00

The validity of this approximation was investigated recently in [36] where comparisons were made with sophisticated eye models, including various anatomical structures and the complex network of blood vessels surrounding the eyeball. It was suggested in [36] that the value of 65 $Wm^{-2}K^{-1}$ was too small. A range between 250 and 300 $Wm^{-2}K^{-1}$ was proposed. These values, however, were found to be effective only when blood acts as a cooling medium instead of a heating source; that is, heat absorption takes place inside the eye, thus causing an increase in temperature beyond the body temperature. For the investigation here, the values of 110, 250, 275, and 300 $Wm^{-2}K^{-1}$ are considered for the purpose of comparison. Table 13.9 shows the numerical results from this analysis.

The results in Table 13.9 indicate that temperature throughout the eye is not significantly affected by changes in the blood convection coefficient even at $h_{bl} = 300$ $Wm^{-2}K^{-1}$. This observation is consistent with the argument given in [36]. When the blood convection coefficient increases from 65 to 300, the corneal temperature increases by a mere 0.41°C.

13.5.6 Effects of Tear Evaporation

Heat loss from tear evaporation for a normal eye was estimated to be 40 Wm^{-2} [9]. This is taken to be the control value. Various other values were given based on experimental measurements [37]. To study the effects of tear evaporation on corneal temperature, values of 20, 70, 100, and 320 Wm^{-2} were considered. The range from 20 to 100 Wm^{-2} corresponds to the evaporation rate under normal conditions, whereas 320 Wm^{-2} represents an abnormal condition in which the lipid layer on the

Table 13.10 Effects of Tear Evaporation on Eye Temperature

$E_{vap}(Wm^{-2})$	Effects of Tear Evaporation					
	Temperature (°C)					
	O	A	B	C	D	E
20	34.43	34.61	35.32	36.25	36.89	36.90
40	34.19	34.39	35.16	36.18	36.88	36.89
70	33.83	34.05	34.93	36.07	36.87	36.88
100	33.47	33.71	34.70	35.97	36.85	36.86
320	30.83	31.26	32.98	35.20	36.75	36.76

tear film is destroyed [38]. The lipid layer is an oily layer that seals and holds the tear film together to prevent excessive evaporation from the corneal surface. Results from this analysis are given in Table 13.10.

For the range of tear evaporation from 20 to 70 Wm^{-2}, the corneal temperature shows very minor changes in temperature (the largest difference is 0.6°C). A smaller value for tear evaporation indicates a lesser cooling effect and thus a higher corneal temperature. At 320 Wm^{-2} a significant drop in corneal temperature is apparent. Dry eye patients have been reported to have higher tear evaporation rates [37, 39]. Based on the results in Table 13.10, the ocular surface temperature may be used as a tool for diagnosing dry eyes.

13.6 Discussion and Conclusions

An axisymmetric boundary element model of the human eye was given here. The optic nerve was omitted in the model. A comparison of the numerical results with those obtained from the finite element model (with the optic nerve) suggests that the optic nerve may not play a very significant role in the temperature distribution in the eye. One of the main advantages of the axisymmetric model is that three-dimensional physical phenomena can be simulated using a two-dimensional analysis. The use of the axisymmetric boundary element method to model the human eye greatly reduces the computer memory required for numerical computations. This became evident when we compared the number of elements required in the present work with the number of elements presented in [2] for a two-dimensional model. In the present work, a total of 193 elements were used, whereas 470 elements were employed in [2].

Temperature distributions at the corneal surface and along the pupillary axis calculated using the axisymmetric boundary element method were compared to those obtained using the three-dimensional finite element model. The two sets of numerical results yielded very good agreement. In addition, the temperature at the center of the cornea was found to lie within the range of corneal temperature measured using infrared thermography. The slight discrepancies between the numerical temperature and the measured temperature may be caused by the differences in the control parameters such as ambient temperature and ambient convection coefficient. Individual variations such as the differences in corneal thickness and age significantly affect the temperature at the corneal surface [15–17]. They were not considered in this study due to the difficulty of incorporating such factors into the mathematical model. Anatomical structures such as the eyelids were also ignored in the model. The presence of eyelids has been found to affect the temperature distribution on the corneal surface [18, 22, 40].

From sensitivity analysis, the ambient temperature, blood temperature, and tear evaporation were found to have significant effects on the corneal temperature. Results from investigations on the effects of ambient convection coefficient, however, were inconclusive. This was due to limitations of the current model, which does not consider the effects of blinking. In the present study, only steady-state temperature was calculated. Blinking induces a time-dependent temperature every 6 to 8 seconds [41]. Each blink of the eyelids spreads warm tears across the corneal surface [42]. Therefore, when the cornea is exposed to forced convection conditions, the cooling process is disrupted by the periodic warming of the cornea due to blinking.

Sensitivity analyses also showed that the temperatures inside a normal eye are not sensitive to changes in the blood convection coefficient. This may not be true when the eye absorbs heat due to exposure to external heating sources such as electromagnetic waves and infrared radiation [10, 36]. Absorption of heat inside the eye causes the temperature to increase beyond the body (blood) temperature [30–32, 43–45]. Thus, blood flow in the choroid becomes a cooling medium to carry away the excessive heat that is absorbed inside the eye. In such situations, the value of the blood convection coefficient may become important because it governs the amount of heat that is transferred away from the eye.

Discontinuous linear elements used in the present work are more accurate than the constant elements used in [2]. Using higher order elements such as the quadratic boundary elements may improve the numerical accuracy of the simulated results. It may, however, be more difficult and more complicated to implement the quadratic boundary elements compared to the discontinuous linear elements and the constant elements. Ultimately, the accuracy of the numerical results depends on how accurate the human eye temperature can be modeled with the linear heat transfer equation in (13.1) using the boundary conditions specified in (13.2) and (13.3).

References

[1] Gaul, L., *Boundary Element Methods for Engineers and Scientists: An Introductory Course with Advanced Topics*, Berlin: Springer, 2003.

[2] Ooi, E. H., W. T. Ang, and E. Y. K. Ng, "Bioheat Transfer in the Human Eye: A Boundary Element Approach," *Eng. Anal. Boundary Elements*, Vol. 31, No. 6, 2007, pp. 494–500.

[3] Ng, E. Y. K., and E. H. Ooi, "FEM Simulation of the Eye Structure with Bioheat Analysis," *Comput. Meth. Programs in Biomedicine*, Vol. 82, No. 3, 2006, pp. 268–276.

[4] COMSOL Multiphysics 3.2, http://www.comsol.com.

[5] Charles, M. W., and N. Brown, "Dimensions of the Human Eye Relevant to Radiation Protection," *Phys. Med. Biol.*, Vol. 20, No. 2, 1975, pp. 202–218.

[6] Fontana, S. T., and R. G. Brubaker, "Volume and Depth of the Anterior Chamber of the Normal Aging Human Eye," *Arch. Ophthalmol.*, Vol. 98, No. 10, 1980, pp. 1803–1808.

[7] Heys, J. J., V. H. Barocas, and M. J. Taravella, "Modeling Passive Mechanical Interaction Between Aqueous Humor and Iris," *J. of Biomech. Eng.*, Vol. 123, No. 6, 2001, pp. 540–547.

[8] Cicekli, U., "Computational Model for Heat Transfer in the Human Eye Using the Finite Element Method," M.Sc. Thesis, Department of Civil & Environmental Engineering, Louisiana State University, 2003.

[9] Scott, J. A., "A Finite Element Model of Heat Transport in the Human Eye," *Phys. Med. Biol.*, Vol. 33, No. 2, 1988, pp. 227–241.

[10] Lagendijk, J. J. W., "A Mathematical Model to Calculate Temperature Distributions in Human and Rabbit Eyes During Hyperthermic Treatment," *Phys. Med. Biol.*, Vol. 27, No. 11, 1982, pp. 1301–1311.

[11] Brebbia, C. A., J. C. F. Telles, and L. C. Wrobel, *Boundary Element Techniques: Theory and Applications in Engineering*, Berlin/Heidelberg: Springer, 1984.

[12]   Katsikadelis, J. T., *Boundary Element Theory and Applications*, Oxford, U.K.: Elsevier Science, 2002.

[13]   Abramowitz, M., and I. Stegun, *Handbook of Mathematical Functions*, New York: Dover, 1970.

[14]   MATLAB 6.5, http://www.mathworks.com/products/matlab.

[15]   Alio, J., and M. Padron, "Influence of Age on the Temperature of the Anterior Segment of the Eye," *Ophthalmic Res.*, Vol. 14, No. 3, 1982, pp. 153–159.

[16]   Morgan, P. B., M. P. Soh, and N. Efron, "Corneal Surface Temperature Decrease with Age," *Contact Lens and Anterior Eye*, Vol. 22, No. 1, 1999, pp. 11–13.

[17]   Morgan, P. B., et al., "Potential Applications of Ocular Thermography," *Optom. Vis. Sci.*, Vol. 70, No. 7, 1993, pp. 568–576.

[18]   Mapstone, R., "Determinants of Ocular Temperature," *Br. J. Ophthalmol.*, Vol. 52, No. 10, 1968, pp. 729–741.

[19]   Schwartz, B., "Environmental Temperature and the Ocular Temperature Gradient," *Arch. Ophthamol.*, Vol. 74, 1965, pp. 237–243.

[20]   Mapstone, R., "Measurement of Corneal Temperature," *Exper. Eye Res.*, Vol. 7, No. 2, 1968, pp. 237–243.

[21]   Fielder, A. R., et al., "Problems with Corneal Arcus," *Trans. of the Ophthalmological Societies of the United Kingdom*, Vol. 101, No. 1, 1981, pp. 22–26.

[22]   Efron, N., G. Young, and N. Brennan, "Ocular Surface Temperature," *Curr. Eye Res.*, Vol. 8, No. 9, 1989, pp. 901–906.

[23]   Girardin, F., et al., "Relationship Between Corneal Temperature and Finger Temperature," *Arch. Ophthamol.*, Vol. 117, No. 2, 1999, pp. 166–169.

[24]   Gugleta, K., S. Orgul, and J. Flammer, "Is Corneal Temperature Correlated with Blood-Flow Velocity in the Ophthalmic Artery?" *Curr. Eye Res.*, Vol. 19, No. 6, 1999, pp. 496–501.

[25]   Purslow, C., J. Wolffsohn, and J. Santodomingo-Rubido, "The Effect of Contact Lens Wear on Dynamic Ocular Surface Temperature," *Contact Lens and Anterior Eye*, Vol. 28, No. 1, 2005, pp. 29–36.

[26]   Amara, E. H., "Numerical Investigations on Thermal Effects of Laser Ocular Media Interaction," *Int. J. Heat Mass Transfer*, Vol. 38, No. 13, 1995, pp. 2479–2488.

[27]   Ridouane, E. H., and A. Campo, "Numerical Computation of the Temperature Evolution in the Human Eye," *Heat Transfer Res.*, Vol. 37, No. 7, 2006, pp. 607–617.

[28]   Siebinga, I., et al., "Age-Related Changes in Local Water and Protein Content of Human Eye Lenses Measured by Raman Microspectroscopy," *Exper. Eye Res.*, Vol. 53, No. 2, 1991, pp. 233–239.

[29]   Spells, K. E., "The Thermal Conductivities of Some Biological Fluids," *Phys. Med. Biol.*, Vol. 5, No. 2, 1960, pp. 139–153.

[30]   Emery, A. F., et al., "Microwave Induced Temperature Rises in Rabbit Eyes in Cataract Research," *J. of Heat Transfer*, Vol. 97, 1975, pp. 123–128.

[31]   Guy, A., et al., "Effect of 2450-MHz Radiation on the Rabbit Eye," *IEEE Trans. on Microwave Theory and Techniques*, Vol. MTT-23, No. 6, 1975 pp. 492–498.

[32]   Scott, J. A., "The Computation of Temperature Rises in the Human Eye Induced by Infrared Radiation," *Phys. Med. Biol.*, Vol. 33, No. 2, 1988, pp. 243–257.

[33]   Neelakantaswamy, P. S., and K. P. Ramakrishnan, "Microwave-Induced Hazardous Nonlinear Thermoelastic Vibrations of the Ocular Lens in the Human Eye," *J. of Biomechanics*, Vol. 12, No. 3, 1979, pp. 205–210.

[34]   Incroprera, F., and D. P. DeWitt, *Fundamentals of Heat and Mass Transfer*, 5th ed., New York: John Wiley & Sons, 2002.

[35]   Pennes, H. H., "Analysis of Tissue and Arterial Blood Temperatures in the Resting Forearm," *J. of Appl. Physiol.*, Vol. 1, No. 2, 1948, pp. 93–122.

[36] Flyckt, V. M. M., B. W. Raaymakers, and J. J. W. Lagendijk, "Modeling the Impact of Blood Flow on the Temperature in the Human Eye and the Orbit: Fixed Heat Transfer Coefficient Versus the Pennes Bioheat Model Versus Discrete Blood Vessels," *Phys. Med. Biol.*, Vol. 51, No. 19, 2006, pp. 5007–5021.

[37] Mathers, W. D., G. Binarao, and A. Petroll, "Ocular Water Evaporation and the Dry Eye: A New Measuring Device," *Cornea*, Vol. 12, No. 4, 1993, pp. 335–340.

[38] Mishima, S., and D. Maurice, "Effects of Normal Evaporation Rate on the Eye," *Exper. Eye Res.*, Vol. 1, 1961, pp. 46–52.

[39] Rolando, M., M. Refojo, and K. Kenyon, "Increased Tear Evaporation in Eyes with Keratoconjunctivitis Sicca," *Arch. Ophthamol.*, Vol. 101, No. 4, 1983, pp. 557–558.

[40] Hill, R. M., and A. J. Leighton, "Temperature Changes of Human Cornea and Tears Under a Contact Lens 2. Effects of Intermediate Lid Apertures and Gaze," *Am. J. Optom. Arch. Am. Acad. Optometry*, Vol. 42, No. 2, 1965, pp. 71–77.

[41] Monster, A. W., H. C. Chan, and D. O'Connor, "Long-Term Patterns in Human Eye Blink Rate," *Biotelemetry and Patient Modeling*, Vol. 5, No. 4, 1978, pp. 206–222.

[42] Mathers, W., "Evaporation from the Ocular Surface," *Exper. Eye Res.*, Vol. 78, No. 3, 2004, pp. 389–394.

[43] Bernardi, P., et al., "SAR Distribution and Temperature Increase in an Anatomical Model of the Human Eye Exposed to the Field Radiated by the User Antenna in a Wireless LAN," *IEEE Trans. on Microwave Theory and Techniques*, Vol. 46, No. 12, 1998, pp. 2074–2082.

[44] Hirata, A., S. Matsuyama, and T. Shiozawa, "Temperature Rises in the Human Eye Exposure to EM Waves in the Frequency Range 0.6–6 GHz," *IEEE Trans. on Electromagnetic Compatibility*, Vol. 44, No. 4, 2000, pp. 594–596.

[45] Hirata, A., "Temperature Increase in Human Eyes Due to Near-Field and Far-Field Exposures at 900 MHz, 1.5 GHz and 1.9 GHz," *IEEE Trans. on Electromagnetic Compatibility*, Vol. 47, No. 1, 2005, pp. 68–76.

CHAPTER 14

Simulation of Aqueous Humor Circulation Inside the Human Eye

E. H. Ooi and Eddie Y. K. Ng

14.1   Introduction

The aqueous humor (AH) inside the human eye consists of the anterior chamber and the posterior chamber. Filling these chambers is the aqueous fluid, which is commonly known as aqueous humor. Rather than being stagnant, which has been assumed in most of the thermal models reported in the literature, the AH flows in a circulatory manner inside the anterior chamber. Clinical observations and mathematical models reported in the literature appear to confirm this. According to Türk in 1906 and 1911 (cited in [1]), the temperature gradient inside the anterior chamber induces buoyant forces that drive the flow of AH. The surface of the anterior chamber that is in contact with the iris is generally warmer than the surface that is in contact with the cornea. Consequently, the warmer AH, which has lower density (due to the Boussinesq approximation), rises and the cooler AH next to the cornea descends, thus creating natural circulation inside the anterior chamber. This phenomenon has since been regarded by many as the primary source of AH flow. Other hypotheses on the flow of AH, such as aqueous secretion from the ciliary processes into the Schlemm's canal, AH mixing during rapid eye movement (REM) sleep, and lens vibration during phakodenesis, have been proposed. Compared to the buoyant driven flow, these factors contributed very little to the circulation of AH inside the anterior chamber [2].

The hydrodynamics of AH flow is poorly understood. Investigations on the AH flow have been limited thus far to computational studies. This is due to the sensitivity of the human eye to touch, which makes experimental studies practically impossible. A number of mathematical models have been proposed, motivated by the need to understand the hydrodynamic behavior of AH [1–5]. In these models, the anterior chamber was modeled as a separate component from the human eye. Nevertheless, these models were able to predict flow and temperature profiles that were in reasonable agreement with each other. According to the findings from these models, the convection currents distort the otherwise symmetrical temperature profile inside the anterior chamber. Heys and Barocas [3], in particular, concluded that the heat transfer across the anterior chamber is not affected by the natural convection of AH, but the temperature profile inside the anterior chamber may change.

The thermal models of the human eye presented in earlier chapters in this book have ignored the presence of AH circulation inside the anterior chamber. Although the predicted corneal surface temperature lies within the range of temperatures obtained using infrared thermography, it may not be safe to dismiss the effects of AH circulation on the corneal surface temperature. On the other hand, the studies conducted in [1–5] only considered the anterior chamber. Therefore, it is difficult to judge whether AH circulation has an impact on the corneal surface temperature. This issue is investigated in this chapter.

A human eye model was developed in two-dimensional Cartesian coordinates (x-y). Heat flow is assumed to be conducted inside the eye model. In addition, AH flow is assumed to take place inside the anterior chamber. The heat conduction equation is used to model the heat flow throughout the entire eye. The Navier-Stokes equation is used to govern the hydrodynamics of AH inside the anterior chamber. The convection of heat resulting from AH circulation is modeled using the energy equation. A constitutive equation is used to couple the Navier-Stokes equation and the energy equation and, in this case, the Boussinesq approximation is chosen. More information is presented in subsequent sections. Comparisons are made on the numerical results obtained for the eye model with and without considering AH circulation inside the anterior chamber. In addition, the model that includes AH circulation is simulated for two orientations of the eye: the vertical orientation, which simulates an upright human position, and the horizontal orientation, which simulates a supine position.

14.2   The Human Eye Model

The human eye model in this chapter follows closely the two-dimensional model presented in Chapter 11. The internal eye components are modeled based on anatomical measurements given in [6, 7]. The retina and the choroid, which are relatively thin, and the iris, which has properties similar to the sclera [8], are combined with the sclera and modeled as one homogeneous region. For the sake of simplicity, the optic nerve is assumed to have properties similar to the sclera and is, thus, combined with the sclera. Six regions are distinguished from the eye model: the cornea, the anterior chamber, the posterior chamber, the lens, the vitreous, and the sclera. Each of these regions is denoted by R_i where $i = 1, 2, 3, 4, 5$, and 6, as labeled in Figure 14.1.

Thermal conductivity values for each eye region can be obtained from the literature and they are listed in Table 14.1. Each region inside the eye model is assumed to be thermally isotropic and homogeneous.

14.2.1   Governing Equations

Three physical phenomena take place in the eye model: heat conduction, heat convection, and fluid flow. As such, three governing equations are needed to described them.

The steady-state heat conduction equation is given by

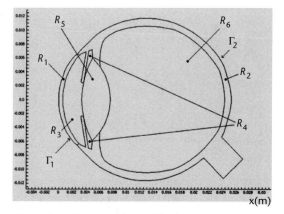

Figure 14.1   The two-dimensional human eye model.

Table 14.1   Thermal Conductivity Values for Each Region of the Human Eye

Region, R	Index, i	Thermal Conductivity, k ($Wm^{-1} K^{-1}$)
Cornea	1	0.58
Sclera	2	1.0042
Anterior chamber	3	0.58
Posterior chamber	4	0.58
Lens	5	0.40
Vitreous	6	0.60

$$\nabla\left(k_i \nabla T_i(x, y)\right) = 0 \quad \text{for } i = 1, 2, 3, 4, 5, \text{ and } 6 \tag{14.1}$$

where k is thermal conductivity and T_i is temperature of region R_i respectively. Heat generated from metabolic activity inside the eye and the blood perfusion inside the iris and ciliary body are neglected when writing (14.1).

The energy equation is used to describe the heat convection process inside the anterior chamber. This is given by

$$u_3 \frac{\partial T_3}{\partial x} + v_3 \frac{\partial T_3}{\partial y} = \frac{k_3}{\rho_3 c_3}\left(\frac{\partial^2 T_3}{\partial x^2} + \frac{\partial^2 T_3}{\partial y^2}\right) \tag{14.2}$$

where u and v are the velocities in the x- and y-directions, respectively; ρ is density; and c is specific heat.

The fluid dynamics of AH is governed by the Navier-Stokes equation. In the x-direction, it is written as

$$\rho_3\left(u_3\frac{\partial u_3}{\partial x}+v_3\frac{\partial u_3}{\partial y}\right)=-\frac{\partial p_3}{\partial x}+\mu_3\left(\frac{\partial^2 u_3}{\partial x^2}+\frac{\partial^2 u_3}{\partial y^2}\right)+\rho_3 g_x \tag{14.3}$$

and in the y-direction, it is given by

$$\rho_3\left(u_3\frac{\partial v_3}{\partial x}+v_3\frac{\partial v_3}{\partial y}\right)=-\frac{\partial p_3}{\partial y}+\mu_3\left(\frac{\partial^2 v_3}{\partial x^2}+\frac{\partial^2 v_3}{\partial y^2}\right)+\rho_3 g_y \tag{14.4}$$

where p is pressure, μ is dynamic viscosity, g_x is the gravitational acceleration in the x-direction, and g_y is the gravitational acceleration in the y-direction.

A constitutive equation is required to couple (14.2), (14.3), and (14.4). In this case, the Boussinesq approximation is chosen, which states that the density of a fluid changes slightly with temperature but negligibly with pressure [9]. The Boussinesq approximation can be written as

$$\rho_3=\rho_{ref}\left[1-\beta_3\left(T_3-T_{ref}\right)\right] \tag{14.5}$$

where ρ_{ref} and T_{ref} are the reference pressure and temperature, respectively, and β_3 is the volumetric expansion coefficient. The subscript 3 appearing in (14.2) to (14.5) refers to the index $i=3$, that is, the anterior chamber. The physical properties of the anterior chamber used in (14.2) to (14.5) are given in Table 14.2.

14.2.2 Boundary Conditions

Each of the governing equations given in the previous section requires boundary conditions to be specified.

14.2.2.1 Heat Conduction

Two boundary conditions are specified for the heat conduction equation. The first boundary condition is written on the corneal surface, Γ_1. If ambient temperature is lower than the corneal surface temperature, heat from inside the eye is transferred away from the corneal surface via convection and radiation. The cooling is aided by the evaporation of the tear film from the corneal surface. Thus, we can write

Table 14.2 Physical Properties of the Anterior Chamber

Property	Value
Density, ρ_3 (kg m^{-3})	998 [10]
Specific heat, c_3 (J kg^{-1} K^{-1})	3,997 [10]
Dynamic viscosity, μ_3 (Nsm^{-2})	7.4×10^{-4} [9]
Gravitational acceleration, g (ms^{-2})	9.81
Volumetric expansion coefficient, β_3 (K^{-1})	3.37×10^{-4} [9]
Reference density, ρ_{ref} (kg m^{-3})	998 [10]
Reference temperature, T_{ref} (°C)	34 [3]

$$-k_1 \frac{\partial T_1}{\partial \mathbf{n}} = h_{amb}\left(T_1 - T_{amb}\right) + \varepsilon\sigma\left(T_1^4 - T_{amb}^4\right) + E_{vap} \quad \text{at } \Gamma_1 \tag{14.6}$$

where the first, second, and last terms on the right-hand side of (14.6) is the heat loss due to convection, radiation, and tear evaporation, respectively; h_{amb} is the ambient convection coefficient; T_{amb} is ambient temperature; ε is corneal emissivity; σ is the Stefan-Boltzmann constant; E_{vap} is the heat loss due to tear evaporation; and $\partial T_1/\partial \mathbf{n}$ is the rate of change of temperature T_1 in the outward normal direction of Γ_1.

The second boundary condition is written on the surface of the sclera, Γ_2. Heat flow from blood at the sclera into the eye is described using a single heat transfer coefficient, which is written as

$$-K_2 \frac{\partial T_2}{\partial \mathbf{n}} = h_{bl}\left(T_1 - T_{bl}\right) \text{ at } \Gamma_2 \tag{14.7}$$

where h_{bl} is the blood convection coefficient, T_{bl} is blood temperature, and $\partial T_2/\partial \mathbf{n}$ is the rate of change of temperature T_2 in the outward normal direction of Γ_2.

14.2.2.2   Heat Convection

The energy equation in (14.2) is written for the anterior chamber. Therefore, the boundary conditions are specified at the boundaries of the anterior chamber. The two boundaries Γ_3 and Γ_4 designated for the anterior chamber are shown in Figure 14.2.

In the anterior chamber models in [1–5], the boundary conditions that were specified on Γ_3 and Γ_4 were basically simplified approximations. For instance, at Γ_4, which is the surface in contact with the iris and lens, a constant temperature of 37°C was specified. This was based on the assumption that the surface Γ_4 is located sufficiently deep inside the eye such that it is reasonable to assume the temperature to be equal to body temperature of 37°C. Thermal models, such as those in [10–15], however, suggested otherwise.

In the current work, we attempt to incorporate the spatial variation in temperature inside the eye. The boundary condition at Γ_3 and Γ_4 is thus, written as

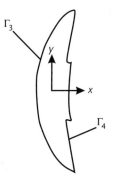

Figure 14.2   Boundaries of the anterior chamber. (*From:* [16]. © 2008 Elsevier. Reprinted with permission.

$$T = \hat{T} \quad \text{at } \Gamma_3 \text{ and } \Gamma_4 \tag{14.8}$$

where $\hat{T}$ is the local temperature obtained after solving (14.1), (14.6), and (14.7).

14.2.3   Aqueous Humor Hydrodynamics

AH hydrodynamics was only assumed to take place inside the anterior chamber [16]. The posterior chamber, which also contains AH, was assumed to remain stagnant. In the actual human eye, AH is produced in ciliary processes in the posterior chamber and flows into the anterior chamber through the very small gap between the iris and the lens. The gap was found to be around $10\,\mu$m [17]. The AH then exits the anterior chamber through the trabecular meshwork into Schlemm's canal [18].

The production of AH and its flow from the ciliary processes into the trabecular meshwork were found to contribute very little to the buoyant driven circulation inside the anterior chamber [2]. It was, therefore, sufficient to treat surfaces Γ_3 and Γ_4 as having a no-slip condition [2]. Regarding the gap between the iris and the lens, although this has been considered in some of the human eye models (e.g., [3, 17]), there were also suggestions that the gap plays no significant role in the circulation of AH due to buoyant forces [17]. In this study, we ignored the gap that exists between the iris and the lens.

At Γ_3 and Γ_4 we specified the no-slip conditions taken from the fluid interface theory, which states that the velocity of a given flow is zero relative to the surface it is in contact with. Mathematically, this is given by

$$\begin{aligned} u &= 0, \\ v &= 0, \quad \text{at } \Gamma_3 \text{ and } \Gamma_4 \end{aligned} \tag{14.9}$$

14.3   Numerical Methodology

The finite element method is used to solve (14.1) through (14.9). To implement this numerical technique, each region of the eye model, such as that illustrated earlier in Figure 14.1 is discretized into triangular elements. Each element is approximated using the Lagrange quadratic approximation. Discretization is refined for regions that are relatively thin such as the cornea and the posterior chamber. In addition, boundaries exposed to a high temperature gradient such as Γ_1 are also refined. Meshing of the eye model resulted in 6,393 triangular elements (Figure 14.3). Numerical experiments indicated a mesh-independent solution. All calculations were done using COMSOL Multiphysics 3.2 [19] on a Pentium 2.41-GHz, 512-Mbyte RAM personal computer.

14.4   Results and Analysis

14.4.1   The Effects of Aqueous Humor Circulation

Two cases are simulated from the human eye model: the first case, without AH circulation, which we denote as the control model, and the second case, with AH circu-

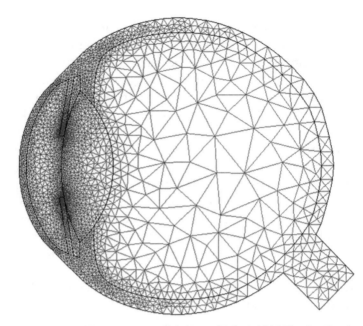

Figure 14.3   The discretized human eye model. (*From:* [16]. © 2008 Elsevier. Reprinted with permission.)

lation, which we denote as the AH model. Figure 14.4 plots the corneal surface temperature against the y-coordinate of points along Γ_1 for the control and AH model.

The corneal surface is coolest at the center [i.e., $(x, y) = (0, 0)$]. The temperature at the center of the corneal surface according to the control model is 33.30°C. At the same location, the AH model predicted a temperature of 33.73°C. In Figure 14.4,

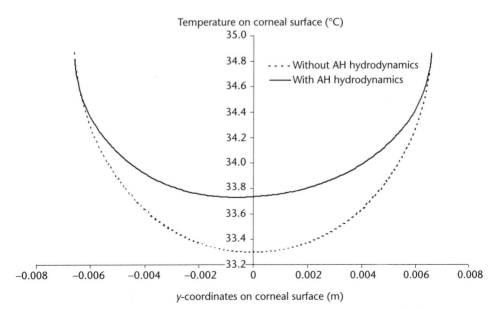

Figure 14.4   Corneal surface temperature for the control and AH models. (*From:* [16]. © 2008 Elsevier. Reprinted with permission.)

the corneal surface temperature predicted by the AH model is found to be higher than the control model at every point except for regions near the edges. Unlike the symmetrical temperature curve of the control model, the temperature curve for the AH model appears to be skewed to the negative y-direction (see Figure 14.2). This implies that the presence of AH circulation inside the anterior chamber is able to distort the temperature profile at the corneal surface.

It may be important to compare temperatures of the control and AH models at other locations inside the eye. To do so, temperature at various points along the pupillary axis ($y = 0$) is plotted. This is shown in Figure 14.5. The region labeled IIC in Figure 14.5 refers to points inside the anterior chamber. The temperature at the cornea and the anterior chamber is found to be affected by the presence of AH circulation. No apparent changes in temperature between the control model and the AH model can be found at points beyond the anterior chamber.

Temperature profiles together with isothermal lines for the control model and the AH model are shown in Figures 14.6 and 14.7, respectively. Both figures are plotted on the same scale. The pupillary axis is denoted by the horizontal line running across the eye model.

The AH model exhibits a warmer cornea and anterior chamber compared to the control model. The isotherms for the control model such as those shown in Figure 14.6 appear to be symmetrical along its pupillary axis except for regions near to the optic nerve. This symmetrical feature is lost when AH circulation is considered inside the anterior chamber (see Figure 14.7).

14.4.2   The Effects of Eye Orientation

Two eye orientations are simulated, namely, the vertical and the horizontal orientations. In the vertical orientation, the gravitational acceleration in the x-direction, g_x, is zero. Consequently, the last term on the right-hand side of (14.3) becomes zero. Similarly, when the eye is in horizontal orientation, component g_y is zero and, there-

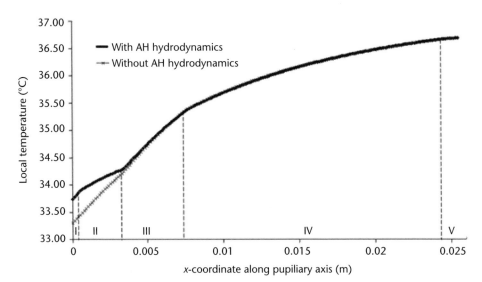

Figure 14.5   Temperature along the pupillary axis at $y = 0$. (*From:* [16]. © 2008 Elsevier. Reprinted with permission.)

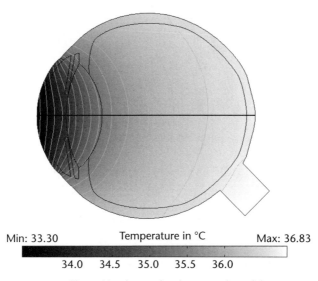

Min: 33.30   Temperature in °C   Max: 36.83

34.0   34.5   35.0   35.5   36.0

Figure 14.6 Temperature profile and isotherms for the control model.

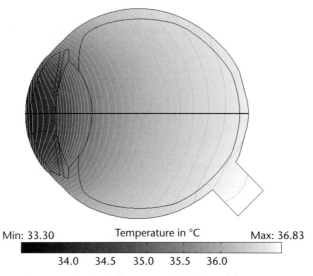

Min: 33.30   Temperature in °C   Max: 36.83

34.0   34.5   35.0   35.5   36.0

Figure 14.7 Temperature profile and isotherms for the AH model.

fore, the last term on the right-hand side of (14.4) becomes zero. To illustrate the changes caused by the different orientations of the eye, we plot the velocity profile and streamlines inside the anterior chamber. This is shown in Figures 14.8 and 14.9 for the vertical and horizontal eye orientation, respectively.

When the eye is in the horizontal orientation, the gravitational effect that is act-ing on the AH is parallel to the direction of the pupillary axis. The symmetrical flow profile shown in Figure 14.8 is very much expected since the geometry of the ante-rior chamber is also symmetrical about the pupillary axis. In the vertical eye orienta-tion, the circulation induced by the buoyant forces and the gravitational effect produces a flow profile such as that shown in Figure 14.9.

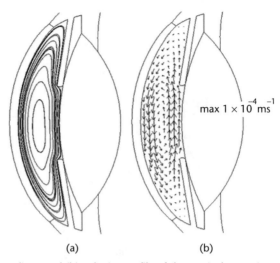

(a)   (b)

Figure 14.8   (a) Streamlines and (b) velocity profile of the vertical eye orientation.

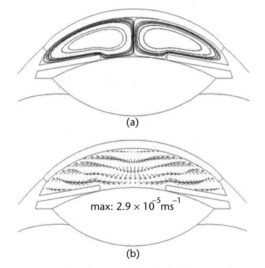

(a)

(b)

Figure 14.9   (a) Streamlines and (b) velocity profile of the horizontal eye orientation.

Temperature along corneal surface Γ_1 is plotted in Figure 14.10 for both the vertical and horizontal eye orientations. Unlike the vertical eye orientation, the horizontal eye orientation does not show any signs of unsymmetrical temperature profile at the corneal surface. This may be attributed to the flow profile such as shown in Figures 14.8 and 14.9.

14.5   Discussion

We have presented a two-dimensional model of the human eye and have successfully simulated the temperature distribution inside the eye with the inclusion of AH circulation inside the anterior chamber. Using the Navier-Stokes equation and the energy

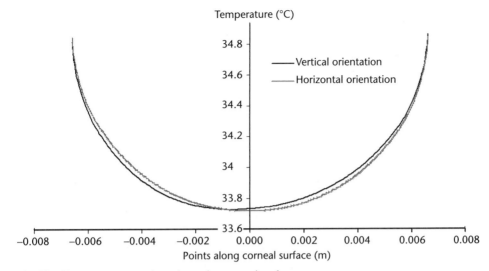

Figure 14.10 Temperature at points along the corneal surface.

equation, coupled with the Boussinesq approximation, we were able to simulate AH flow that is similar to that described in [1–5].

When AH circulation was included as part of the thermophysical phenomena inside the eye, the temperature at the cornea and the anterior chamber appeared to be warmer. This is perhaps caused by the convective currents induced by the circulation of AH. If we look at the temperature distribution inside the human eye (see Figure 14.6), we observe a temperature difference of approximately 1°C between the posterior surface of the cornea and the surface of the lens and iris. This creates a thermal gradient across the anterior chamber that is necessary to induce buoyant forces on the AH. The convective currents created by the circulation of AH may aid the transfer of heat from blood flow at the sclera, thus causing a higher temperature at both the cornea and anterior chamber.

The asymmetrical temperature profile at the corneal surface may be explained by observing the flow profile of AH (see Figure 14.8). The anterior chamber can be separated into the positive y and the negative y section. The warmer AH rises at the back of the anterior chamber and descends next to the surface that is in contact with the cornea. Heat, which is carried through convective flow, is diffused to the cornea as the AH descends. Therefore, it is reasonable to assume that the temperature of AH at the positive y section is greater than the temperature of AH at the negative y section. Consequently, the corneal surface temperature may also have a greater temperature at the positive y section compared to the negative y section.

Changes in eye orientation are found to have no significant effect on the temperature at the corneal surface. Temperature profile at the corneal surface for the horizontal eye orientation showed a more symmetrical profile compared to the vertical eye orientation. This slight difference may not be significant in a practical sense since IR thermography may not be able to capture it. The maximum circulation velocity of the AH in the vertical eye orientation is 1×10^{-4} m s^{-1} (see Figure 14.8). This is of the same order of magnitude as the results predicted by the theoretical model presented in [1]. In the horizontal eye orientation, the maximum circulation velocity is 2.8×10^{-5} m s^{-1} (see Figure 14.9). We can thus conclude that the flow of

AH is greater in the vertical eye orientation compared to the horizontal eye orientation.

Despite the current work showing an increase in cornea and anterior chamber temperature when AH circulation is present, this result should not be mistaken as to suggest that the AH circulation is important in human eye heat transfer; particularly when no experimental validations have been conducted. One of the difficulties in carrying out experimental validations is the lack of a proper experimental method for investigating the effects of AH circulation on the temperature distribution inside the human eye. Comparison between the calculated temperatures at the corneal surface and measured temperatures found in the literature may not be reliable, particularly when the human eye is sensitive to factors such as ambient temperature, ambient convection coefficient, and tear evaporation [10, 12].

Perhaps one of the best ways to deal with this problem is to obtain a large database of corneal surface temperatures for individuals with no abnormal ocular conditions and under a control environment. Information extracted from this database could then be used as a benchmark to validate the calculated results from the models of this study for the eye both with and without AH hydrodynamics.

One of the limitations of the eye model in the present work is the use of a two-dimensional geometry to represent the actual human eye. By doing so, we have neglected any possible thermal variation that may take place in the direction perpendicular to the $(x-y)$ eye model (see Figure 14.1). The two-dimensional model may also overestimate the flow of AH inside the anterior chamber due to the absence of a complete corneal and scleral boundary. Although a three-dimensional model best simulates the actual eye, efforts may be hampered by the requirement for large computer memory. Anatomical and physiological improvements such as the eyelids and the effects of blinking should also be considered in the eye model since these features were found to have significant effects on the corneal surface. Despite these limitations, the information provided by the two-dimensional eye model in this study may help researchers and ophthalmologists gain primary understanding on the thermal interactions between AH circulation inside the anterior chamber and other regions inside the eye.

14.6 Conclusions

The temperature distribution inside a two-dimensional human eye model has been simulated with and without the effects of AH circulation. A higher temperature in the cornea and anterior chamber is found when AH circulation is present inside the anterior chamber. When the AH circulation is not present, a symmetrical temperature profile along the pupillary axis is observed. This feature is destroyed when AH circulation is considered part of the thermophysical processes that take place inside the human eye.

Changes in eye orientation from vertical to horizontal do not produce any significant changes on the corneal temperature. Because no experimental validations were performed, no conclusions can be drawn based on the results of this study regarding the importance of AH circulation on human eye heat transfer. This issue, along with the development of a three-dimensional model, is currently being investigated to further improve our knowledge on the thermal interactions between AH circulation and heat transfer in the human eye.

References

[1]   Canning, C. R., et al., "Fluid Flow in the Anterior Chamber of a Human Eye," *IMA J. Math. Appl. Med. Biol.*, Vol. 19, No. 1, 2002, pp. 31–60.

[2]   Fitt, A. D., and G. Gonzalez, "Fluid Mechanics of the Human Eye: Aqueous Humor Flow in the Anterior Chamber," *Bull. Mathematical Biol.*, Vol. 68, No. 1, 2006, pp. 53–71.

[3]   Heys, J. J., and V. H. Barocas, "A Boussinesq Model of Natural Convection in the Human Eye and the Formation of Krukenberg's Spindle," *Ann. Biomed. Eng.*, Vol. 30, No. 3, 2002, pp. 392–401.

[4]   Avtar, R., and R. Srivastava, "Modelling the Flow of Aqueous Humor in Anterior Chamber of the Eye," *Appl. Math. Computation,* Vol. 181, No. 2, 2006, pp. 1336–1348.

[5]   Kumar, S., et al., "Numerical Solution of Ocular Fluid Dynamics in a Rabbit Eye: Parametric Effects," *Ann. Biomed. Eng.,* Vol. 34, No. 3, 2006, pp. 530–544.

[6]   Fontana, S. T., and R. F. Brubaker, "Volume and Depth of the Anterior Chamber of the Normal Aging Human Eye," *Arch. Ophthalmol.,* Vol. 98, No. 10, 1980, pp. 1803–1808.

[7]   Charles, M. W., and N. Brown, "Dimensions of the Human Eye Relevant to Radiation Protection," *Phys. Med. Biol.*, Vol. 20, No. 2, 1975, pp. 202–218.

[8]   Cicekli, U., "Computational Model for Heat Transfer in the Human Eye Using the Finite Element Method," M.S. Thesis, Department of Civil and Environmental Engineering, Louisiana State University, 2003.

[9]   Incroprera, F., and D. P. DeWitt, *Fundamentals of Heat and Mass Transfer*, 5th ed., New York: John Wiley & Sons, 2002.

[10] Scott, J. A., "A Finite Element Model of Heat Transport in the Human Eye," *Phys. Med. Biol.*, Vol. 33, No. 2, 1988, pp. 227–241.

[11] Amara, E. H., "Numerical Investigations on Thermal Effects of Laser Ocular Media Interaction," *Int. J. Heat and Mass Transfer,* Vol. 38, No. 13, 1995, pp. 2479–2488.

[12] Ng, E. Y. K., and E. H. Ooi, "FEM Simulation of the Eye Structure with Bioheat Analysis," *Comput. Meth. Programs in Biomedicine*, Vol. 82, No. 3, 2006, pp. 268–276.

[13] Ridouane, E. H., and A. Campo, "Numerical Computation of the Temperature Evolution in the Human Eye," *Heat Transfer Res.*, Vol. 37, No. 7, 2006, pp. 607–617.

[14] Ng, E. Y. K., and E. H. Ooi, "Ocular Surface Temperature: A 3D FEM Prediction Using Bioheat Equation," *Comput. Biol. Med.*, Vol. 37, No. 6, 2007, pp. 829–835.

[15] Ooi, E. H., W. T. Ang, and E. Y. K. Ng, " Bioheat Transfer in the Human Eye: A Boundary Element Approach," *Eng. Anal. Boundary Elements*, Vol. 31, No. 6, 2007, pp. 494–500.

[16] Ooi, E. H., and E. Y. K. Ng, "Simulation of Aqueous Humor Hydrodynamics in Human Eye Heat Transfer," *Computers in Biology and Medicine*, Vol. 38, No. 2, 2008, pp. 252–262.

[17] Heys, J. J., V. H. Barocas, and M. J. Taravella, "Modeling Passive Mechanical Interaction Between Aqueous Humor and Iris," *J. of Biomechanical Eng.*, Vol. 123, No. 6, 2001, pp. 540–547.

[18] Remington, L. A., *Clinical Anatomy of the Visual System,* 2nd ed., Boston, MA: Elsevier Butterworth Heinemann, 2005.

[19] Comsol, http://www.comsol.com.

CHAPTER 15

Clinical Implications for Thermography in the Eye World: A Short History of Clinical Ocular Thermography

Christine Purslow

15.1 Introduction

Ocular thermography, as we currently know it, is a technique that allows for noninvasive temperature assessment of the anterior eye and surrounding adnexa at both high spatial and temporal resolutions. However, attempts to measure ocular temperature, according to Holmberg, go back more than 100 years—the first recorded attempt was by Dohnberg in 1875 [1]. The desire to record eye temperature comes from fundamental physiological principles: All metabolic processes within the human body produce heat, which must be dissipated, and an excess of heat often indicates disease. The potential applications for ocular thermography are widely acknowledged within the literature [2–6], but the actual technique employed to measure ocular temperature has been driven by the prevailing technology, most recently with the application of infrared thermal imaging methods taken from the world of engineering.

15.1.1 Early Techniques of Ocular Temperature Assessment

It was not until the 1970s that radiated ocular temperature was assessed using infrared thermal imaging. Prior to this development, all techniques involved direct contact with the eye to some degree. This division in time was due to the available instrumentation and mimicked similar developments occurring in the measurement of human body temperature (Figure 15.1). A summary of early ocular temperature studies cited by Dohnberg (1952) is given in Table 15.1. The results vary greatly, probably due to a combination of factors: the differing methods employed, which part of the eye was used, whether anesthetic was employed, and what temperature the room was, and so forth.

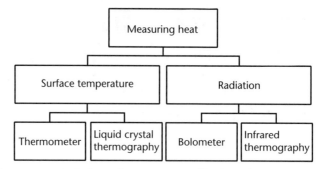

Figure 15.1   Measurement techniques for assessment of body and eye temperature.

Table 15.1   Earliest Recorded Attempts to Measure Ocular Temperature

Date	Author(s)	Method	Results (Mean Temperature, °C)
1875	Dohnberg	Specially adapted glass mercury thermometer placed in *conjunctival sac*; registered higher temperatures in acute keratitis	36.5–36.7
1877	Galezowski	Mercury thermometer in normal and inflamed eyes	36.4 (norm) 36.9–37.1 (inflamed)
1893	Silex	Thermoelement in the inferior fornix; temperatures in acute iritis higher	35.55 (norm) 37.11 (iritis)
1894	Giese	Thermoelement	35.72 (norm) 38.19 (iritis)
1900	Hertel	Mercury thermometer	35.65 (norm) 36.95 (iritis)
1942	Kirisawa	1.0-mm thermoelement in conjunctiva	34.50
1942	Kirisawa	0.2-mm thermoelement in conjunctiva	36.34
1952	Holmberg	0.2-mm thermoelement in conjunctiva	2.5°C lower than oral temperature

Several investigators since have attempted to make ocular temperature measurements using small thermocouples or thermistors (Table 15.2) [7–12].

These contact, sometimes invasive techniques have some obvious disadvantages:

1. The contact lens methods of Kinn and Tell [13] and Hill and Leighton [7, 8] are clearly not representative of a cornea in air, and complex modeling would be required to determine any correction factors. The color responses from Kinn and Tell's liquid crystal lens device were also difficult to interpret, due to poor resolution.

2. Where the cornea is perforated, the work is obviously limited to animals, and even where only the corneal surface is being examined, there is a possibility of trauma.

3. The use of topical anesthetics influences the temperature recorded in two ways: the temperature of the drug itself and the reflex tearing its instillation involves.

Table 15.2 Contact Methods of Ocular Thermometry

Date	Author	Method	Application
1950*	Stoll and Hardy	Contact	Observations of ocular temperature
1952*	Braendstrup	Contact	Observations of ocular temperature
1962*	Schwartz and Feller	Contact	Observations of ocular temperature
1965	Hill and Leighton [7, 8]	Thermistor attached to scleral contact lens	Observations of changes in tear film temperature behind scleral lens
1965	Schwartz [9]	Thermistor in hypodermic needle inserted in eyes of rabbits	Demonstrated ocular gradients in rabbits
1970	Kolstad [10]	Thermistor in glass probe placed directly on anesthetized cornea	Observations of corneal sensitivity at low temperatures
1972	Fatt and Forester [11]	Thermistor in metallic probe inserted into agar jelly and rabbit eyes	Demonstrated the errors that can arise when using a metallic probe
1973	Freeman and Fatt [12]	Thermistor inserted into rabbit eyes to measure surface and intracorneal temperature	Observations of effect of environment on eye temperature in rabbits
1973	Kinn and Tell [13]	Liquid crystal layer in scleral contact lens that is heat sensitive	Measurements of surface temperature
1975	Hørven [14]	Thermistor probe placed on cornea	Corneal surface temperature in normal eyes and in eye disease
1977	Rosenbluth and Fatt [15]	Fine-wire thermocouple inserted into rabbit eyes	Exploring sources of error in this contact method
1982	Auker et al. [16]	Flat, circular thermistor probe	Scleral and conjunctival temperature assessed as a measure of choroidal blood flow

*Cited by Mapstone [17].

4. With any probes containing a thermistor, the actual temperature measured is somewhere between that of the air and the cornea, as it is exposed to both. Even when the air-exposed part of the probe is sheathed, temperature gradients still exist along the probe [11, 15]. Schwartz [18] did respond to this criticism by claiming that the heat of the experimenter's fingers diminished any temperature gradient.

5. It has also been demonstrated that the probe acts as a cooling fin, increasing the available surface area for heat conduction away from the eye, particularly when the probe is used at minimal depth, or when large-gauge needles are used [11, 15]. This has the effect of recorded temperatures actually being lower than the true corneal temperature.

6. All of these methods only measure temperature at one specific point at any one time, a measurement that is of limited value, considering the intrinsic hemodynamics. In 1952, Braendstrup had already demonstrated that a temperature gradient was present across the ocular surface.

7. Contact measurements of temperature vary with the pressure of application of the probe [17].

However, thermistors have continued to be used by some researchers especially with animal work and in work involving contact lenses. Auker et al. [16] used a circular, flat thermistor on cats and monkeys to investigate whether scleral and conjunctival temperatures could be used as measures of choroidal blood flow—previously this comparison had been made with intraocular temperatures. Holden and Sweeney [19] used thermistors and oxygen sensors embedded in a ring designed to fit into the human palpebral aperture to directly measure the oxygen tension and temperature of the upper palpebral conjunctiva, representing closed eye conditions. Martin and Fatt [20] attempted to model the heat transfer at the cornea/contact lens interface. They used thermistors between thin contact lenses in their work.

15.1.2 Noncontact Ocular Temperature Measurement

Contact thermometry of the human eye has largely been rejected in favor of the more practical and, potentially, more accurate techniques of noncontact infrared radiometry.

A radiometric method was first employed by Zeiss in 1930 [17], but the ideas were really explored by Mapstone in the 1960s and presented in a series of four articles [17, 21–23].

Mapstone used a *bolometer* (Figure 15.2). A bolometer employs the principle that infrared radiation incident on an electrical conductor produces a change in resistance (through a change in temperature), which can be amplified and converted directly into a temperature reading. It is sensitive to infrared radiation in the range between 1 and 25 μm and is calibrated against a blackbody. The bolometer was positioned 0.5 cm from the cornea (lids retracted if necessary), so that the area measured had a diameter of 10 mm. Measurements were taken in between blinks, and after 15 minutes of room adaptation, the room was kept between 18°C and 26°C.

Figure 15.2 Mapstone's bolometer setup. (*From:* [21]. © 1968 British Journal of Opthamology. Reprinted with permission.)

15.1.3 Converting Radiation Readings to Temperatures

To convert radiation readings to true temperatures, it is necessary to know how close the tear film and cornea are to the blackbody ideal. If the transmission/absorption characteristics of the ocular tissues are known, then the radiation characteristics can be deduced by Kirchoff's law, which states that the absorption and emission characteristics of a body are equal at any given temperature, and by the application of the Stefan-Boltzmann equation.

Water is an effective absorber of infrared radiation, so it is logical that the high water content of tears, cornea, and lens should ensure high absorption characteristics. The cornea transmits visible radiation, but at the 1.4-μm infrared level, transmission begins to fall off rapidly. Above 1.8 μm, transmission is only a few percent and is absent above 2.3 μm [24]. Thus, the cornea can be considered an efficient *absorber* (and, therefore, *radiator*) of infrared radiation in this portion of the spectrum, and to have radiation *emissivity* close to that of a blackbody [25]. For example, if the radiation emitted by the eye has a spectrum curve between 1 and 30 μm with a peak at 9 μm, then the blackbody that has a similar curve is known to be at 32°C (Figure 15.3).

Exactly how close the cornea is to a perfect blackbody is important: Mapstone reasoned that the true temperature of the cornea ought to be same as the body core (37°C) and, therefore, if the Stefan-Boltzmann equation is applied, the cornea would have an emissivity of 0.97. Because the cornea will be generally cooler than body temperature, the emissivity will actually lie between 0.97 and 1 [17]. An emissivity value of 0.98 is generally accepted by scientists for both the skin and ocular surface.

Mapstone also recognized that the tear film must play an important role in the measured temperature: He regarded the tear film and cornea as one continuous water phase, with both behaving as blackbodies. He suggested that the bolometer recorded a transient equilibrium between the tear film and cornea, where heat exchange would take place following a blink to restore this equilibrium. It has been established that the media of the human eye absorb and emit infrared radiation efficiently, in a similar way to water [26]; and the tear film, with its high water content, should also be an efficient absorber/radiator. Hamano's work [27] suggests that a tear film thickness of 20 μm or more will exhibit 100% absorption and emission of infrared radiation in the region of 7.5 to 15 μm (i.e., the cornea will have no direct influence on measured radiation).

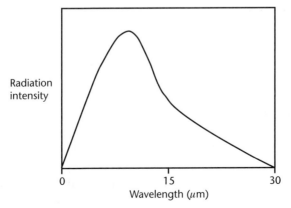

Figure 15.3 Emission spectrum of blackbody at 32°C (peak at 9 μm).

However, tear film thickness is still under debate [28]: If the minimum values are selected from research [29], Hamano's work suggests an absorption of around 80%.

Further studies have therefore suggested that the measured temperature is essentially that of the tears [4, 27, 30], and only where the tears are absent will the radiation detected be that from the cornea itself. Ocular surface temperature measured by infrared thermography has been shown to be highly correlated with tear film stability and body temperature [31]. The tear film is a dynamic structure, and changes in its thickness, composition, and evaporation rate will alter the measured temperature. However, in view of the current dispute over tear film thickness, it is perhaps more reasonable to predict that as the tear film decreases in thickness, the posterior structure, that is, the cornea, will have increasing contribution to the radiated temperature that is measured—a dynamic situation with every blink.

Mapstone considered the deeper ocular structures to contribute little to the emergent radiation from the anterior eye: All wavelengths above 3 μm radiated by posterior tissues will be absorbed by the lens, cornea, and tear film, leaving only a small portion to effectively contribute. The vitreous is opaque to radiation beyond 1.4 μm. However, that is not to say that measured radiation is unaffected by its surroundings; it will be influenced by internal and external factors [32], including the following:

- Heat transfer from adjacent structures occurs by conduction and convection [2].
- Blinking spreads warm tears over the cornea and conjunctiva [21].
- The relatively cooler environment causes the surface to cool in an effort to maintain equilibrium [32, 33].
- As tears thin after a blink, the surface will appear to cool [34, 35]. It has been shown that rapid cooling of the tear film in dry eye conditions appears to be related to reduced tear film stability [36] and an increased rate of evaporation.
- Changes in blood flow to the eye [2] or head [21, 37, 38].

It is perhaps more appropriate to call the radiated temperature ocular surface temperature (OST), rather than assign it to any particular structure in the anterior eye. In most studies, the OST in the central area, overlying the cornea and limbal areas, has been the main focus.

Table 15.3 displays a summary of research using infrared ocular thermometry from the 1968–1988 period, including Mapstone's work. These methods still had the disadvantages of being very close to the eye (inducing reflex tearing) and only recording temperatures over a small corneal area [39–47].

15.2 Ocular Thermography

By 1970, Mapstone had progressed to utilizing an infrared camera that produced a thermal picture, rather than a reading on a dial. This had the advantages of being entirely noninvasive and allowed the surface thermal pattern over an area to be seen. Measurement and color-coded display of eye temperature is referred to as *ocular thermography* [2].

Since the 1950s, medical thermography has been recognized as a useful clinical tool, but it was not until the 1970s that *ocular* thermography was explored. The reasons for this were largely practical because early infrared cameras could not cope with the small temperature gradients involved, magnification was limited, and scanning times were too long. Table 15.4 summarizes previous applications of ocular thermography with these "first generation" cameras.

A second generation of infrared cameras was used extensively by Morgan and coworkers at UMIST in the 1990s, among other researchers (Table 15.5). The 6T62 Thermo Tracer (NEC San-ei Instruments, Japan) had a germanium window set before the detector unit that was sensitive to wavelengths of 8 to 13 μm. The optical scanning device employed mirrors to scan the surface from top to bottom and left to right; scan times varied between 0.25, 0.5, and 1 second, but spatial resolution suffered at anything faster than 1 Hz. The focusing of the system was controlled by adjusting the position of the objective lens. It had a working distance of 40 cm, but a close-up lens could be employed to allow a more detailed view; this had a focal range of 37 to 53 mm, and a maximum magnification of up to ×30 using a zoom facility. A chopper device created an alternating signal to the detector [an alloy of cadmium, mercury, and tellurium (CMT)], which was cooled by liquid nitrogen. The thermal image was displayed in 8-bit color coding (256 colors) on a 30-cm monitor. Morgan's experiments were conducted within a remote-controlled thermal cubicle containing subject and camera; temperature was recorded from five areas across the ocular surface, and the typical thermal sensitivity employed was 0.2°C [4, 48].

Temperature changes across the ocular surface are small and rapidly changing: The latest generation of thermal cameras has adequate spatial and, perhaps more importantly, temporal resolution, allowing their application to be extended in the world of tear film research. They no longer need to be cooled by liquid nitrogen and are very portable, being only the size of a video camera. Typical spatial resolution is 50 μm, with a temperature resolution of 0.08°C, all at 60 Hz. The author has used such cameras to examine tear flow dynamics, tear film stability, and contact lens wear [31, 49, 50].

15.2.1 Typical OST

The observation of a characteristic thermal profile has been noted by most studies of ocular temperature using infrared thermography. Values in the literature indicate normal OST ranges from 32.9°C to 36°C, but for some studies this is not the central temperature, rather an average across the corneal area. In normal eyes, the limbal area has been shown to be warmer than the corneal center by 0.45°C to 1.0°C, with the nasal areas (limbal/conjunctival/canthus) warmer than the corresponding temporal areas [4, 43, 44, 47]. A thermogram represents areas of equal temperature by similar color: "Corneal" isotherms appear elliptically shaped with a long horizontal axis. It has been observed that they become more circular when the eyelids are retracted [4, 47] (Figure 15.4).

Such a temperature distribution across the ocular surface might be expected because the temperature is being differentially influenced by differing physical characteristics of the anterior eye including the following:

Table 15.3 Early Studies Using Infrared Thermometry

Reference	Application	Method	Subjects	Conclusions
[17]	To measure normal corneal temperature.	Bolometer: subject room adapted for 15 minutes; late afternoon; room temperature 18°–26°C; lids retracted; blink between measures; area measured 10 mm in diameter.	140 normal corneas from 70 subjects	Mean temperature 34.8°C Range 33.2°–36.0°C
[21]	To examine factors that determine corneal temperature and cause differences between the eyes.	Bolometer: as above plus 1. Lids closed for 5 minutes to assess effect of closed lids 2. Lids retracted for 5 minutes with topical anesthetic 3. Irrigation with and without blinking allowed 4. Four subjects with unilateral anterior uveitis 5. Five subjects with carotid artery stenosis 6. Four subjects with malignant choroidal melanoma	1–5 subjects	1. Lid closure causes mean rise of 1.5°C. 2. Lid retraction causes mean drop of 1.1°C. 3. Blinking helps restore normal corneal temperature. 4. Interocular temperature differences decline as condition improves. 5. Cornea of affected side had lower temperature than that of opposite side. 6. Insignificant difference between affected eye and unaffected eye. => Abnormal temperature differences between two eyes are a response to different aqueous temperatures, not to external or posterior eye factors.
[22]	To examine normal patterns in cornea and periorbital skin.	Bolometer: 15 minutes room adaptation; late afternoon; room temperature 18°–27°C; measurements made over four areas: cornea, medial forehead, lateral forehead, and lower lid.	70 normal subjects	Normal differences between eyes are no greater than 0.4°C–0.6°C. Normal subjects either have no temperature difference or one side is consistently hotter/colder than the other. Environmental temperature changes affect periorbital areas and cornea by similar amounts.
[23]	To examine thermal patterns in anterior uveitis.	Same protocol as above.	53 unselected subjects with unilateral anterior uveitis	An increase in corneal and periorbital temperature is seen on the affected side. A positive correlation is seen between maximum corneal temperature and duration of ciliary injection.
[39]	To examine the influence of age, sex, race, relative humidity, oral temperature, room illumination, and time of day on ocular temperature and forehead temperature.	Stoll–Hardy radiometer.	131 subjects	Ocular temperature highly correlates with forehead temperature, but independent of all other variables except time of day and age of subject.
[15]	To explore the sources of error in using intraocular thermocouples on rabbits by comparing readings with those from a Dermo–Therm unit.	Dermo–Therm unit (similar to Mapstone's bolometer), held 1 cm from the cornea; area measuring 3 mm in diameter.	Rabbits	Errors exist when using probes due to heat conduction losses from the probe and trauma to the eye.
[30]	To look at the mean corneal temperature and compare bare corneas with those covered by hard and soft contact lenses. Temperature was also measured after the lens was removed.	Dermo–Therm bolometer: area measuring 4 mm in diameter; sensitivity of 0.25°C.	6 subjects for bare cornea (25–55 years); another 3 wearing hard lenses, and 3 wearing soft lenses	Bare corneas range was 33°C–36°C. Eye wearing hard contact lens (lower thermal conductivity) was 0.5°C–1.5°C below that of same eye without lens. Eye wearing soft contact lens was never more than 0.5°C below that of same eye without lens.
[40]	To investigate the relationship between bulbar conjunctival hyperemia and temperature.	C-600M isotherm noncontact infrared bolometer (Linear Lab, United States); use of hypertonic saline to induce redness.	18 subjects (12 male, 6 female, ages 23 ± 4 years)	Hypertonicity and resulting blepharospasm induce redness and temperature increases. An increase in one grade of hyperemia corresponded with an increase in temperature of 0.15°C, but intersubject variability was high, precluding the use of a single measurement for diagnostic purposes.

Table 15.4 Applications of Ocular Thermography: First Generation Cameras

Reference	Application	Method	Subjects	Conclusions
[2]	To examine thermographic patterns in normal, ischemic, and hyperemic ocular states.	Bofors infrared camera system; thermal picture registered on an oscilloscope.	30 subjects	Thermal patterns around the eye are described.
[32]	To examine normal eyes and eyes with pathology.	Bofors system and AGA Thermovision.	38 subjects with unilateral or bilateral eye disease; 41 healthy subjects	In cases of anterior and posterior eye disease, the affected eye was on average warmer than the fellow eye. It was felt that technology needed to improve before it could become a useful clinical tool.
[25, 41]	1. Observations of corneal temperature during cold stress. 2. Observations of corneal temperature in humans and rabbits (who blink less frequently than humans)	AGA Thermovision 680 system.	20 subjects; 78 rabbit eyes; 12 human eyes	1. A fall in corneal temperature with cold stress may be correlated with the thickness of the cornea and the depth of the anterior chamber. 2. Corneal temperature correlates strongly with body temperature in both rabbits and humans; blinking is followed by a rise in temperature; a human cornea cools more rapidly than does a rabbit's, which was postulated to be due to the larger surface area or possibly a shallower anterior chamber.
[42]	A study of normal and laser-injured corneas in rabbits.	Radiometer from Barnes Engineering Company; measurements taken at baseline and on subsequent days after laser insult to cornea; 7 mm diameter area measured.	172 rabbit corneas	Rise in corneal temperature after laser injury observed.
[43]	Examined thermographic patterns in relation to corneal arcus.	AGA Thermovision 680 system; purpose-built draft-free room.	Unclear	Arcus begins in the warmest part of the eye, and in cases of unilateral arcus, it is the warmer eye that is involved.
[44, 45]	Examined normal variations in the orbito-ocular region and the influence of age on corneal temperature.	AGA Thermovision 680 system—102B. Absolute temperatures a five points measured: cornea center, limbus, sclera, and outer and inner canthus.	96 subjects	Asymmetry between the eyes featured in 57% of subjects (difference of 0.51 ± 0.06°C)—shows fallacy of considering thermographic asymmetry as criteria of abnormality. Corneal, limbal, scleral, and outer canthus temperatures decrease significantly with age, whereas inner canthus temperatures show very little variation (outer canthus shows most). No significant difference between sexes or sides was found ($P > 0.05$). Limbus was 0.6°C warmer than corneal center.
[46]	To relate stages of wound healing and inflammatory responses to temperature in rabbits.	DRGX thermodigital camera.		Thermograms indicate stages of wound healing and intra-ocular inflammatory responses, in limbal incisions in rabbits.
[47]	To explore the corneal thermal profile and the temporal stability of temperature.	Thermo Tracer 6T61 (NEC San-ei Instruments, Japan); 6 × 4.5-cm area imaged at 15 cm distance from cornea.	21 subjects (12 male, 6 female, ages 31 ± 10 years)	Advent of superior technology. Recognition of ellipsoidal isotherms with major axis horizontal, concentric about a temperature apex that is slightly inferior to the geometric corneal center. Observed decrease in temperature following a blink, possibly due to tear evaporation. Limitations due to inability to provide accurate anatomical localization and high cost.

Table 15.5 Applications of Ocular Thermography in Second Generation Cameras

Reference	Application	Method	Subjects	Conclusions
[2]	To examine thermographic patterns in normal, ischaemic, and hyperaemic ocular states.	Bofors infrared camera system; thermal picture registered on an oscilloscope.	30 subjects	Thermal patterns around the eye are described.
[32]	To examine normal eyes and eyes with pathology.	Bofors system and AGA Thermovision.	38 subjects with unilateral or bilateral eye disease; 41 healthy subjects	In cases of anterior and posterior eye disease, the affected eye was on average warmer than the fellow eye. It was felt that technology needed to improve before it could become a useful clinical tool.
[25, 41]	1. Observations of corneal temperature during cold stress. 2. Observations of corneal temperature in humans and rabbits (who blink less frequently than humans).	AGA Thermovision 680 system	20 subjects: 78 rabbit eyes; 12 human eyes	1. A fall in corneal temperature with cold stress may be correlated with the thickness of the cornea and the depth of the anterior chamber. 2. Corneal temperature correlates strongly to body temperature in both rabbits and humans; blinking is followed by a rise in temperature; a human cornea cools more rapidly than a rabbit cornea—postulated to be due to larger surface area or possibly shallower anterior chamber?
[42]	A study of normal and laser-injured corneas in rabbits.	Radiometer from Barnes Engineering Company; measurements taken at baseline and subsequent days after laser insult to cornea; 7-mm diameters are measured.	172 rabbit corneas	Rise in corneal temperature after laser injury observed.
[43]	Examined thermographic patterns in relation to corneal arcus.	AGA Thermovision 680 system; purpose-built draught-free room.	Unclear	Arcus begins in the warmest part of the eye, and in cases of unilateral arcus, it is the warmer eye that is involved.

Table 15.5 (continued)

Reference	Application	Method	Subjects	Conclusions
[44, 45]	Examined normal variations in the orbito-ocular region and the influence of age on corneal temperature.	AGA Thermovision 680 system—102B. Absolute temperatures of five points measured—cornea center, limbus, sclera, and outer and inner canthus.	96 subjects	Asymmetry between the eyes is featured in 57% of subjects (difference of $0.51 \pm 0.06°C$) and shows the fallacy of considering thermographic asymmetry as criteria of abnormality.
				Corneal, limbal, scleral, and outer canthus temperatures decrease significantly with age, whereas inner canthus temperatures show very little variation (outer canthus shows most).
				No significant difference between sexes or sides was found ($P > 0.05$).
				Limbus $0.6°C$ warmer than corneal center.
[46]	To relate stages of wound healing and inflammatory responses to temperature in rabbits.	DRGX thermo-digital camera		Thermograms indicate stages of wound healing and intra-ocular inflammatory responses in limbal incisions in rabbits.
[47]	To explore the corneal thermal profile and the temporal stability of temperature.	Thermo Tracer 6T61 (NEC San-ei Instruments, Japan); 6´ 4.5 cm area imaged at 15-cm distance from cornea.	21 subjects (12 males, 6 females, aged 31 ± 10 years)	Advent of superior technology.
				Recognition of ellipsoidal isotherms with major axis horizontal, concentric about a temperature apex that is slightly inferior to the geometric corneal center.
				Observed decrease in temperature following a blink, possibly due to tear evaporation?
				Limitations due to inability to provide accurate anatomical localization and high cost.

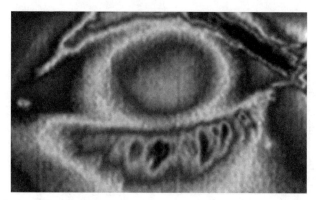

Figure 15.4 Typical ocular thermogram.

- An avascular cornea is surrounded by a vascularized limbal area [4, 43, 47].
- The anterior chamber is naturally shallower in the periphery compared to the center. It has been suggested that eyes with deeper anterior chambers demonstrate less influence from body temperature on OST [25, 41].
- Steeper corneas exhibit a significantly steeper gradient across the cornea [4]. This may be because steeper corneas are more exposed and lose radiation to the environment more readily.
- The palpebral aperture accounts for the elliptical nature of the isotherms and the position of the thermal apex [47].
- The central cornea (4 mm) averages 0.534 mm in thickness, compared to the peripheral average of 0.672 mm, and the area superior to the center is reported to be marginally thicker (10 μm) [67]. Morgan reported a significant decrease in OST with increasing corneal thickness [48].
- Blinking causes tear fluid to accumulate initially centrally, followed by rapid dispersal to the canthi [68]. Temperature changes appear to reflect this tear flow and drainage after a blink [49].

No significant interocular difference in OST has been reported in normal eyes [14, 44, 62]. Mapstone found differences between normal eyes no greater than 0.4°C to 0.6°C [22]. Alio and Padron [44] noted some degree of asymmetry (central OST 0.5°C ± 0.06°C) in 51% of their subjects (n = 96). In a study of 98 normal subjects, Morgan [4] found no significant difference in eye temperature with respect to the right versus the left eye, or the first eye versus the second, and it was established that 95% of the subjects had an interocular difference within 0.53°C.

15.2.2   The Influence of the Environment on OST

OST has been shown to vary significantly throughout the day. There is general agreement that it increases throughout the day, even if changes in room temperature are accounted for [9, 39, 48, 62]. This may reflect the diurnal variation in body temperature: A positive correlation between body temperature and OST has been reported in the literature [4, 9, 12, 14, 21, 25, 31, 39, 60], and room temperature is similarly correlated with OST [9, 10, 12, 14, 25, 53], demonstrating a typical rise of

0.15°C to 0.2°C per degree rise in room temperature [21, 48]. Korb et al. [69] have shown that lipid layer thickness can double if humidity doubles, but the effects of humidity are unclear. Over a narrow range it has been shown to have minimal effect on OST [9, 12]. Increased air flow within a room has been shown to increase heat transfer and therefore decrease OST [12]. Most studies have therefore tried to control the environment with respect to these factors. A period of room adaptation prior to measurement of OST has been implemented in all studies, but the appropriate duration and nature of this period has not been fully established [4, 25].

15.2.3 Individual Variations in OST

There is conflicting evidence about the effect of age on the tear film [69], but it is generally agreed that volume, evaporation, and lipid layer thickness are constant in the normal eye with increasing age, despite changes in tear production and stability: the tear film becomes less stable with age [70]. A negative correlation between age and OST has been reported in the literature [14, 45, 48, 60], with an average decrease of 0.015°C/year [47]. No significant effect of gender or race on OST has been observed in previous studies [4, 39, 45]. Observed interocular differences between eyes are not significant (<0.6°C), [14, 22, 44, 62], but present in more than half of subjects [4, 44].

Blinking spreads warm tears over the ocular surface, and the globe will rotate upward to align the anterior eye to the vascular palpebral conjunctiva [71]. Heat transfer from the tear film to the environment occurs immediately and a decrease in OST over time is observed after blinking [8, 47]. The tear film destabilizes after a blink, which may explain the continued reduction in OST, but the exact mechanism by which tears destabilize is not fully understood. Efron et al. [47] demonstrated that subjects whose corneas cooled more slowly had the capacity to avoid blinking for longer periods, suggesting greater tear stability and/or increased tear film thickness. The stimulus to blink is still under debate: The change in OST following a blink has been suggested as a stimulus to blinking [53], and the cornea and conjunctiva have been found to be sensitive to a cooling stimulus, such as tear film thinning [63, 66].

Ocular temperature has been shown to increase when the eye is closed [7, 17, 20, 30]. Mapstone recorded a 1.5°C rise in OST after 5 minutes of lid closure, and a similar decrease when blinking was prevented for long periods [21].

There have been few reports in the literature observing the temperature of the anterior eye during contact lens wear, and these are discussed in Chapter 16.

15.3 Ocular Disease and Ocular Temperature

As has been demonstrated, OST increases when blood flow to the anterior eye is increased or when anterior uveitis is present [4, 21, 23, 42]. Results from studies with carotid artery stenosis subjects show that there is a significant negative correlation between ocular temperature and degree of stenosis [14, 21, 38]. Mapstone boldly deduced that if an abnormal temperature difference between two eyes exists, it must result from differences in blood supply to the anterior segments. He found a positive correlation between maximum OST and duration of ciliary injection. Efron and colleagues [40] induced hyperemia in the bulbar conjunctiva and found a positive correlation between OST and grade of redness (the McMonnies scale), corresponding to 0.5°C for a three-grade change in redness. In a small, observational study, Fielder and colleagues [43] noted that arcus begins in the warmer parts of the "cornea" (superiorly and inferiorly), and if arcus was unilateral it was the warmer eye that was affected. It was suggested by the authors that increased capillary permeability (in response to or as result of increased temperature) may explain their findings. Thermography has been used to examine the temperature effects of postherpetic neuralgia; the eye of the affected side was significantly colder, but skin temperature was not [37, 72]. Morgan explained that the increase in sympathetic innervation thought to occur with this condition was the likely mechanism of lower OST [48].

There are conflicting results from the few studies that have examined the effect of posterior eye conditions on OST. Some studies showed no correlation between choroidal abnormalities and OST [21], whereas others [32] show a 60% correlation, seemingly irrespective of tumor position. Gugleta et al. [61] suggested that retrobulbar hemodynamics influence OST, and it has been argued that peripheral blood flow may show parallel changes with blood flow variation in the eye [73]. However, it is most likely that any metabolic heat produced by a choroidal tumor will be dissipated by the retina (as a vascular area of high metabolic activity) and, as a result, the thermal gradient will be minimal and thus less likely to affect radiated OST.

A change in ocular temperature has been described as an indicator of the stages of wound healing and correlates well with inflammatory responses in rabbits [46]. It has been demonstrated that cooling eyes postoperatively in cataract surgery reduces inflammation and improves comfort [51, 52]. More contemporary is the use of thermography during refractive surgery. Betney and colleagues [55] showed that OST increased during PRK surgery, to levels at which proteins can denature, but did not find that ablation depth or duration of procedure correlated with OST. This contrasts with the findings of Maldonado-Codina et al. [65] who demonstrated that corneas undergoing larger treatments were subject to greater rises in OST for longer periods of time.

Most recently, the author has been able to use a high-resolution thermal camera to record facial asymmetry in a large cohort of subjects, demonstrating a mean temperature difference between right and left corneas of 0.06 ± 0.23°C (95% confidence interval; -0.01°C to 0.13°C; $p = 0.07$), further suggesting an application for ocular thermography to examine corneal disease.

15.4 Summary

It is apparent from the literature that the potential use of OST in monitoring ocular physiology is well recognized. As early as 1970, it was acknowledged that technology was the key to progress in this area [32], and that improvements in sensitivity and equipment costs would be necessary to make this technique a useful research or clinical tool [40]. The latest generation of cameras offers improvements in terms of portability, cost, and temporal analysis, following the parallel move in medicine toward dynamic thermography [74] as a more useful diagnostic tool. The tear film is a dynamic structure and is continuously affected by eyelid movement, thinning and evaporation, and ocular surface properties. The ability to measure OST in real time offers great opportunities for progress in monitoring ocular physiology, particularly in areas such as dry eyes and contact lens wear.

References

[1] Holmberg, A., "The Temperature of the Eye During the Application of Hot Packs, and After Milk Injections," *Acta Ophthalmologica*, Vol. 30, No. 4, 1952, pp. 347–364.

[2] Mapstone, R., "Ocular Thermography," *Br. J. Ophthalmol.*, Vol. 54, 1970, pp. 751–754.

[3] Keeney, A. H., and P. Guibor, "Thermography and Ophthalmology," *Trans. Am. Acad. Ophthalmol. Otolaryngol.*, Vol. 74, No. 5, 1970, pp. 1032–1043.

[4] Morgan, P. B., et al., "Potential Applications of Ocular Thermography," *Optometry Vis. Sci.*, Vol. 70, 1993, pp. 568–576.

[5] Purslow, C., and J. S. Wolffsohn, "Ocular Surface Temperature: A Review," *Eye and Contact Lens*, Vol. 31, No. 3, 2005, pp. 117–123.

[6] Yang, W. J., and P. P. Yang, "Literature Survey on Biomedical Applications of Thermography," *Bio-Medical Mater. Eng.*, Vol. 2, No. 1, 1992, pp. 7–18.

[7] Hill, R. M., and A. J. Leighton, "Temperature Changes of a Human Cornea and Tears Under a Contact Lens: 1. The Relaxed Open Eye, and the Natural and Forced Closed Eye Conditions," *Am. J. Optometry Arch. Am. Acad. Optometry*, Vol. 42, 1965, pp. 9–16.

[8] Hill, R. M., and A. J. Leighton, "Temperature Changes of Human Cornea and Tears Under a Contact Lens: 2. Effects of Intermediate Lid Apertures and Gaze," *Am. J. Optometry Arch. Am. Acad. Optometry*, 1965.

[9] Schwartz, B., "Environmental Temperature and the Ocular Temperature Gradient," *Arch. Ophthalmol.*, Vol. 74, 1965, pp. 237–243.

[10] Kolstad, A., "Corneal Sensitivity by Low Temperatures," *Acta Ophthalmologica*, Vol. 48, No. 4, 1970, pp. 789–793.

[11] Fatt, I., and J. F. Forester, "Errors in Eye Tissue Temperature Measurements When Using a Metallic Probe," *Exper. Eye Res.*, Vol. 14, 1972, pp. 270–276.

[12] Freeman, R. D., and I. Fatt, "Environmental Influences on Ocular Temperature," *Invest. Ophthalmol.*, Vol. 12, No. 8, 1973, pp. 596–602.

[13] Kinn, J. B., and R. A. Tell, "A Liquid-Crystal Contact Lens Device for Measurement of Corneal Temperature," *IEEE Trans. on Biomedical Engineering*, Vol. 20, 1973, pp. 387–388.

[14] Horven, I., "Corneal Temperature in Normal Subjects and Arterial Occlusive Disease," *Acta Ophthalmologica*, Vol. 53, 1975, pp. 863–874.

[15] Rosenbluth, R. F., and I. Fatt, "Temperature Measurements in the Eye," *Exper. Eye Res.*, Vol. 25, 1977, pp. 325–341.

[16] Auker, C. R., et al., "Choroidal Blood Flow. I. Ocular Tissue Temperature as a Measure of Flow," *Arch. Ophthalmol.*, Vol. 100, No. 8, 1982, pp. 1323–1326.

[17] Mapstone, R., "Measurement of Corneal Temperature," *Exper. Eye Res.*, Vol. 7, 1968, pp. 237–243.

[18] Schwartz, B., "Letter to the Editor: The Measurement of Ocular Temperature," *Exper. Eye Res.*, Vol. 17, 1973, pp. 385–386.

[19] Holden, B. A., and D. F. Sweeney, "The Oxygen Tension and Temperature of the Superior Palpebral Conjunctiva," *Acta Ophthalmologica*, Vol. 63, No. 1, 1985, pp. 100–103.

[20] Martin, D., and I. Fatt, "The Presence of a Contact Lens Induces a Very Small Increase in the Anterior Corneal Surface Temperature," *Acta Ophthalmologica*, Vol. 64, 1986, pp. 512–518.

[21]   Mapstone, R., "Determinants of Ocular Temperature," *Br. J. Ophthalmol.*, Vol. 52, 1968, pp. 729–741.

[22]   Mapstone, R., "Normal Thermal Patterns in Cornea and Periorbital Skin," *Br. J. Ophthalmol.*, Vol. 52, 1968, pp. 818–827.

[23]   Mapstone, R., "Corneal Thermal Patterns in Anterior Uveitis," *Br. J. Ophthalmol.*, Vol. 52, 1968, pp. 917–921.

[24]   Lerman, S., *Radiant Energy and the Eye,* New York: Macmillan, 1980.

[25]   Rysa, P., and J. Sarvaranta, "Corneal Temperature in Man and Rabbit. Observations Made Using an Infrared Camera and a Cold Chamber," *Acta Ophthalmologica*, Vol. 52, 1974, pp. 810–816.

[26]   van den Berg, T. J. T. P., and H. Spekreijse, "Near Infrared Absorption in the Human Eye Media," *Vis. Res.*, Vol. 37, No. 2, 1997, pp. 249–253.

[27]   Hamano, H., S. Minami, and Y. Sugimori, "Experiments in Thermometry of the Anterior Portion of the Eye Wearing a Contact Lens by Means of Infra-Red Thermometer," *Contacto*, Vol. 13, No. 2, 1969, pp. 12–22.

[28]   Craig, J. P., "Structure and Function of the Preocular Tear Film," in *The Tear Film: Structure, Function and Clinical Examination*, D. Korb, (ed.), Boston, MA: Butterworth-Heinemann, 2002, pp. 18–44.

[29]   King-Smith, P. E., et al., "The Thickness of the Human Precorneal Tear Film: Evidence from Reflection Spectra," *Invest. Ophthalmol. Vis. Sci.*, Vol. 41, No. 11, 2000, pp. 3348–3359.

[30]   Fatt, I., and J. Chaston, "Temperature of a Contact Lens on the Eye," *Int. Contact Lens Clinic*, Vol. 7, 1980, pp. 195–198.

[31]   Purslow, C., and J. S. Wolffsohn, "The Relation Between Physical Properties of the Anterior Eye and Ocular Surface Temperature," *Optometry Vis. Sci.*, Vol. 84, No. 3, 2007.

[32]   Wachtmeister, L., "Thermography in the Diagnosis of Diseases of the Eye and the Appraisal of Therapeutic Effects: A Preliminary Report," *Acta Ophthalmologica*, Vol. 48, 1970, pp. 945–958.

[33]   Serway. R. A., and R. J. Beichner, *Physics for Scientists and Engineers,* 5th ed., London, U.K.: Brooks/Cole, 2000.

[34]   Morgan, P. B., A. B. Tullo, and N. Efron, "Infrared Thermography of the Tear Film in Dry Eye," *Eye,* Vol. 9, 1995, pp. 615–618.

[35]   Craig, J. P., et al., "The Role of Tear Physiology in Ocular Surface Temperature," *Eye,* Vol. 14, 2000, pp. 635–641.

[36]   Craig, J. P., "Importance of the Lipid Layer in Human Tear Film Stability and Evaporation," *Optometry Vis. Sci.*, Vol. 74, No. 1, 1997, pp. 8–13.

[37]   Cardona, G., et al., "Ocular and Skin Temperature in Ophthalmic Postherpetic Neuralgia," *Pain Clinic*, Vol. 9, No. 2, 1996, pp. 145–150.

[38]   Morgan, P. B., et al., "Ocular Temperature in Carotid Artery Stenosis," *Optometry Vis. Sci.*, Vol. 76, No. 12, 1999, pp. 850–854.

[39]   Schwartz, B., S. Packer, and S. C. Himmelstein, "Ocular Thermoradiometry," *Invest. Ophthalmol.*, Vol. 7, No. 2, 1968, p. 231.

[40]   Efron, N., et al., "Temperature of the Hyperaemic Bulbar Conjunctiva," *Curr. Eye Res.*, Vol. 7, No. 6, 1988, pp. 615–618.

[41]   Rysa, P., and J. Sarvaranta, "Thermography of the Eye During Cold Stress," *Acta Ophthalmologica*, Vol. 123 (Supplement), 1973, pp. 234–239.

[42]   Mikesell, G. W. J., "Corneal Temperatures—A Study of Normal and Laser-Injured Corneas in the Dutch Belted Rabbit," *Am. J. Physiolog. Opt.*, Vol. 55, No. 2, 1978, pp. 108–115.

[43]   Fielder, A. R., et al., "Problems with Corneal Arcus," *Trans. Ophthalmological Soc. UK*, Vol. 101, 1981, pp. 22–26.

[44]   Alio, J., and M. Padron, "Normal Variations in the Thermographic Pattern of the Orbito-Ocular Region," *Diagnostic Imaging*, Vol. 51, 1982, pp. 93–98.

[45] Alio, J., and M. Padron, "Influence of Age on the Temperature of the Anterior Segment of the Eye," *Ophthalmic Res.,* Vol. 14, 1982, pp. 153–159.

[46] Coles, W. H., et al., "Ocular Surface Temperature (Ocular Thermography) as a Predictor of Corneal Wound Healing," *Invest. Ophthalmol. Vis. Sci.,* Vol. 29 (Supplement), 1988, p. 313.

[47] Efron, N., G. Young, and N. Brennan, "Ocular Surface Temperature," *Curr. Eye Res.,* Vol. 8, No. 9, 1989, pp. 901–906.

[48] Morgan, P. B., "Ocular Thermography in Health and Disease," in *Optometry,* Manchester, U.K.: University of Manchester, 1994.

[49] Purslow, C., "Dynamic Thermography of the Anterior Eye," in *Optometry,* Birmingham: Aston University, 2005, p. 241.

[50] Purslow, C., J. S. Wolffsohn, and J. Santodomingo-Rubido, "The Effect of Contact Lens Wear on Dynamic Ocular Surface Temperature," *Contact Lens and Anterior Eye,* Vol. 28, No. 1, 2005, pp. 29–36.

[51] Fujishima, H., et al., "Quantitative Evaluation of Post-Surgical Inflammation by Infrared Radiation Thermometer and Laser Flare-Cell Meter," *J. of Cataract Refractive Surg.,* Vol. 20, No. 4, 1994, pp. 451–454.

[52] Fujishima, H., et al., "Increased Comfort and Decreased Inflammation of the Eye by Cooling after Cataract Surgery," *Am. J. Ophthalmol.,* Vol. 119, No. 3, 1995, pp. 301–306.

[53] Hata, S., et al., "Corneal Temperature and Inter-Blinking Time," *Invest. Ophthalmol. Vis. Sci.,* Vol. 35, No. 4, 1994, p. S999.

[54] Morgan, P. B., A. B. Tullo, and N. Efron, "Ocular Surface Cooling in Dry Eye—A Pilot Study," *J. Br. Contact Lens Assoc.,* Vol. 19, No. 1, 1996, pp. 7–10.

[55] Betney, S., et al., "Corneal Temperature Changes During Photorefractive Keratectomy," *Cornea,* Vol. 16, No. 2, 1997, pp. 158–161.

[56] Schrage, N. F., et al., "Temperature Changes of the Cornea by Applying an Eye Bandage," *Ophthalmologe,* Vol. 94, No. 7, 1997, pp. 492–495.

[57] Mori, A., et al., "Use of High-Speed, High-Resolution Thermography to Evaluate the Tear Film Layer," *Am. J. Ophthalmol.,* Vol. 124, No. 6, 1997, pp. 729–735.

[58] Fujishima, H., et al., "Effects of Artificial Tear Temperature on Corneal Sensation and Subjective Comfort," *Cornea,* Vol. 16, No. 6, 1997, pp. 630–634.

[59] Du Toit, R., et al., "Diurnal Variation of Corneal Thickness, Sensitivity and Temperature," *Optometry Vis. Sci.,* Vol. 75 (Supplement), 1998, p. 82.

[60] Girardin, F., et al., "Relationship Between Corneal Temperature and Finger Temperature," *Arch. Ophthalmol.,* Vol. 117, 1999, pp. 166–169.

[61] Gugleta, K., S. Orgul, and J. Flammer, "Is Corneal Temperature Correlated with Blood-Flow Velocity in the Ophthalmic Artery?" *Curr. Eye Res.,* Vol. 19, No. 6, 1999, pp. 496–501.

[62] Kocak, I., S. Orgul, and J. Flammer, "Variability in the Measurement of Corneal Temperature Using a Non-Contact Infrared Thermometer," *Ophthalmologica,* Vol. 213, 1999, pp. 345–349.

[63] Murphy, P. J., et al., "Corneal Surface Temperature Change as a Mode of Stimulation of the Non-Contact Corneal Aesthesiometer," *Cornea,* Vol. 18, No. 3, 1999, pp. 333–342.

[64] Morgan, P. B., M. P. Soh, and N. Efron, "Corneal Surface Temperature Decreases with Age," *Contact Lens and Anterior Eye,* Vol. 22, No. 1, 1999, pp. 11–13.

[65] Maldonado-Codina, C., P. B. Morgan, and N. Efron, "Thermal Consequences of Photorefractive Keratectomy," *Cornea,* Vol. 20, No. 5, 2001, pp. 509–515.

[66] Murphy, P. J., et al., "The Minimum Stimulus Energy Required to Produce a Cooling Sensation in the Human Cornea," *Ophthalmic Physiol. Opt.,* Vol. 21, No. 5, 2001, pp. 407–410.

[67] Doughty, M. J., and M. L. Zaman, "Human Corneal Thickness and Its Impact on Intraocular Pressure Measures: A Review and Meta-Analysis Approach," *Survey Ophthalmol.,* Vol. 44, No. 5, 2000, pp. 367–408.

[68] Fatt, I., "Architecture of the Lid-Cornea Junction," *CLAO J.*, Vol. 18, No. 3, 1992, pp. 187–192.

[69] Korb, D. R., et al., *The Tear Film: Structure, Function and Clinical Examination*, Boston, MA: Butterworth-Heinemann, 2002.

[70] Patel, S., and J. C. Farrell, "Age-Related Changes in Pre-Corneal Tear Film Stability," *Optometry Vis. Sci.*, Vol. 66, No. 3, 1989, pp. 175–178.

[71] Vandersteen, J., R. M. Steinman, and H. Collewijn, "The Dynamics of Lid and Eye-Movements Associated with Blinking and Eye Closure," *Experientia*, Vol. 40, No. 11, 1984, p. 1298.

[72] Gispets, J., et al., "Central Thickness of Hydrogel Contact Lenses as a Predictor of Success When Fitting Patients with Tear Deficiency," *Contact Lens and Anterior Eye*, Vol. 25, No. 2, 2002, pp. 89–94.

[73] Guthauser, U., J. Flammer, and F. Mahler, "The Relationship Between Digital and Ocular Vasospasm," *Graefe's Arch. Clin. Exper. Ophthalmol.*, Vol. 226, 1988, pp. 224–226.

[74] Anbar, M., "Clinical Thermal Imaging Today," *IEEE Eng. Med. Biol.*, Vol. 17, No. 4, 1998, pp. 25–33.

CHAPTER 16

Temperature Measurement of the Anterior Eye During Hydrogel Contact Lens Wear

Christine Purslow

This chapter presents the results of a clinical study utilizing dynamic ocular thermography to determine the temperature changes of the anterior eye during and immediately after wearing hydrogel contact lenses, by use of noncontact infrared thermography.

A noncontact infrared camera (Thermo Tracer TH7102MX, NEC San-ei, Japan) was used to record the ocular surface temperature in 32 subjects (mean age: 21.5 ± 1.5 years) wearing Lotrafilcon-A contact lenses on a daily wear basis (LDW; $n = 8$); Balafilcon-A contact lenses on a daily wear (BDW; $n = 8$); Etafilcon-A contact lenses on a daily disposable basis (EDW); and no lenses (CONTROLS; $n = 8$). All contact lens wearers had been wearing lenses for more than 1 year. The ocular surface temperature was measured continuously for 8 seconds after a blink, as the subjects held their eyes open, following a minimum of 2 hours of wear, and immediately following lens removal. Custom-designed software (LabVIEW, National Instruments, Austin, Texas) allowed dynamic quantitative objective analysis of computerized thermal images of the anterior eye. Absolute temperature, changes in temperature after blinking, and the dynamics of temperature changes were calculated.

Ocular surface temperature (OST) immediately following contact lens wear was significantly greater compared to nonlens wearers ($35.0 \pm 1.1°C$ versus $36.7 \pm 1.6°C$; $p < 0.01$). Temperature on top of the lens was highly correlated ($r = 0.97$) to, but lower than ($-0.51 \pm 0.4°C$) that beneath the lens. The OSTs of subjects who had worn silicone hydrogel lenses were significantly greater than those of subjects who had worn Etafilcon A contact lenses ($p < 0.005$). Ocular surface cooling following a blink was not significantly affected by contact lens wear ($p = 0.32$).

OST is greater with hydrogel contact lenses in situ and significantly greater in subjects who have just removed silicone hydrogel contact lenses. The reasons why and the possible clinical significance are discussed.

16.1   Introduction

Contact lens wearers will often use terms such as *hot* or *burning* to describe their symptoms, and *coolness* can be associated with relief from discomfort [1]. This is particularly true for patients with dry eye [2–4]. Conjunctival hyperemia has been suggested to be a measure of the ocular response to contact lens wear, whether symptomatic or otherwise [5–9], and changes in blood flow to the sclera or conjunctiva produce changes in local temperature [10]. Grading this response has attracted much interest from researchers and clinicians alike [11, 12]. There are, however, many occasions when ocular responses to contact lens wear may be initially minimal, particularly with the advent of silicone hydrogel contact lenses [13]. The ability to measure OST using noncontact, dynamic thermography has potential importance to research and, ultimately, to clinical situations in monitoring ocular physiology, particularly in response to contact lens wear.

There have been few reports in the literature observing the temperature of the anterior eye during contact lens wear, and most of these studies have used contact methods of temperature measurement (Table 16.1). It is difficult to compare such studies because it has been shown that corneal temperature is influenced by many factors including environmental conditions [14–16], lid position [17], ocular health [18–20], age of subject [21–24], and point of measurement on the cornea [23, 25, 26]. These factors may have varied between studies; in addition, different measurement techniques were used.

It might be anticipated that OST is influenced by contact lens wear because short-term changes occur in the tear film when a contact lens is inserted, but the long-term changes are not clear from the literature. A contact lens disrupts the integrity of the lipid layer and increases tear film evaporation [27]. This increase in tear evaporation has not been found to be related to the water content of a soft lens [28]. The physical properties of the tear film above and beneath the contact lens will influence OST. Volume and mixing of the post-lens tear film (PoLTF) depends on factors such as blinking, lens edge profile, tear production and drainage, and lens material [29]. For the pre-lens tear film (PLTF), a thin lipid layer will form over a hydrogel lens at best, but with a rigid lens that moves more, the lipid layer will be absent [30].

The purpose of this study was to examine the influence of contact lens wear on OST, including the dynamic changes in OST after a blink.

16.2   Method

16.2.1   Study Population

The subject sample consisted of the right eye of 32 young, healthy volunteers. Subject group details are given in Table 16.2. Contact lens wearers had all worn lenses to the specified modality for at least 1 year prior to the start of the study, and had a well-fitting contact lens with no history of ocular inflammation. Subjects were excluded if they showed any sign of ocular inflammation or ill health or were taking any medication at the time of presentation. Subjects were examined on a slit lamp prior to recording, and asked to refrain from rubbing their eyes prior to and during the session. Full ethical approval from the university was granted prior to com-

Table 16.1 Previous Studies Examining the Effect of Contact Lenses on Eye Temperature

Date	Reference	Subject Numbers	Method	Results
1965	Hill and Leighton [15, 46]	8 eyes	Thermistors under scleral contact lenses. Studied effect of eye closure and palpebral aperture.	Eye slightly cooler with scleral lens in place, but not significant (31.3°C). Eye closure produced significant increase in temperature (33.3°C). Temperature stabilizes after 4 minutes. Vertical apertures < 6 mm produced significant increases in temperature.
1969	Hamano et al. [45]	10 eyes	Use of a noncontact radiometer and rigid contact lenses.	Less than 0.5°C difference was observed between eyes with and without rigid contact lenses. Open eye: 32.0°C; closed eye: 34.2°C.
1980	Fatt and Chaston [14]	6 eyes	Measured temperature with and without hard and soft contact lenses. Use of a noncontact bolometer. Use of a thermistor in fornix under closed-eye conditions.	Temperature wearing hard contact lenses was 0.5°C–1.5°C lower than that of bare cornea. Temperature wearing soft contact lenses was never more than 0.5°C lower than that of bare cornea. Observed increased temperature under closed-eye conditions, especially when a contact lens is worn.
1986	Martin and Fatt [16]	13 eyes	Describes modeling and experiments to investigate the surface temperature beneath hydrogel contact lenses. Thermistors sandwiched between thin hydrogel contact lenses.	Demonstrated small flow of heat from eye to environment. Exhibited insignificant increases in anterior corneal surface temperature beneath a contact lens predicted by model. Increase in temperature observed under closed-eye conditions (34.5°C compared to 36.2°C).
1991	Montoro et al. [47]	38 subjects, including 19 contact lens wearers	Observations of the effects of eye rubbing and contact lens wear. Use of a scanning infrared camera (true thermography) and customized software for objective analysis of the digital thermal image.	Observed that some subjects with increased lacrimation showed a significant cooling in temperature. Observed a minimal, transient effect of eye rubbing, concurrent with a small effect on fellow eye. Observed an irregular thermal pattern in contact lens wearers.

Table 16.2 Details of Subjects Used in the Study

Group	Contact Lens Wear	Number	Gender		Mean Age ± SD (years)	Age Range (years)	Range of Refractive Error (MSE)
Control	None	8	3 male	5 female	23.5 ± 2.8	19–28	+0.75 to –7.50
LDW	Lotrafilcon A daily wear	8	2 male	6 female	20.8 ± 1.5	18–24	+1.50 to –7.00
BDW	Balafilcon A daily wear	8	4 male	4 female	20.0 ± 1.7	18–23	–0.75 to –6.50
EDW	Etafilcon A daily wear	8	4 male	4 female	21.6 ± 2.1	24–30	–0.50 to –6.00

mencement of the study, and informed consent was obtained from every subject. All procedures conformed to the tenets of the Declaration of Helsinki.

16.2.2 Study Design and Procedure

Subjects attended randomly from each group over the course of the study, midmorning, after at least 2 of hours contact lens wear. Temperature measurements took place within a controlled environment, with room temperature maintained at $21.4 \pm 0.5°C$ and humidity at $39 \pm 1\%$. Room adaptation of 20 minutes was found to be necessary as a result of the authors' preliminary studies using this camera. A self-calibrating, infrared camera system was used for measurements (Thermo Tracer TH7102MX, NEC San-ei, Japan) of OST across the palpebral aperture. It allows dynamic thermal imaging at 60 Hz, with thermal and spatial resolutions of $0.08°C$ and 50 μm, respectively, at a close-up focus of 60 mm. The camera was positioned with respect to lower lid and canthi for each subject, so that the geometric center of the cornea could be aligned with the system (Figure 16.1).

Each subject was asked to blink and then hold the eye open for 8 seconds. This was repeated five times (8 seconds was selected as an easily achievable target, without causing reflex tearing in subjects). Contact lens wearers were then instructed to remove their right contact lens as gently as possible, and five further readings were taken. To remove the contact lens, it was digitally decentered onto the sclera and pinched off the eye. The physical effect of manipulating the lens in this way has previously been found by the author to have an insignificant effect on the measured temperature (average temperature change = $0.18°C$; $p > 0.10$). Every effort was made to minimize the delay between removing the contact lens and recording temperature.

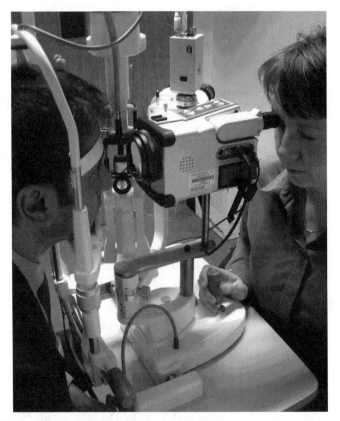

Figure 16.1 Thermal camera setups.

16.2.3 Data Acquisition and Analysis

The processing within the camera allows up to 10 point settings for recording static temperature but to permit greater flexibility, monochrome thermal images are analyzed at 60 Hz using a purpose-designed computer program (LabVIEW and Vision Software, National Instruments). The program recorded data from 23 points across the anterior eye, and for ease of analysis this data is grouped via a spreadsheet into 5 regions: central, superior, inferior, nasal and temporal (Figure 16.2).

For each recording, the point of blink (maximum temperature as lid lowers) was identified during the analysis, and temperature changes analyzed for 8 seconds following the blink. The initial temperature, the temperature decrease over 8 seconds, and the time taken to reach one-quarter, one-half, and three-quarters of the decrease were calculated.

The null hypothesis was that contact lens wear would have no effect on OST or the dynamic changes in OST following a blink. Analysis of variance was used to compare group means, and any significant differences were evaluated with Scheffe's post-hoc testing to examine differences among the groups. Pearson's correlation coefficient was used to investigate the relationship between ocular surface temperature with and without contact lenses, in the contact lens–wearing groups. All analyses were performed using Excel, version XP (Microsoft Corporation, Redmond, Washington) and StatView, version 5 (SAS Institute Inc., Cary, North Carolina).

16.3 Results

Average surface temperature on top of the hydrogel contact lenses ($36.2 \pm 1.6°C$) was greater than the mean ocular surface temperature (OST) of the non-lens-wearing control group ($35.0 \pm 1.1°C$; $p = 0.05$; Figure 16.3).

Temperature beneath the contact lens was highly correlated ($r = 0.970$; $p < 0.0001$) to that on top of the contact lens in all five areas across the ocular surface (Figure 16.4).

OST immediately following contact lens wear was significantly greater when compared to controls ($35.0 \pm 1.1°C$ versus $36.7 \pm 1.6°C$; $p < 0.01$) in all areas across the anterior eye surface. Figure 16.5 shows the mean OST for all subject groups; differences between noncontact lens wearers (controls) and those who had been wear-

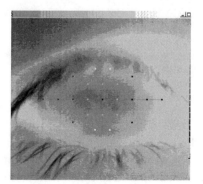

Figure 16.2 Data collection points.

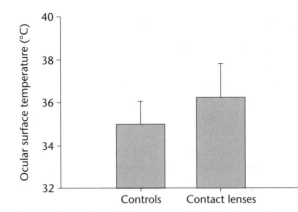

Figure 16.3 Mean temperature of controls compared to subjects wearing contact lenses (error bars = 1 SD).

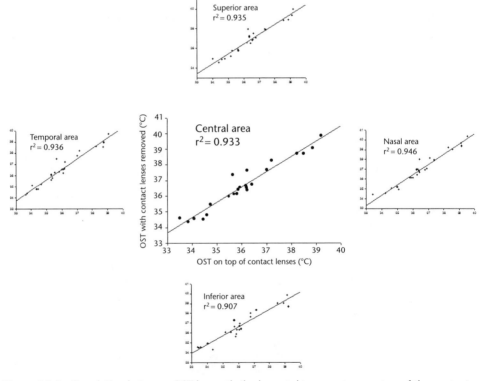

Figure 16.4 Correlation between OST beneath the lens and temperature on top of the contact lens for five regions across the ocular surface.

ing Etafilcon A (EDW) contact lenses were insignificant (Scheffe; $p = 0.86$), but significant when silicone hydrogel contact lenses had been worn (Scheffe; $p < 0.005$). No significant differences were found between subjects who had worn Lotrafilcon A contact lenses (LDW) compared to those who had worn Balafilcon A contact lenses (BDW) (Scheffe; $p = 0.93$).

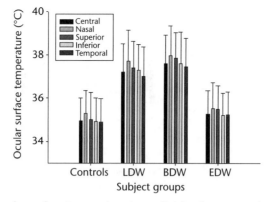

Figure 16.5   Mean ocular surface temperature immediately after contact lens wear compared to noncontact-lens wearers for five regions across the ocular surface (error bars = 1 SD).

OST decreased after a blink for all subjects, but no significant differences were found between controls and groups that had been wearing contact lenses (Figure 16.6; $p = 0.32$).

The rate of decline in the OST after a blink was calculated by measuring the time taken to reach one-quarter, one-half, and three-quarters of the total decrease; no significant differences were found between controls and subjects who had been wearing contact lenses ($p = 0.35$).

16.4   Discussion

Previous studies have observed insignificant differences in measured surface temperature during contact lens wear compared to noncontact-lens wearers [31–33]. The superior sensitivity of the current instrumentation and greater subject numbers in this study may explain the significant results seen here. The results concur with the model proposed by Martin and Fatt [34]: that there is a flow of heat from the eye to the environment that, due to the insulating properties of the contact lens material, will result in increases in surface temperature beneath a contact lens. The hydrogel contact lenses used in this study will be less efficient than the cornea and the tear film in terms of the absorption and emission of thermal radiation from the anterior eye, due to their relative percentage water content (after Lerman [35]). This accumulation of temperature beneath the lens will be transferred to the lens surface due to tear exchange and conduction/convection through the material. At the lens surface radiation will occur (due to the temperature gradient), causing a temperature differential between PoLTF and PLTF. This would explain the strong correlation between the recorded temperatures on top of the contact lenses and OST.

Variables affecting this model of thermodynamics will be water content of the contact lens, post-lens tear circulation and tear mixing (which is influenced by many factors including base curve radius [29] and lens diameter [36]), and lens fit. This study observed significantly greater OST beneath silicone hydrogel contact lenses compared to Etafilcon A lenses. These lenses have lower water content and signifi-

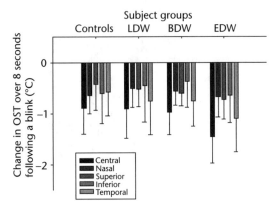

Figure 16.6 Decrease in OST for 8 seconds after a blink immediately after contact lens wear compared to noncontact-lens wearers for five regions across the ocular surface (error bars = 1 SD).

cantly less volume of PoLTF, but more tear mixing due to their high modulus of elasticity [29, 37]. The stiffness of these materials is also a contributing factor to the tarsal conjunctival hyperemia and localized papillary conjunctivitis that can be an adverse response [13]. Bulbar conjunctival hyperemia has previously been to shown to be strongly correlated with OST [38].

Further work is needed to verify the relative influence of each of these physical characteristics on OST. Differences among lens types may also vary according to wearing time. In this study all measures were taken after 2 hours of wear. Polse and colleagues have indicated that wearing time affects post-lens tear thickness under a soft contact lens [37].

The finding that changes in OST following a blink were not significantly affected by contact lens wear or between different between lens types may be a result of the wide variability in this part of the data. Wearing contact lenses has been shown to increase evaporation rates and decrease thinning time, but no significant differences have been observed between Balafilcon and Etafilcon [39].

16.4.1 Possible Clinical Relevance

Inadequate tear mixing is thought to play a part in adverse responses during extended wear of hydrogel contact lenses [40], and the closed-eye environment is known to cause an increase in OST [31, 32, 34, 41]. It may be that such increases in OST are implicated here. Silicone hydrogel lenses are being increasingly used as therapeutic contact lenses, for corneal protection and/or therapy [42, 43], including postrefractive surgery in which it has been shown that the eye already reaches high temperatures [44, 45]. Bacterial binding and protein deposition differ between organisms, lens materials [46], and modalities of wear [47]: in vitro cultures of bacteria have been shown to adhere in the greatest numbers if grown under low temperature (25°C compared to 37°C) [47], and oxygen transmissibility has been shown by some studies to be inversely related to binding of *Pseudomonas aeruginosa* [48], although others have found no difference [49]. Further work is needed to investigate the relevance of OST to such biochemical processes.

16.5 Conclusions

OST increases when contact lenses are worn, but the relative influence of the likely contributory factors (physical properties of the lens material, PoLTF mixing and volume, lid hyperemia) needs further investigation. This novel, noninvasive technique offers greater understanding of ocular physiology related to contact lens wear.

References

[1]   Du Toit, R., et al., "The Effects of Six Months of Contact Lens Wear on the Tear Film, Ocular Surfaces, and Symptoms of Presbyopes," *Optom. Vis. Sci.*, Vol. 78, No. 6, 2001, pp. 455–462.

[2]   Begley, C. G., et al., "Use of the Dry Eye Questionnaire to Measure Symptoms of Ocular Irritation in Patients with Aqueous Deficient Dry Eye," *Cornea*, Vol. 21, No. 7, 2002, pp. 664–670.

[3]   Bandeen-Roche, K., et al., "Self-Reported Assessment of Dry Eye in a Population-Based Setting," *Ophthalmol. Vis. Sci.*, Vol. 38, No. 12, 1997, pp. 2469–2475.

[4]   McMonnies, C. W., "Key Questions in a Dry Eye History," *J. of Am. Optometric Assoc.*, 1986, pp. 512–517.

[5]   Owen, C. G., F. W. Fitzke, and E. G. Woodward, "A New Computer Assisted Objective Method for Quantifying Vascular Changes of the Bulbar Conjunctivae," *Ophthalmic Physiol. Opt.*, Vol. 16, No. 5, 1996, pp. 430–437.

[6]   McMonnies, C. W., and A. Chapman-Davies, "Assessment of Conjunctival Hyperemia in Contact Lens Wearers. Part 1," *Am. J. Optometry Physiol. Opt.*, Vol. 64, No. 4, 1987, pp. 246–250.

[7]   McMonnies, C. W., A. Chapman-Davies, and B. A. Holden, "The Vascular Response to Contact Lens Wear," *Am. J. Optometry Physiol. Opt.*, Vol. 59, No. 10, 1982, pp. 795–799.

[8]   Holden, B. A., "The Ocular Response to Contact Lens Wear," *Optometry Vis. Sci*, Vol. 66, 1989, pp. 717–733.

[9]   Bron, A. J., S. Mengher, and C. C. Davey, "The Normal Conjunctiva and Its Responses to Inflammation," *Trans. Ophthalmol. Soc. UK*, Vol. 104, 1985, pp. 424–435.

[10] Mapstone, R., "Determinants of Ocular Temperature," *Br. J. Ophthalmol.*, Vol. 52, 1968, pp. 729–741.

[11] Efron, N., "Grading Scales for Contact Lens Complications," *Ophthalmic Physiol. Opt.*, Vol. 18, No. 2, 1998, pp. 182–186.

[12] Wolffsohn, J. S., and C. Purslow, "Clinical Monitoring of Ocular Physiology Using Digital Image Analysis," *Contact Lens and Anterior Eye*, Vol. 26, 2003, pp. 27–35.

[13] Dumbleton, K., "Adverse Events with Silicone Hydrogel Continuous Wear," *Contact Lens and Anterior Eye*, Vol. 25, No. 3, 2002, pp. 137–146.

[14] Schwartz, B., "Environmental Temperature and the Ocular Temperature Gradient," *Arch. Ophthalmol.*, Vol. 74, 1965, pp. 237–243.

[15] Freeman, R. D., and I. Fatt, "Environmental Influences on Ocular Temperature," *Invest. Ophthalmol.*, Vol. 12, No. 8, 1973, pp. 596–602.

[16] Rysa, P., and J. Sarvaranta, "Corneal Temperature in Man and Rabbit. Observations Made Using an Infrared Camera and a Cold Chamber," *Acta Ophthalmologica*, Vol. 52, 1974, pp. 810–816.

[17] Hill, R. M., and A. J. Leighton, "Temperature Changes of Human Cornea and Tears Under a Contact Lens 2. Effects of Intermediate Lid Apertures and Gaze," *Am. J. Optometry Arch. Am. Acad. Optometry*, 1965.

[18] Morgan, P. B, "Ocular Thermography in Health and Disease," *Optometry*, 1994.

[19] Mapstone, R., "Corneal Thermal Patterns in Anterior Uveitis," *Br. J. Ophthalmol.*, Vol. 52, 1968, pp. 917–921.

[20] Mikesell, G. W. J., "Corneal Temperatures—A Study of Normal and Laser-Injured Corneas in the Dutch Belted Rabbit," *Am. J. Physiol. Opt.*, Vol. 55, No. 2, 1978, pp. 108–115.

[21] Horven, I., "Corneal Temperature in Normal Subjects and Arterial Occlusive Disease," *Acta Ophthalmologica*, Vol. 53, 1975, pp. 863–874.

[22] Morgan, P. B., M. P. Soh, and N. Efron, "Corneal Surface Temperature Decreases with Age," *Contact Lens and Anterior Eye*, Vol. 22, No. 1, 1999, pp. 11–13.

[23] Alio, J., and M. Padron, "Influence of Age on the Temperature of the Anterior Segment of the Eye," *Ophthalmic Res.*, Vol. 14, 1982, pp. 153–159.

[24] Girardin, F., et al., "Relationship Between Corneal Temperature and Finger Temperature," *Arch. Ophthalmol.*, Vol. 117, 1999, pp. 166–169.

[25] Efron, N., G. Young, and N. Brennan, "Ocular Surface Temperature," *Curr. Eye Res.*, Vol. 8, No. 9, 1989, pp. 901–906.

[26] Fielder, A. R., et al., "Problems with Corneal Arcus," *Trans. Ophthalmol. Soc. UK*, Vol. 101, 1981, pp. 22–26.

[27] Korb, D. R., et al., *The Tear Film: Structure, Function and Clinical Examination*, Oxford, U.K.: Butterworth-Heinemann, 2002.

[28] Cedarstaff, T. H., and A. Tomlinson, "A Comparative Study of Tear Evaporation Rates and Water Content of Soft Contact Lenses," *Am. J. Optometry Physiol. Opt.*, Vol. 60, 1983, pp. 167–174.

[29] Lin, M. C., Y. Q. Chen, and K. A. Polse, "The Effects of Ocular and Lens Parameters on the Postlens Tear Thickness," *Eye & Contact Lens*, Vol. 29, No. 1S, 2003, pp. S33–S36.

[30] Korb, D. R., "Tear Film–Contact Lens Interactions," *Advances in Experimental Medicine and Biology*, Vol. 350, 1994, pp. 403–410.

[31] Fatt, I., Chaston, J., "Temperature of a Contact Lens on the Eye," *Int. Contact Lens Clinic*, Vol. 7, 1980, pp. 195–198.

[32] Hill, R. M., and A. J. Leighton, "Temperature Changes of a Human Cornea and Tears Under a Contact Lens 1. The Relaxed Open Eye, and the Natural and Forced Closed Eye Conditions," *Am. J. Optometry Arch. Am. Acad. Optometry*, Vol. 42, 1965a, pp. 9–16.

[33] Hamano, H., S. Minami, and Y. Sugimori, "Experiments in Thermometry of the Anterior Portion of the Eye Wearing a Contact Lens by Means of Infra-Red Thermometer," *Contacto, Vol. 13, No. 2, 1969, pp. 12–22.*

[34] Martin, D., and I. Fatt, "The Presence of a Contact Lens Induces a Very Small Increase in the Anterior Corneal Surface Temperature," *Acta Ophthalmologica*, Vol. 64, 1986, pp. 512–518.

[35] Lerman, S., *Radiant Energy and the Eye*, New York: Macmillan, 1980.

[36] McNamara, N. A., et al., "Tear Mixing Under a Soft Contact Lens: Effects of Lens Diameter," *Am. J. Ophthalmol.*, Vol. 127, No. 6, 1999, pp. 659–665.

[37] Polse, K., M. Lin, and S. Han, "Wearing Time Affects Post-Lens Tear Thickness under a Soft Contact Lens," *ARVO Meeting Abstracts*, Vol. 43, No. 12, 2002, p. 970.

[38] Efron, N., et al., "Temperature of the Hyperaemic Bulbar Conjunctiva," *Curr. Eye Res.*, Vol. 7, No. 6, 1988, pp. 615–618.

[39] Thai, L., M. Doane, and A. Tomlinson, "Effect of Different Soft Contact Lens Materials on the Tear Film," *Invest. Ophthalmol. Vis. Sci.*, Vol. 43, No. 12, 2002, p. 3083.

[40] Miller, K. L., K. A. Polse, and C. J. Radke, "Fenestrations Enhance Tear Mixing Under Silicone-Hydrogel Contact Lenses," *Invest. Ophthalmol. Vis. Sci.*, Vol. 44, No. 1, 2003, pp. 60–67.

[41]   Mapstone, R., "Measurement of Corneal Temperature," *Exper. Eye Res.*, Vol. 7, 1968, pp. 237–243.

[42]   Kanpolat, A., and O. Ucakhan, "Therapeutic Use of Focus® Night & Day™ Contact Lenses," *Cornea*, Vol. 22, No. 8, 2003, pp. 726–734.

[43]   Montero, J., J. Sparholt, and R. Mely, "Retrospective Case Series of Therapeutic Applications of a Lotrafilcon a Silicone Hydrogel Soft Contact Lens," *Eye & Contact Lens*, Vol. 29, No. 1S, 2003, pp. S54–S56.

[44]   Betney, S., et al., "Corneal Temperature Changes During Photorefractive Keratectomy," *Cornea*, Vol. 16, No. 2, 1997, pp. 158–161.

[45]   Maldonado-Codina, C., P. B. Morgan, and N. Efron, "Thermal Consequences of Photorefractive Keratectomy," *Cornea*, Vol. 20, No. 5, 2001, pp. 509–515.

[46]   Willcox, M. D. P., et al., "Bacterial Interactions with Contact Lenses; Effects of Lens Material, Lens Wear and Microbial Physiology," *Biomaterials*, Vol. 22, No. 24, 2001, pp. 3235–??.

[47]   Cavanagh, H. D., et al., "Effects of Daily and Overnight Wear of a Novel Hyper Oxygen-Transmissible Soft Contact Lens on Bacterial Binding and Corneal Epithelium: A 13-Month Clinical Trial," *Ophthalmology*, Vol. 109, No. 11, 2002, pp. 1957–1969.

[48]   Ren, D. H., et al., "The Relationship between Contact Lens Oxygen Permeability and Binding of *Pseudomonas aeruginosa* to Human Corneal Cells After Overnight and Extended Wear," *The CLAO Journal*, Vol. 25, No. 2, 1999, pp. 80–96.

[49]   Borazjani, R. N., B. Levy, and D. G. Ahearn, "Relative Primary Adhesion of *Pseudomonas aeruginosa*, *Serratia marcescens* and *Staphylococcus aureus* to Hema-Type Contact Lenses and an Extended Wear Silicone Hydrogel Contact Lens of High Oxygen Permeability," *Contact Lens and Anterior Eye*, Vol. 27, No. 1, 2004, pp. 3–8.

CHAPTER 17

Variations of Ocular Surface Temperature with Different Age Groups

Eddie Y. K Ng, J. H. Tan, E. H. Ooi, Caroline Chee, and Rajendra Acharya

17.1 Introduction

The OST has been of great interest since the 1800s and is still being researched by many scientists today. Before IR thermography was available, OST was measured by placing thermal sensors (mercury thermometer, thermistor, and thermocouple) onto the surface of the cornea, and the readings were obtained when steady state was achieved. The earliest reported work done by Dohnberg (cited by Holmberg [1]) in 1875 involved the use of a mercury bulb thermometer to measure temperatures at the conjunctival sacs of humans. A more comprehensive work was reported by Holmberg [1] who found that OST was 2.51 ± 0.13°C lower than oral temperature. Use of the early thermal sensors required direct contact with the ocular surface, and if intraocular temperatures were to be measured, penetration of a needle probe into the eye was performed, which was usually preceded by the administration of anesthesia. This method resulted in measurement errors because heat from the eye is constantly transferred away through the needle probe, which in this case behaves as a pin fin [2]. Such procedures are now limited to animal studies due to the trauma faced by the test subject and the possibility of damaging the eye.

The pioneer of noninvasive OST measurement was Mapstone [3] who used a bolometer to measure the amount of IR radiation from the corneal surface before converting the measured radiation into temperature using the Stefan-Boltzmann equation. The use of a bolometer did not involve any contact between the measuring device and the corneal surface. The ability to obtain instantaneous results also made the noninvasive measuring technique preferable. Mapstone [3] reported a mean OST of 34.8°C based on 70 subjects. Although the bolometer was superior to the invasive measuring techniques, it failed to show the topographical variation of the temperature within the eye.

During the late 1950s, IR thermography, which was originally used by the military, was adapted for use in the medical field [4]. Later, in the early 1970s, IR thermography began to gain popularity in ophthalmology as a device to measure OST. Mapstone [5] was the first to report on this work using a primitive version of an IR camera in which the captured temperature profiles were displayed on a black-and-white monitor. Although primitive, IR cameras were able to show the topographical variations in OST and this agreed with Mapstone's earlier reported works [3, 5–8]. A similar IR camera was later used by Rysä and Sarvaranta [9, 10] to measure the OST changes during cold stress and by Fielder et al. [11] who reported that the corneal temperature is 3.6°C below the aural temperature, which is higher than the observation made by Holmberg [1]. A summary of the characteristics and the advantages and disadvantages of both the contact and noncontact eye temperature measurements is given in Table 17.1.

The advancement of IR technology resulted in the development of more efficient real-time IR cameras and display units that were able to present measured temperatures in a color-coded display. Further developments in IR technology led to a new breed of IR camera that is small and portable with high spatial and thermal resolutions and can obtain real-time temperature readings. This encouraged many ophthalmologists to conduct research over the years for the purpose of understanding the OST. Among the more notable works using IR thermography are those by Efron et al. [12], Morgan et al. [13, 14], Craig et al. [15], and Purslow et al. [16] as tabulated in Table 17.2 [17–20].

17.1.1 Variations of Ocular Surface Temperature with Age

The OST is sensitive and is affected by many parameters such as environmental conditions [6, 21–23], blinking [6, 9, 10, 12, 24–26], age [18, 19, 27–29], pathophysiology (vascular and nonvascular diseases), and the presence of contact lens [16, 17, 24, 25, 30]. Factors such as environmental conditions (environment temperature and humidity) and blinking and contact lens wear are parameters that can be controlled during an OST measuring session (although the effects of long-term contact lens wear are not yet fully understood). However, the age and pathophysiological factors cannot be controlled.

It is generally agreed that the OST is negatively correlated with age although the degree of correlation varies between reports. This decrease was attributed to atherosclerosis, which is a degenerative condition that reduces the caliber of all major arteries beyond middle age [29]. A decrease of 0.01°C per year in OST was reported by Morgan et al. [29]. The work done by Alio and Padron [18] on the correlation

Table 17.1 Characteristics of Invasive and Noninvasive Eye Temperature Measurements

Invasive/Contact Measurement	*Noninvasive/Noncontact Measurement*
Earliest reported work: Dohnberg, 1875	Pioneered by Mapstone [3]
Requires direct contact	No contact involved
Equipment: thermometer, thermistor, thermocouple	Equipment: bolometer, IR thermography
Penetration for intraocular temperature	Leading method in OST measurements
Limited to animal studies	
Disadvantages	
Trauma and damage are possible.	Method is unable to produce intraocular temperatures.
The reading is not instantaneous.	
Anasthesia is required.	
Only allows local measurement.	
Measurement error up to 6°C results because probe acts as a heat sink [2].	

Table 17.2 Summary of OST from Various Literatures Using the Noncontact Method

Author(s)	Mean Temperature (°C)	Technique
Mapstone [3]	34.80	Bolometer
Rysä and Sarvaranta [9, 10]	34.80	IR thermography
Fatt and Chaston [17]	34.50	IR thermography
Alio and Padron [18]	32.90	IR thermography
Fielder et al. [11]	33.40	IR thermography
Efron et al. [12]	34.30	IR thermography
Morgan et al. [13]	33.50	IR thermography
Morgan et al. [14]	31.94	IR thermography
Girardin et al. [19]	33.70	IR thermography
Gugleta et al. [20]	33.65	IR thermography
Craig et al. [15]	33.82	IR thermography
Purslow et al. [16]	35.00	IR thermography

between OST and age was not conclusive. However, based on the graph presented by them, the decrease in OST can be roughly estimated to be within 0.023°C per year. Both reports conducted measurements using IR thermography.

The age factor was also studied using contact thermometry, as was done by Isenberg and Green [28] with a reported decrease of 0.02°C per year. Hørven [27] reported a similar decrease (0.023°C per year) using a thermoelectrical measuring method. Both reports were in good agreement with the results obtained by Alio and Padron [18] using IR thermography. Note, however, that the result in [18] discussed here is merely an estimation from the graph presented in the paper. Morgan et al. [29] pointed out that the measuring techniques used in [27, 28] required the administration of anesthesia and this could be the reason for the deviation of results.

In the present study, the effects of age on the variation of OST are investigated. The purpose of this study is to gain a better understanding on the decrease in the OST with increasing age. In addition, the temperature variations between the left and right eye are further investigated. The effectiveness of IR thermography in obtaining the OST is examined in the latter part of this chapter.

17.1.2 Clinical Applications

Understanding the OST has its importance in clinical applications. Hippocrates, a Greek physician in 400 B.C., reported that in any part of the body that experiences an excess of heat or cold, disease is there to be found. This statement—although dated centuries ago—is still being applied today. A typical example is the measurement of body temperature by clinical doctors to determine if a person is feverish and infectious. The advent of IR thermography has led to medical applications for detecting human body abnormalities [31–35]. Among the highly successful fields are its use in detecting rheumatism [4], vascular disorders [36, 37], and breast cancers [37–42].

A report by Beutelspacher [43] suggested the use of IR thermography in identifying ocular diseases. Earlier works by Holmberg [1] had discovered that the eye temperature is higher in patients with various eye inflammations compared to normal ones. Holmberg [1] also cited much earlier works by Dohnberg, Galezowski, Silex, Giese, Hertel, and Kirisawa where temperatures for acute conjunctivitis, keratitis, and iritis were higher than in normal cases. During the measurement of the OST using an IR camera, environmental conditions can be controlled. Blinking of eyelids could also be induced or prevented at the will of the patient. The variation of OST with the age of an individual is an uncontrollable factor, however. It is therefore important to gain a strong understanding of the effects of increasing age on the OST, particularly in the diagnosis of a particular pathophysiological disease. If OST is dependent on the age of an individual, as is reported in the literature [18, 19, 27–29], then a poor understanding of this issue may result in a clinician arriving at an incorrect diagnosis.

17.2 Methodology

17.2.1 Equipment Specifications

The VarioCAM (Figure 17.1; JENOPTIK Laser, Optik, Systeme GmbH, http://www.jenoptik-los.com or http://www.InfraTec.co.uk) together with a proper chin rest was used to obtain ocular surface thermograms of test subjects. The VarioCAM is a portable device that uses an uncooled microbolometer detector (focal plane array). It has a measurement accuracy of $\pm 2\%$ and a spatial resolution of 320×320 pixels. At 30°C, it has a temperature resolution of 0.08K. Calibration of the machine is unnecessary because the VarioCAM comes with a built-in, self-calibrated system.

The use of the VarioCAM requires input for the value of effective emissivity, which is critical to ensure that temperature measurements are accurate [44]. Due to the absence of an actual living human cornea as the testing material, the effective emissivity used in this study was estimated using the reference emitter technique [44]. The subsequent section briefly highlights the steps involved in estimating the effective emissivity.

17.2.2 Reference Emitter Technique

The reference emitter technique estimates the emissivity of a given material based on a sample material. In this study, polymethyl methacrylate (PMMA) was used as the

Figure 17.1 The VarioCAM used to obtain ocular surface thermograms.

sample material. Two reasons prompted the selection of PMMA. First, it is practically impossible to prepare an actual cornea as the sample material and, second, PMMA was once used to manufacture hard contact lens. The use of PMMA as the sample material will thus best simulate the properties of the cornea.

Two sets of sample material were molded into hemispheres with a radius of 50 mm (to be large enough to contain several spot sizes or instantaneous fields of view of the camera). The first sample was sprayed with black paint and covered with a substance of known high emissivity, whereas the second sample was maintained at its normal condition. Both samples were placed side by side and heated to 37°C uniformly from room temperature (25°C) with an electrical heater. The VarioCAM emissivity was set to the emissivity of the black paint (1.0), and the temperature of the first sample on reaching 37°C was recorded. The VarioCAM was then immediately pointed to the second, unpainted material, and the emissivity input to the VarioCAM was adjusted until the temperature recorded from the second sample reached the temperature recorded from the first sample. The final adjusted emissivity value of 0.978 was used throughout this study.

17.2.3 Test Subjects

A total of 48 individuals, 24 males and 24 females, were used as pilot test subjects. The subjects ranged in age from 20 to 59 with an average of 38.9 ± 12.2 years (mean $\pm$ SD) (Table 17.3). Selection procedures were controlled and based on the following criteria:

1. No current eye disease;
2. No history of mild or serious eye disease;
3. No history of ocular or facial surgery;
4. Generally in good health;
5. Free of any recent usage of tear drops or medications.

Subjects were brought to a room where temperature was maintained at $25 \pm 1°C$ with a mean humidity of 84% throughout the entire session. External factors that could cause discrepancies in measurements such as reflection, sunlight, and air flow (natural and artificial) were prevented from entering the room. Prior to the measurement, subjects were given a minimum period of 20 minutes to allow thermal equilibrium with the environment. During this adaptation period, subjects refrained from:

Table 17.3 Age Groups of the Pilot Test Subjects

Age Group	Male	Female	Total
20–29	6	6	12
30–39	6	6	12
40–49	6	6	12
50–59	6	6	12
Total	24	24	48

1. Consuming any hot or cold beverages that might cause changes to the overall body temperature;
2. Rubbing the eyes;
3. Washing of face or eyes;
4. Using eyedrops (e.g., lubricant);
5. Exercising or walking around;
6. Being stressed.

For every measurement, each subject was seated and VarioCAM mounted onto a stand where the camera lens and the subject's eye (with chin rest) were maintained at a constant distance of 15 cm. (The camera was aimed normal to the target surface because the effective emissivity of a target surface is partially dependent on the surface texture [45].) During the focusing of the VarioCAM's lens, the subject was asked to naturally close his or her eyes. Once the focusing was done, the subject was asked to open his or her eyes, and the temperature was captured after 3 seconds following the prolonged opening of the eyes. These steps were repeated two more times for (averaging of the three attempts) each eye. A total of six thermographic images for each subject are the result.

17.3   Results and Discussion

A sample of the left ocular surface thermogram obtained for a 22-year-old male participant is shown in Figure 17.2. The white circle depicted by the "C11" label is the area of interest (AOI) in the present study, which is termed the central corneal temperature (CCT). The built-in software of the VarioCAM scans the area covered by circle C11 to calculate the average temperature across the circular area.

From Figure 17.2, we can see that the cooler region is located at the center of the cornea and the temperature increases away to the limbus. This agrees well with most works obtained from the literature. The central corneal region is furthest from the heating source of the limbal vasculature and thus having the lowest temperature on the ocular surface. The elliptical isotherm observed in Figure 17.2 agrees well with the observations made by Efron et al. [12] and Morgan et al. [13]. An interesting observation here is the nasal area (the left side of Figure 17.2), which is shown on the left side of Figure 17.2. At the nasal region, the temperature appeared to be high and

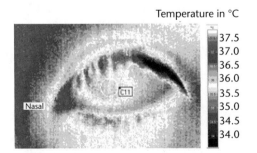

Figure 17.2 A typical left ocular surface thermogram for a 22-year-old male.

almost similar to the body core temperature. Such an observation had been reported in a recent report by Ng et al. [46] on 750 random samples of febrile and nonfebrile subjects.

From the recorded thermograms on all subjects, the average CCT is 35.07 ± 0.71°C and 35.18 ± 0.63°C for the right and left eye, respectively. The slightly higher temperature compared to reports from the literature [3, 11, 12, 29] may be caused by the closure of the subjects' eyelids during focusing of the VarioCAM. Closure of eyelids was found to increase the OST [6, 9, 10, 12, 24–26]. Although measurements were taken 3 seconds after reopening of eyelids, this time could be insufficient for OST to reach thermal equilibrium.

A statistical analysis was conducted on the significance of differences between the right and left eyes and the differences were found to be significant (student t test for independent variables: $t = 3.555$, $\alpha = 0.01$). The left eye is, on average, 0.11°C higher compared to the right eye. This contradicts previously reported work [13]. Figure 17.3 plots the correlation between the CCT for the left and right eye. Correlation between the CCT for the left and right eye is high ($R^2 = 0.7644$).

Ng et al. [46] reported that the left aural temperature is slightly higher (not more than 1°C) than that of the right ear. The present study shows that the left eye is also at a slightly higher temperature than the right (0.11°C). The reason behind this slight increase for both the ocular and aural temperatures could probably be the same. One possible explanation for the increase in the left-sided temperature could be the different functions of the brain in the left and right hemispheres.

According to Nicolas Cherbuin [47], when one side of the brain hemisphere is activated, blood that flows into the brain increases. Since the temperature of the blood which flows into the brain is cooler compared to the blood that is already in the brain, a drop in brain temperature occurs. During the measurement of OST, the subjects were instructed to close their eyes during the camera's lens focusing process

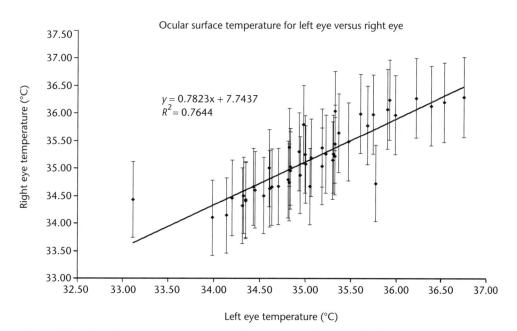

Figure 17.3   Correlation of the central corneal temperature between the left and right eyes.

and gaze into the VarioCAM's lens soon after. This action would provoke the right side of the brain (the right brain is known to be responsible for visual and action), thus reducing the temperature. Although there is still no evidence that the OST correlates with brain temperature, this may explain that this could be true, and it explains the higher OST for the left eye. More investigation on this issue should be carried out to further understand the response of OST with brain activity.

Figure 17.4 shows the changes of left eye CCT with the subject's age. A negative correlation was found and agrees qualitatively with reports by Hørven [27], Alio and Padron [18], Isenberg and Green [28], Morgan et al. [29], and Girardin et al. [19]. The decrease in temperature is 0.0383°C per year and is higher compared to results by other researchers mentioned earlier. Statistical analyses show that the linear regression approximation is adequate ($t = 7.175$, $\alpha = 0.01$). The decrease is statistically different from zero ($\alpha = 0.01$), which agrees with results by Morgan et al. [29]. The decrease of OST with age was postulated by Morgan et al. to be caused by atherosclerosis, a degenerative condition in which the ability of all major arteries to function properly is reduced on reaching middle age.

By taking 40 years as the threshold for middle age [29] and analyzing the data, it can be shown that the rate of decrease of left eye OST with age is higher for older subjects (> 40 years) compared to the younger ones (< 40 years) (−0.0376°C versus −0.0035 per year). This, along with the correlation coefficient R^2 is shown graphically in Figure 17.5.

The actual cause for the decrease in OST with age is not well understood. Atherosclerosis, as suggested by Morgan et al. [29], is generally accepted as the cause. A more recent work by Burhan et al. [48] reported that the blood flow inside the internal carotid artery (main artery where the ophthalmic artery branches off from) decreases when human age increases. Since the source of heat for the eye is from the flow of blood circulating the choroidal layer and the iris, reduction in ocular blood

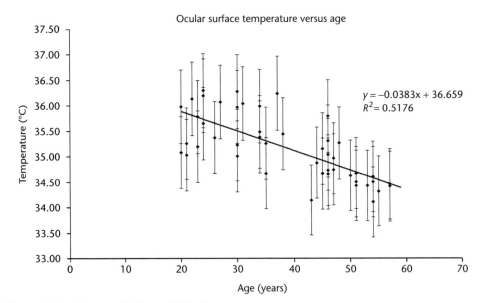

Figure 17.4 Changes of left eye CCT with age.

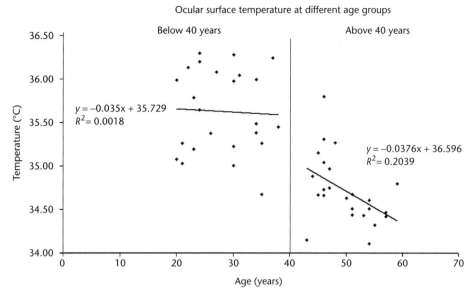

Figure 17.5   Relationship between left eye CCTs with different age ranges.

flow would result in a decrease in OST [6, 27]. This would then enhance the hypothesis given by Morgan et al.

Other factors such as changes in dimension and properties inside the human eye could also explain the decrease in OST. The water content inside the human lens was found to increase as individuals get older [49]. Changes in the lens water content ultimately alter its thermal properties, such as thermal conductivity, specific heat, and density. This, in turn, leads to changes in the overall eye temperature distribution.

The depth of the anterior chamber was also found to be dependent on an individual's age [50, 51] where a decrease is observed with increasing age. Rysa and Sarvaranta [9] found that the rate of change in temperature increases across the anterior chamber when its depth decreases. This consequently points out that the OST would be lower for smaller anterior chamber depth. Dimensions of the lens such as its radii curvature and its true lens thickness were also reported to change with age [50, 52, 53]. Lens radii were found to decrease with age, whereas true lens thickness increases. As a person gets older, the depth of the anterior chamber decreases. Alteration in dimension could thus also cause changes to the eye temperature distribution.

The relationship between OST and age may also be attributed to the decrease in tissue metabolism when age increases. Lower tissue metabolism would suggest that less heat is being expelled or transferred from the body. Therefore, as metabolism for human decreases with age (this is in accordance with the Harris-Benedict equation for calculating basal metabolic rate), the temperature throughout the body should decrease as well. In brief, accurately pinpointing the cause for the changes in OST with age would require a thorough investigation on the physiological changes of the eye as a person gets older.

17.3.1   Effectiveness of Infrared Cameras in Ocular Surface Thermography

In the current study, 288 measurements were taken from 48 test subjects. It is important to ensure that the measured temperatures produce consistent results throughout the entire thermal scanning session and within the certain specified range. Figure 17.6 illustrates the reproducibility of the measured temperatures shown as the differences between the left and right eye temperatures. (The reproducibility of both the instrument and physiological assumptions was checked and established by comparing paired left-right readings of the eyes.) The graph exhibits a trend similar to the normal distribution curve, in which the majority of the measured temperatures showed differences in the range between 0°C and 0.2°C. In fact, we can see that a near-zero difference was obtained for the highest number of subjects. The results shown in Figure 17.6 suggest that the IR camera used gives an acceptable reproducible measurement throughout the 288 recordings.

17.4   Conclusions

The OST was found to decrease with age at a rate of 0.0383°C per year. The reason behind this may be atherosclerosis. Other factors that could cause this decrease were suggested and need to be further investigated for verification. The OST for the left eye was, on average, higher than that of the right eye. The use of IR thermography has proven to be an efficient and reliable method for obtaining measurements of the OST. Future work should use more advanced software such as automatic edge detection or other more advanced image processing approaches to locate objectively and better the central corneal surface temperature.

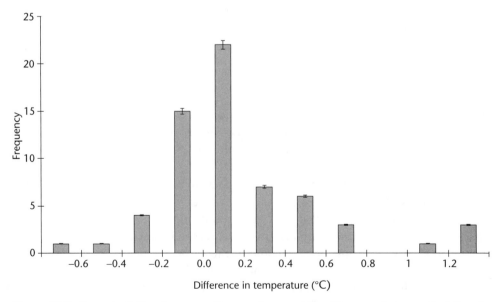

Figure 17.6   Reproducibility of measured temperatures and the differences between the left and right eyes.

Acknowledgments

The authors would like to thank Z. J. Tan Nanyang Technological University's B. Eng. Final Year Project and Eric Er of Eetarp Engineering for setting up the VarioCAM system. We also thank Dr. Manjunath Gupta (Department of Ophthalmology, National University Hospital of Singapore) for his interest in this initial phase of study regarding the age factor in OST with the healthy control subjects screened in Nanyang Technological University.

References

[1] Holmberg, A., "The Temperature of the Eye During Application of Hot Packs and After Milk Injections," *Acta Ophthalmologica*, Vol. 30, No. 4, 1952, pp. 347–364.

[2] Fatt, I., and J. F. Forester, "Errors in Eye Tissue Temperature Measurements When Using a Metallic Probe," *Exper. Eye Res.*, Vol. 14, No. 3, 1972, pp. 270–276.

[3] Mapstone, R., "Measurement of Corneal Temperature," *Exper. Eye Res.*, Vol. 7, No. 2, 1968, pp. 237–243.

[4] Ring, E. F. J., "Progress in the Measurement of Human Body Temperature," *IEEE Eng. Med. Biol.*, Vol. 17, No. 4, 1998, pp. 19–24.

[5] Mapstone, R., "Ocular Thermography," *Br. J. Ophthalmol.*, Vol. 54, No. 11, 1970, pp. 751–754.

[6] Mapstone, R., "Determinants of Ocular Temperature," *Br. J. Ophthalmol.*, Vol. 52, No. 10, 1968, pp. 729–741.

[7] Mapstone, R., "Normal Thermal Patterns in Cornea and Periorbital Skin," *Br. J. Ophthalmol.*, Vol. 52, No. 11, 1968, pp. 818–827.

[8] Mapstone, R., "Corneal Thermal Patterns in Anterior Uveitis," *Br. J. Ophthalmol.*, Vol. 52, No. 12, 1968, pp. 917–921.

[9] Rysa, P., and J. Sarvaranta, "Thermography of the Eye During Cold Stress," *Acta Ophthalmologica*, Vol. 123, 1973, pp. 234–239.

[10] Rysä, P., and J. Savaranta, "Corneal Temperature in Man and Rabbit. Observations Made Using an Infra-Red Camera and a Cold Chamber," *Acta Ophthalmologica*, Vol. 52, No. 6, 1974, pp. 810–816.

[11] Fielder, A. R., et al., "Problems with Corneal Arcus," *Trans. Ophtalmological Soc. UK*, Vol. 101, No. 1, 1981, pp. 22–26.

[12] Efron, N., G. Young, and N. Brennan, "Ocular Surface Temperature," *Curr. Eye Res.*, Vol. 8, No. 9, 1989, pp. 901–906.

[13] Morgan, P. B., et al., "Potential Applications of Ocular Thermography," *Optometry Vis. Sci.*, Vol. 70, No. 7, 1993, pp. 568–576.

[14] Morgan, P. B., A. B. Tullo, and N. Efron, "Infrared Thermography of the Tear Film in Dry Eye," *Eye*, Vol. 9, No. 5, 1995, pp. 615–618.

[15] Craig, J. P., et al., "The Role of Tear Physiology in Ocular Surface Temperature," *Eye*, Vol. 14, No. 4, 2000, pp. 635–641.

[16] Purslow, C., J. Wolffsohn, and J. Santodomingo-Rubido, "The Effect of Contact Lens Wear on Dynamic Ocular Surface Temperature," *Contact Lens and Anterior Eye*, Vol. 28, No. 1, 2005, pp. 29–36.

[17] Fatt, I., and J. Chaston, "Temperature of a Contact Lens on the Eye," *Int. Contact Lens Clinic*, Vol. 7, 1980, pp. 195–198.

[18] Alio, J., and M. Padron, "Influence of Age on the Temperature of the Anterior Segment of the Eye," *Ophthalmic Res.*, Vol. 14, No. 3, 1982, pp. 153–159.

[19] Girardin, F., et al., "Relationship Between Corneal Temperature and Finger Temperature," *Arch. Ophthalmol.*, Vol. 117, No. 2, 1999, pp. 166–169.

[20] Gugleta, K., S. Orgul, and J. Flammer, "Is Corneal Temperature Correlated with Blood-Flow Velocity in the Ophthalmic Artery?" *Curr. Eye Res.*, Vol. 19, No. 6, 1999, pp. 496–501.

[21] Schwartz, B., "Environmental Temperature and the Ocular Temperature Gradient," *Arch. Ophthalmol.*, Vol. 74, 1965, pp. 237–243.

[22] Freeman, R. D., and I. Fatt, "Environmental Influences on Ocular Temperature," *Invest. Ophthalmol. Vis. Sci.*, Vol. 12, No. 8, 1973, pp. 596–602.

[23] Kocak, I., S. Orgul, and J. Flammer, "Variability in the Measurement of Corneal Temperature Using a Non-Contact Infrared Thermometer," *Ophthalmologica*, Vol. 213, No. 6, 1999, pp. 345–349.

[24] Hill, R. M., and A. J. Leighton, "Temperature Changes of a Human Cornea and Tears Under a Contact Lens: 1. The Relaxed Open Eye and the Natural and Forced Closed Eye Conditions," *Am. J. Optometry Arch. Am. Acad. Optometry*, Vol. 42, No. 2, 1965, pp. 9–16.

[25] Hill, R. M., and A. J. Leighton, "Temperature Changes of Human Cornea and Tears Under a Contact Lens: 2. Effects of Intermediate Lid Apertures and Ga57ze," *Am. J. Optometry Arch. Am. Acad. Optometry*, Vol. 42, No. 2, 1965, pp. 71–77.

[26] Hata, S., et al., "Corneal Temperature and Inter-Blinking Time," *Invest. Ophthalmol. Vis. Sci.*, Vol. 35, 1994, p. S999.

[27] Hørven, I., "Corneal Temperature in Normal Subjects and Arterial Occlusive Disease," *Acta Ophthalmologica*, Vol. 53, No. 6, 1975, pp. 863–874.

[28] Isenberg, S. J., and B. F. Green, "Changes in Conjunctival Oxygen Tension and Temperature with Advancing Age," *Critical Care Med.*, Vol. 13, No. 3, 1985, pp. 683–685.

[29] Morgan, P. B., M. P. Soh, and N. Efron, "Corneal Surface Temperature Decrease with Age," *Contact Lens and Anterior Eye*, Vol. 22, No. 1, 1999, pp. 11–13.

[30] Dixon, J., and L. Blackwood, "Thermal Variations of the Human Eye," *Trans. Am. Ophthalmological Soc.*, Vol. 89, 1991, pp. 183–193.

[31] Ng, E. Y.-K., G. J. L. Kaw, and W. M. Chang, "Analysis of IR Thermal Imager for Mass Blind Fever Screening," *Microvascular Res.*, Vol. 68, No. 2, 2004, pp. 104–109.

[32] Ng, E. Y. -K., C. Chong, and G. J. L. Kaw, "Classification of Human Facial and Aural Temperature Using Neural Networks and IR Fever Scanner: A Responsible Second Look," *J. of Mechanics in Medicine and Biology*, Vol. 5, No. 1, 2005, pp. 165–190.

[33] Ng, E. Y. -K., and G. J. L. Kaw, "IR Scanners as Fever Monitoring Devices: Physics, Physiology and Clinical Accuracy," in *Biomedical Engineering Handbook*, N. Diakides, (ed.), Boca Raton, FL: CRC Press, 2006.

[34] Ng, E. Y. K, "Is Thermal Scanner Losing Its Bite in Mass Screening of Fever Due to SARS?" *Medical Phys.*, Vol. 32, No. 1, 2005, pp. 93–97.

[35] Harding, J. R., "Investigating Deep Venous Thrombosis with Infrared Imaging," *IEEE Eng. Med. Biol.*, Vol. 17, No. 4, 1998, pp. 43–46.

[36] Fushimi, H., et al., "Peripheral Vascular Reactions to Smoking Profound Vasoconstriction by Atherosclerosis," *Diabetes Res. Clinical Practice*, Vol. 42, No. 1, 1998, pp. 29–34.

[37] Kerseyling, J. R., et al., "Infrared Imaging of the Breast: Initial Reappraisal Using High Resolution Digital Technology in Successive Cases of Stage I and Stage II Breast Cancer," *Breast J.*, Vol. 4, No. 4, 1998, pp. 245–251.

[38] Ng, E. Y. K., Y. Chen, and L. N. Ung, "Computerized Breast Thermography: Study of Image Segmentation and Temperature Cyclic Variations," *J. of Medical Eng. Technol.*, Vol. 25, No. 1, 2001, pp. 12–16.

[39] Ng, E. Y.-K., et al., "Computerized Detection of Breast Cancer with Artificial Intelligence and Thermograms," *J. of Medical Eng. Technol.*, Vol. 26, No. 4, pp. 152–157.

[40] Fok, S. C., E. Y.-K. Ng, and K. Tai, "Early Detection and Visualization of Breast Tumor with Thermogram and Neural Network," *J. of Mechanics in Medicine and Biology,* Vol. 2, No. 2, 2002, pp. 185–196.

[41] Ng, E. Y. -K., and S. C. Fok, "A Framework for Early Discovery of Breast Tumor Using Thermography with Artificial Neural Network," *Breast J.,* Vol. 9, No. 4, 2003, pp. 341–343.

[42] Ng, E. Y.-K., and Y. Chen, "Segmentation of Breast Thermogram: Improved Boundary Detection with Modified Snake Algorithm," *J. of Mechanics in Medicine and Biology,* Vol. 6, No. 2, 2006, pp. 123–136.

[43] Beutelspacher, S, "Thermography Shows 'Enormous Promise' for Diagnosis and Treatement of Eye Diseases," http://www.escrs.org/eurotimes/March2003/thermo.asp#top, 2003.

[44] Kaplan, H., *ASNT Level III Study Guide: Infrared and Thermal Testing Method,* Columbus, OH: American Society for Nondestructive Testing, 2001.

[45] Hatcher, D. J., and J. A. D'Andrea, "Effects on Thermography Due to Curvature of the Porcine Eye," *InfraMation: The Thermographers Conference,* Pub. ITC 092-A-2003-08-15, Infrared Training Center, 2003, http://www.infraredtraining.com/store/infra2003.asp.

[46] Ng, E. Y. K., W. Muljo, and B. S. Wong, "Study of Facial Skin and Aural Temperature," *IEEE Eng. Med. Biol.,* Vol. 25, No. 3, 2006, pp. 68–74.

[47] Skatssoon, J., "Health & Medical News: Cool Ears Give the Brain Away," http://www.abc.net.au/science/news/health/HealthRepublish_1328120.htm, March 23, 2005.

[48] Burhan, Y., B. Erdoðmuþ, and A. Tugay, "Cerebral Blood Flow Measurements of the Extracranial Carotid and Vertebral Arteries with Doppler Ultrasonography in Healthy Adults," *Diagnostic and Interventional Radiology (Ankara, Turkey),* Vol. 11, No. 4, 2005, pp. 195–198.

[49] Siebinga, I., et al., "Age-Related Changes in Local Water and Protein Content of Human Eye Lenses Measured by Raman Microspectroscopy," *Exper. Eye Res.,* Vol. 53, No. 2, 1991, pp. 233–239.

[50] Charles, M. W., and N. Brown, "Dimensions of the Human Eye Relevant to Radiation Protection," *Phys. Med. Biol.,* Vol. 20, No. 2, 1975, pp. 202–218.

[51] Fontana, S. T., and R. F. Brubaker, "Volume and Depth of the Anterior Chamber of the Normal Aging Human Eye," *Arch. Ophthalmol.,* Vol. 98, No. 10, 1980, pp. 1803–1808.

[52] Lowe, R. F., and B. A. Clark, "Posterior Corneal Curvature: Correlations in Normal Eyes and in Eyes Involved with Primary Angle-Closure Glaucoma," *Br. J. Ophthalmol.,* Vol. 57, No. 7, 1973, pp. 464–470.

[53] Lowe, R. F., and B. A. Clark, "Radius of Curvature of the Anterior Lens Surface: Correlations in Normal Eyes and in Eyes Involved with Primary Angle-Closure Glaucoma," *Br. J. Ophthalmol.,* Vol. 57, No. 7, 1973, pp. 471–474.

About the Editors

Rajendra Acharya U is a visiting faculty member at Ngee Ann Polytechnic, Singapore. He received his doctorate from the National Institute of Technology, Karnataka, Surathkal, India. He served as an assistant professor in the Department of Bio-Medical Engineering, Manipal Institute of Technology, Manipal, India, until 2001. He served as a technical consultant for Larsen and Tourbo (L&T), Mysore, India, and Honeywell India Software Operations, Bangalore, India, in 2000. He has published more than 47 papers in refereed international journals and more than 25 papers in national and international conference proceedings. Recently, he edited the book *Advances in Cardiac Signal Processing* (Springer-Verlag, 2007). He is a reviewer for more than 15 international journals and served as a scientific chair for the Third Biomedical Society Singapore Meeting, Singapore, held in May 2004. He was also the chair of an invited session for the Fifteenth International Conference on Mechanics in Medicine and Biology in Singapore in 2006. His résumé was selected for Marquis *Who's Who in Medicine and Healthcare* in 2006. His major interests are biomedical signal processing, bioimaging, data mining, visualization, and biophysics for better health care design, delivery, and therapy. He is a member of the IEEE and BMESI.

Eddie Y. K. Ng worked as a marine engineer for the Exxon Tankers Fleet from 1983 to 1985. He obtained a B.Eng. from the University of Newcastle upon Tyne, England. He pursued his Ph.D. at Queens' College and Whittle Research Laboratory at Cambridge University with a Cambridge Commonwealth Scholarship. Dr. Ng obtained a postgraduate diploma in teaching higher education from NIE-NTU in 1995. His main areas of research are thermal imaging, human physiology, biomedical engineering, computational turbomachinery aerodynamics, microscale cooling problems, and CFD/CHT. He is an associate professor at Nanyang Technological University in the School of Mechanical and Aerospace Engineering and adjunct Singapore National University Hospital scientist. Dr. Ng has published more than 250 papers in refereed international journals, 70 international conference proceedings, and textbook chapters. He has also coedited the book *Cardiac Pumping and Perfusion Engineering* (World Scientific Press, 2007) and coauthored the book *Compressor Instability with Integral Methods* (Springer-Verlag, 2007). Dr. Ng is the editor-in-chief for the *Journal of Mechanics in Medicine and Biology* and the associate editor for the *International Journal of Rotating Machinery, Computational Fluid Dynamics Journal*, and the *Chinese Journal of Medicine*. He is an invited keynote lecture speaker for many international scientific conferences and workshops.

Jasjit S. Suri has spent more than 20 years in imaging sciences; in the last 15 years he has worked in the field of medical imaging modalities and their fusion. He has published more than 200 technical papers on body imaging related to modalities such as magnetic resonance imaging, computed tomography, X-ray, ultrasound, PET, SPECT, elastography, and molecular imaging. While working for more than a decade for companies such as Siemens Research, Philips Research, Fischer Research, and Eigen Divisions in the capacity of scientist, senior director of R&D, and chief technology officer, Dr. Suri has submitted more than 30 patents to the U.S. Patent Office, covering the areas of medical imaging modalities. Dr. Suri has also written 14 collaborative books on body imaging for areas such as cardiology, neurology, pathology, mammography, angiography, atherosclerosis/plaque imaging, ophthalmology, molecular imaging, PDE/level set, deformable models, and performance evaluation of computer vision systems. These books covered medical image segmentation, image and volume registration, cardiac signal processing, and the physics of medical imaging modalities and emerging applications of medical imaging technologies. He is a lifetime member of the following research engineering societies: Tau Beta Pi, Eta Kappa Nu, and Sigma Xi; a member of the New York Academy of Sciences, the Engineering in Medicine and Biology Society, the American Association of Physics in Medicine, SPIE, and ACM; and a senior member of the IEEE. Dr. Suri is on the editorial board or is a reviewer for several international journals, including *Real Time Imaging, Pattern Analysis and Applications, Engineering in Medicine and Biology Magazine, Radiology, Journal of Computer Assisted Tomography, IEEE Transactions of Information Technology in Biomedicine,* and the IASTED Imaging Board. He has also chaired biomedical imaging tracks at several international conferences and has given more than 40 international presentations and seminars. Dr. Suri has been listed in *Who's Who* 10 times in a row and was a recipient of the President's Gold Medal in 1980; he has received more than 50 scholarly and extracurricular awards during his career. He is also a Fellow of the American Institute of Medical and Biological Engineering and ABI. He is a visiting faculty member at several schools: the Department of Computer Sciences, University of Exeter, Exeter, United Kingdom; the Department of Computer Sciences, University of Barcelona, Spain; the Department of Computer Sciences, Kent State University, Ohio; and the director of the medical imaging division, Jebra Wellness Technologies. Dr. Suri is on the board of directors of Biomedical Technologies Inc. and Cardiac Health Inc. He received a B.S. in computer engineering with distinction from Maulana Azad College of Technology, Bhopal, India, an M.S. in computer sciences (in conjunction with the Neurology Division, School of Medicine) from the University of Illinois, Chicago; a Ph.D. in electrical engineering (in conjunction with the Cardiology Division, School of Medicine) from the University of Washington, Seattle; and an M.B.A. from Weatherhead School of Management (in conjunction with the Department of Biomedical Engineering), Case Western Reserve University, Cleveland, Ohio. His major interests are in the fields of biomedical imaging (from pixel to molecular to fusion), pharmaceutical imaging, invasive and noninvasive biomedical devices, displays and image guided surgeries, computer vision, graphics and bioimaging, software engineering, and biophysics for better health care design, delivery, and therapy.

List of Contributors

A. Abu-Dalhoum
Computer Sceince Department
Jordan University
Amman 11942
Jordan

Rajendra Acharya U
Department of Electronics and Computer
Engineering
Ngee Ann Polytechnic
Singapore
E-mail: aru@np.edu.sg

M. Al-Rawi
Computer Science Department
Jordan University
Amman 11942
Jordan
E-mail: rawi@ju.edu.jo

Whye-Teong Ang
School of Mechanical and Aerospace
Engineering
College of Engineering
Nanyang Technological University
50 Nanyang Avenue
Singapore 639798

Fábio J. Ayres
Department of Electrical and Computer
Engineering
University of Calgary
2500 University Drive
N.W. Calgary
Alberta, Canada T2N 1N4

Madhusudhanan Balasubramanian
Hamilton Glaucoma Center
Department of Ophthalmology
University of California, San Diego
La Jolla, CA 92093
E-mail: madhu@glaucoma.ucsd.edu

Christopher Bowd
Hamilton Glaucoma Center
Department of Ophthalmology
University of California, San Diego
La Jolla, CA 92093

Roger W. Beuerman
Singapore Eye Research Institute
Department of Ophthalmology
Yong Loo Lin School of Medicine
National University of Singapore
Singapore
School of Biomedical and Chemical
Engineering
Nanyang Technical University
Singapore
LSU Eye Center

LSU Health Sciences Center
New Orleans, LA 70112

Vinod Chandran
Queensland University of Technology
Australia

Caroline Chee
Eye NUH/Surgery Centre
National University Hospital
Department of Opthamology
Singapore

Johnny Chee
Department of Electronics and Computer
Engineering
Ngee Ann Polytechnic
Singapore
E-mail: chj@np.edu.sg

Chua Kuang Chua
Department of Electronics and Computer
Engineering
Ngee Ann Polytechnic
Singapore

Kenneth Er
Department of Electronics and Computer
Engineering
Ngee Ann Polytechnic
Singapore

Peyman Eshghzadeh-Zanjani
Department of Electrical and Computer
Engineering
University of Calgary
2500 University Drive
N.W. Calgary
Alberta, Canada T2N 1N4

Gracielynn Flores
Department of Electronics and Computer
Engineering
Ngee Ann Polytechnic
Singapore

Manjunath Gupta
Eye NUH/Surgery Centre
Department of Opthamology
National University Hospital
Singapore

S. Sitharama Iyengar
Department of Computer Science
Louisana State University
Baton Rouge, LA 70803

H. Karajeh
Computer Sceince Department
Jordan University
Amman 11942
Jordan

Lim Choo Min
Department of Electronics and Computer
Engineering
Ngee Ann Polytechnic
Singapore

Jagadish Nayak
Department of E&C
Manipal Institute of Technology
Manipal-India, 5761204

Eddie Y. K. Ng
Adjunct NUH Scientist
Office of Biomedical Research
National University Hospital of Singapore
Associate Professor
School of Mechanical and Aerospace
Engineering
College of Engineering
Nanyang Technological University
Singapore
E-mail: mykng@ntu.edu.sg

Faraz Oloumi
Department of Electrical and Computer
Engineering
University of Calgary
2500 University Drive
N.W. Calgary
Alberta, Canada T2N 1N4

Foad Oloumi
Department of Electrical and Computer
Engineering
University of Calgary
2500 University Drive
N.W. Calgary
Alberta, Canada T2N 1N4

E. H. Ooi
School of Mechanical and Aerospace
Engineering
College of Engineering
Nanyang Technological University
50 Nanyang Avenue
Singapore 639798

A. Louise Perkins
Computer Science
University of Southern Mississippi
Hattiesburg, MS 39406

Christine Purslow
School of Optometry & Vision Sciences
Anterior Eye Group
Cardiff University
Redwood Building
King Edward VII Avenue
Cardiff, Wales, CF10 3NB
United Kingdom
E-mail: purslowc@Cardiff.ac.uk

Rangaraj M. Rangayyan
Department of Electrical and Computer
Engineering
University of Calgary
2500 University Drive
N.W. Calgary
Alberta, Canada T2N 1N4
Contact: ranga@ucalgary.ca

P. Subbanna Bhat
Department of E&C Eng.
National Institute of Technology Karnataka
Surathkal, India, 574157

Jasjit S. Suri
Research Professor
Biomedical Research Institute
Pocatello, Idaho, United States
Chief Technology Officer
Biomedical Technologies, Inc.
Colorado, United States
E-mail: suri0256@msn.com

William Tan
Department of Electronics and Computer
Engineering
Ngee Ann Polytechnic
Singapore

Melissa Tan Yan Jun
Department of Electronics and Computer
Engineering
Ngee Ann Polytechnic
Singapore

Wenwei Yu
Graduate School of Medical System
Engineering
Chiba University
Japan 263-8522

Wong Li Yun
Department of Electrical Engineering
National University of Singapore
Singapore

Linda M. Zangwill
Hamilton Glaucoma Center
Department of Ophthalmology
University of California, San Diego
La Jolla, CA 92093
E-mail: zangwill@glaucoma.ucsd.edu

Index

Related Titles from Artech House

Biomolecular Computation for Bionanotechnology, Jian-Qin Liu and
  Katsunori Shimohara

Electrotherapeutic Devices: Principles, Design, and Applications, George D. O'Clock

Fundamentals and Applications of Microfluidics, Second Edition, Nam-Trung and
  Steven T. Wereley

Matching Pursuit and Unification in EEG Analysis, Piotr Durka

Micro and Nano Manipulations for Biomedical Applications, Tachung C. Yih and
  Ilie Talpasanu, editors

Microfluidics for Biotechnology, Jean Berthier and Pascal Silberzan

For further information on these and other Artech House titles, including
previously considered out-of-print books now available through our
In-Print-Forever® (IPF®) program, contact:

Artech House	Artech House
685 Canton Street	46 Gillingham Street
Norwood, MA 02062	London SW1V 1AH UK
Phone: 781-769-9750	Phone: +44 (0)20 7596-8750
Fax: 781-769-6334	Fax: +44 (0)20 7630-0166
e-mail: artech@artechhouse.com	e-mail: artech-uk@artechhouse.com

Find us on the World Wide Web at: www.artechhouse.com

RETURN TO: **FONG OPTOMETRY LIBRARY**

490 Minor Hall · 642-1020

LOAN PERIOD 1 MONTH	1	2	3
	4	5	6

All books may be recalled after 7 days.
Renewals may be requested by phone or, using GLADIS,
type **inv** followed by your patron ID number.

DUE AS STAMPED BELOW.

This book will be held
in OPTOMETRY LIBRARY
until AUG 22 2008

5/21/2010

FORM NO. DD 17
5M 9-02

UNIVERSITY OF CALIFORNIA, BERKELEY
Berkeley, California 94720–6000